Modelling Pulsar Wind Nebulae

Astrophysics and Space Science Library

More information about this series at http://www.springer.com/series/5664

Diego F. Torres
Editor

Modelling Pulsar Wind Nebulae

Springer

Sant Cugat Forum on Astrophysics

Editor
Diego F. Torres
ICREA & Institute of Space Sciences
(IEEC-CSIC) Campus UAB
Barcelona, Spain

ISSN 0067-0057 ISSN 2214-7985 (electronic)
Astrophysics and Space Science Library
ISBN 978-3-319-63030-4 ISBN 978-3-319-63031-1 (eBook)
DOI 10.1007/978-3-319-63031-1

Library of Congress Control Number: 2017954960

Cover illustration: This composite image of the Crab Nebula, a supernova remnant, was assembled by combining data from five telescopes spanning nearly the entire breadth of the electromagnetic spectrum: the Very Large Array, the Spitzer Space Telescope, the Hubble Space Telescope, the XMM-Newton Observatory, and the Chandra X-ray Observatory. Credit: G. Dubner (IAFE, CONICET-University of Buenos Aires) et al.; A. Loll et al.; T. Temim et al.; F. Seward et al.; NASA, ESA,VLA/NRAO/AUI/NSF; Chandra/CXC; Spitzer/JPL-Caltech; XMM-Newton/ESO; and Hubble/STScI

Printed on acid-free paper

This Springer imprint is published by Springer Nature
The registered company is Springer International Publishing AG
The registered company address is: Gewerbestrasse 11, 6330 Cham, Switzerland

Preface

This book contains a collection of reviews of specific aspects of the modelling of nebulae by some of the experts in the field, intending to give a broad and comprehensive overview on the topic. Particularly, the book pays special attention to those areas for which reviews were not yet published. The aim of this book is then to provide a reference point, useful for researchers in the field of pulsar wind nebulae and beyond.

The book assesses, among others, the following questions: What kind of models do we already have, and what kinds of models are needed to reach a more profound understanding of nebulae? Can models be combined? Which are the most promising avenues for unifying model classes? Can they be made versatile enough to interpret observations of hundreds of sources? To what extent are the results from different radiative models comparable? What key features are they missing? Up to what extent are time-dependent models without spatial information reliable/useful? Are hybrid hadronic/leptonic models necessary for modelling nebulae in general? What is the best case for a hadronic-dominated nebula? How can we differentiate hadronic from leptonic nebulae at an observational level? What is the impact of hybrid models and how can they be observationally tested? How do we move forward: What features are the models missing to account for the forthcoming data?

This book was conceived at the 5th Sant Cugat Forum on Astrophysics session, "Modelling Nebulae", held on June 14–17, 2016. This workshop joined together an international group of experts with the aim of assessing the theoretical state of the art in modelling pulsar wind nebulae. It did so in view of the current and forthcoming observational data, which were reviewed.

The Sant Cugat Forum on Astrophysics[1] is a logistic and scientific framework designed to host international meetings of scientists. The forum workshops focus on a specific aspect of astrophysics and have a duration of 3–4 days.

[1]http://www.ice.csic.es/research/forum.

Since such a reference book was currently lacking in the literature, the Scientific Organizing Committee believed this volume could be a welcomed addition to it and helped in planning its contents.

The Scientific Organizing Committee was formed by

Elena Amato (INAF/Obs. Arcetri, Italy)
Aya Bamba (Aoyama University, Japan)
Emma de Oña Wilhelmi (IEEC-CSIC, Spain)
Joseph Gelfand (NYUAD, United Arab Emirates)
Simon Johnston (CSIRO, Australia)
Serguei Komissarov (Leeds, UK)
Anatoly Spitkovsky (Princeton, USA)
Diego F. Torres (ICREA/IEEC-CSIC, Spain)

As main organizer of the workshop and editor of this book, it is my pleasure to thank them all, together with acknowledging the help received from the Sant Cugat City Hall, ICREA, and IEEC. I thank as well, of course, all speakers and attendees to the workshop for providing such a lively setting for discussion and learning. I hope some of that spirit is reflected in these pages and, in turn, that they provide a useful entry reading to a topic so active and diverse.

Sant Cugat, Spain
March 2017

Diego F. Torres

Contents

Chapter 1
Radio Properties of Pulsar Wind Nebulae

Roland Kothes

Abstract Fast rotating neutron stars produce a steady energetic wind of relativistic particles and magnetic field that is released into the expanding ejecta of a core-collapse supernova explosion. This pulsar wind creates a nebula, in which the relativistic particles travel through the magnetic field producing synchrotron emission observable from radio to the γ-ray band. In the radio band, smooth, highly linearly polarized emission arises from well ordered magnetic fields that can be characterized by a power-law distribution of flux, with typical spectral indices similar to optically thin thermal emission from HII regions. In this review I will summarize the development of the radio structure of pulsar wind nebulae with respect to the different phases of their evolution. I will present the characteristics of the radio continuum emission that we observe, such as spectral behaviour and the presence of cooling breaks in the radio band. I will derive magnetic field configurations and discuss their impact on overall PWN models.

1.1 Introduction

On the 4th of July in the year AD 1054, ancient Chinese and Japanese astronomers discovered a new "guest star" in the constellation Taurus (Stephenson and Green 2002). There also may have been sightings in Europe as early as April 11, 1054 (Collins et al. 1999). This guest star remained visible for almost 2 years before it disappeared to the naked eye. The remnant of this supernova explosion (SN) is the well-known Crab Nebula, which also made it as M 1 into the Messier catalogue of optical nebulae, after being miss-classified as Halley's comet by Charles Messier. As one of the brightest sources in the radio sky the radio emission of the Crab Nebula was discovered very early in the history of radio astronomy by Bolton and Stanley (1949). In 1967 Pacini (1967) proposed a fast rotating neutron star as the central engine for the Crab Nebula. Pulsating radio emission from this neutron star, the pulsar B 0531+21, was discovered by Staelin and Reifenstein (1968), shortly

R. Kothes (✉)
Herzberg Programs in Astronomy and Astrophysics, Dominion Radio Astrophysical Observatory, National Research Council Canada, P.O. Box 248, Penticton, BC V2A 6J9, Canada
e-mail: Roland.Kothes@nrc-cnrc.gc.ca

D.F. Torres (ed.), *Modelling Pulsar Wind Nebulae*, Astrophysics and Space Science Library 446, DOI 10.1007/978-3-319-63031-1_1

after the very first discovery of a pulsar by Hewish et al. (1968). The Crab Nebula remains an astrophysical laboratory of immense value to study the development of a young neutron star and the synchrotron nebula generated by the relativistic wind it injects into its surroundings.

Over the last decades, after the discovery of the Crab Nebula as a radio source, numerous radio pulsar wind nebulae (PWNe), also known as filled-centre, plerion-type or crab-like supernova remnants (SNRs), have been identified and catalogued (e.g. Green 2014; Ferrand and Safi-Harb 2012). Most of them exhibit the common characteristics of radio PWNe such as a centre-filled morphology, a flat radio synchrotron spectrum, highly linearly polarized radio continuum emission and a fast rotating neutron star as their central engine. Their appearance reflects the characteristics of the central energy source and their radio continuum spectrum reflects the energy spectrum of the injected relativistic particles. Studies of pulsar wind nebulae in the radio waveband serve to probe the spectral and morphological distribution of the relativistic particles, constraining in turn the nature of the central engines and the processes by which energy is transferred to their surroundings.

I summarize basic properties of pulsars and their wind nebulae in Sect. 1.2. This includes a discussion about the evolution of the nebula and its radio emission in the context of overall supernova remnant development. In Sect. 1.3 I describe the properties of radio continuum emission from pulsar wind nebulae, specifically spectral characteristics and the presence of breaks in the radio spectrum. In Sect. 1.4 I investigate magnetic field configurations inside pulsar wind nebulae derived from radio polarimetric observations and compare their characteristics with theoretical expectations.

1.2 Background Information

The supernova explosion of a massive star leaves a rapidly rotating neutron star behind which produces a steady energetic wind of magnetic field and highly relativistic particles. This wind is released into the interior of the expanding shockwave of the core-collapse supernova explosion. The relativistic particles interact with the magnetic field producing a synchrotron emitting nebula. The characteristics of such a pulsar wind nebula depend mainly on the properties of the central neutron star, a pulsar and the density and structure of the medium this relativistic wind is expanding into. Early on this ambient medium would be the supernova ejecta. In a later stage the PWN may interact with the reverse shock of the supernova remnant or, after it leaves its host SNR, with the ambient interstellar medium.

1.2.1 Pulsars

A pulsar, short for **pulsa**ting source of **r**adio emission, is a highly magnetized fast rotating neutron star. Jocelyn Bell and Anthony Hewish discovered the very first pulsar, now known to us as PSR B 1919+21, in 1967 during a survey of radio sources (Hewish et al. 1968). Pulsars are characterized by their period P and their period derivative $\dot{P} = \frac{dP}{dt}$. Those characteristics of the pulsar can be easily determined by timing observations over a long timescale. Rotational periods found for pulsars range from a little more than 1 ms up to a few seconds. Period derivatives between 10^{-21} and 10^{-12} s s^{-1} have been observed.

Surface magnetic fields and age of a pulsar can be inferred from P and $\dot{P}$. Assuming that the magnetic field is dipolar, and therefore the pulsar is spinning down by magnetic dipole radiation (Pacini 1967; Ostriker and Gunn 1969), the surface magnetic field can be estimated by:

$$B_{\rm PSR} = 3.2 \times 10^{19} \sqrt{P \dot{P}} \text{ (G)}. \tag{1.1}$$

For pulsars a wide range of magnetic field strengths $B_{\rm PSR}$ have been detected, from as low as 10^8 G for the so-called millisecond pulsars to 10^{15} G for magnetars. Most "normal" pulsars have surface magnetic fields between 10^{12} and 10^{14} G.

The characteristic age $\tau_{\rm ch}$ of a pulsar is defined by:

$$\tau_{\rm ch} = \frac{P}{2\dot{P}}. \tag{1.2}$$

Assuming that the intrinsic period of the pulsar P_0 is significantly shorter than the currently observed period P and the pulsar is spinning down by magnetic dipole radiation the characteristic age $\tau_{\rm ch}$ is comparable to the real age t.

The pulsar produces a steady wind of magnetic field and relativistic particles. The energy source for that wind is the slow dissipation of rotational energy. The current rotational energy loss rate or "spin down luminosity" of the pulsar is given by:

$$\dot{E}_{\rm c} = 4\pi^2 I \frac{\dot{P}}{P^3}. \tag{1.3}$$

Here, I is the moment of inertia of the neutron star ($I = 10^{45}$ g cm^{-2} for typical values for the radius and mass of the pulsar of 10 km and 1.4 $M_\odot$, respectively). $\dot{E}_{\rm c}$ values for pulsars between about 10^{28} and 4.5×10^{38} erg s^{-1} for the Crab pulsar have been observed. According to Gotthelf (2004) a minimum critical rotational energy loss rate $\dot{E}_{\rm crit} \geq 10^{36}$ erg s^{-1} is required to produce a prominent radio pulsar wind nebula. This may be the reason why only about 15% of the supernova remnants listed in Green (2014) contain a radio pulsar wind nebula even though more than 80% are supposed to be the result of a core-collapse supernova explosion, which leaves a neutron star behind.

For a constant braking index n the evolution of the spin down luminosity can be described by (e.g. Pacini and Salvati 1973):

$$\dot{E} = \frac{\dot{E}_0}{\left(1 + \frac{t}{\tau_0}\right)^{\frac{n+1}{n-1}}} \tag{1.4}$$

Here $\dot{E}_0$ and τ_0 are the pulsar's intrinsic values for the spin down luminosity and the characteristic age. Most measured values for the braking index n have been found between 2 and 3, but typically $n = 3$ is assumed. This would indicate a pure dipolar field on the pulsar's surface. If we do assume a dipolar magnetic field and know the pulsar's true age t, we can estimate the total energy released by the pulsar into the nebula by integrating the energy loss rate over its lifetime:

$$E_{\rm tot} = \int_0^t \dot{E}\, dt = \int_{\tau_0'}^{\tau_{\rm ch}} \dot{E}_0 \left(\frac{\tau_0}{\tau}\right)^2 d\tau = \frac{\dot{E}_{\rm c} \tau_{\rm ch} t}{\tau_{\rm ch} - t} \tag{1.5}$$

For the Crab pulsar this results in a total energy release of $E_{\rm crab} = 5.8 \times 10^{49}$ erg since the supernova explosion in AD 1054. For the PWN 3C 58 and its pulsar J 0205+6449 this results in a total energy content of $E_{\rm 3C\,58} = 8.4 \times 10^{47}$ erg for the pulsar wind nebula. The only historical supernova that produced an observable shell-type remnant with a PWN inside, G11.2–0.3 and its pulsar J 1811–1925, contains a synchrotron nebula with an integrated energy of $E_{\rm G11.2-0.3} = 3.5 \times 10^{47}$ erg inside.

1.2.2 *Pulsar Wind Nebulae*

A central rapidly rotating neutron star, created in a supernova explosion, produces a steady highly energetic wind of relativistic particles and magnetic field that is released into the ejecta of a core-collapse supernova explosion (see Fig. 1.1). For details about the generation of this wind I refer to Gaensler and Slane (2006) and references therein, since this is beyond the scope of this article.

A wind termination shock is formed through deceleration of the pulsar wind as it interacts with the more slowly moving ejecta of the SN. At this termination shock electron-positron pairs are re-accelerated to relativistic velocities. If we assume equality between the wind ram pressure and the internal pressure of the PWN at the termination shock, its radius can be expressed by:

$$r_{\rm t} = \sqrt{\frac{\dot{E}}{4\,\pi\, c\, P_{\rm PWN}}}, \tag{1.6}$$

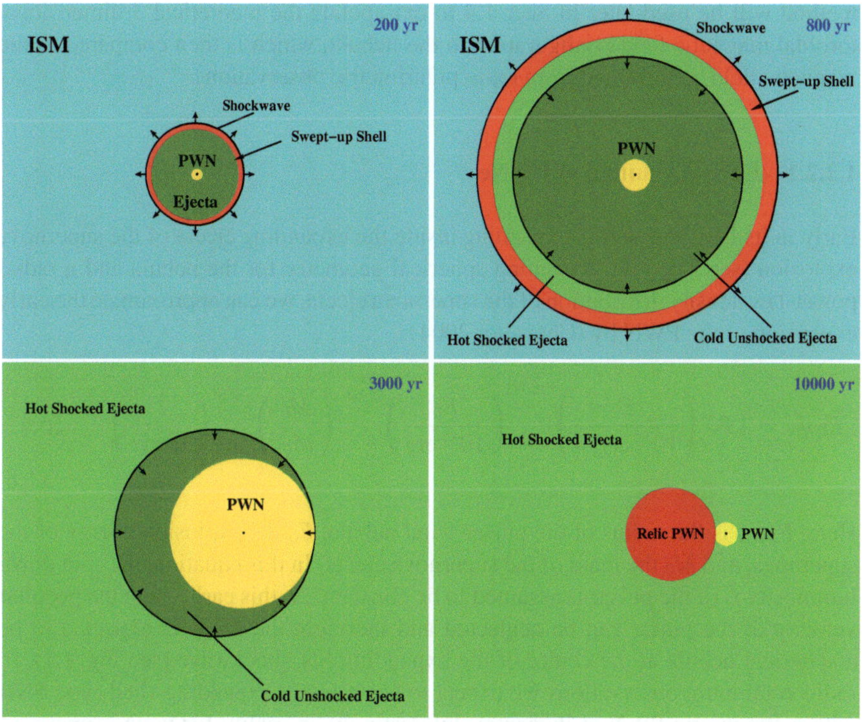

Fig. 1.1 Sketch of the different phases of PWN evolution. *Top left*: Shortly after the explosion the PWN is expanding inside the ejecta. *Top right*: Due to the interaction with the ambient medium, the supernova shockwave sends a reverse shock back into the interior which shocks and heats up the ejecta. *Bottom left*: The reverse shock first collides with the expanding PWN, which starts a period of interaction in which the pulsar wind bubble oscillates. *Bottom right*: The reverse shock has dissipated and the pulsar again can power an expanding PWN. It also may have left a relic PWN behind. The scales are based on the following parameters: $\dot{E} = 10^{37}$ erg s^{-1}, $E_0 = 10^{51}$ erg, $M_0 = 5\,M_\odot$, $n_0 = 1$ cm^{-3}, and $v_{PSR} = 227$ km s^{-1} (for more details see text)

the pressure inside the PWN P_{PWN} can be estimated via:

$$P_{PWN} = \frac{B_{PWN}^2}{8\,\pi}. \tag{1.7}$$

B_{PWN} represents the internal magnetic field of the pulsar wind nebula.

Some PWNe display toroidal and/or jet-like components close to the central pulsar in high resolution X-ray observations. An axis-symmetric wind creates those torus-like features in the equatorial plane of the pulsar's spin axis along with jets that are collimated along the axis. The size of those features, of few tenth of a pc, place them near the pulsar wind termination shock. The morphology observed in the jets and torii reflects the orientation of the pulsar spin axis, which can be determined by fitting 3-dimensional models to the jet-tours systems (Ng and Romani 2004). This method will be used later in Sect. 1.4 to extrapolate the theoretical 3-dimensional toroidal magnetic field configuration in the nebula, which is then compared to the magnetic field, determined from radio polarimetric observations.

1.2.2.1 Early Evolution of PWNe

Early in its life the PWN is expanding inside the expanding ejecta of the supernova explosion (see Fig. 1.1). Assuming spherical geometry for the nebula and a radial power-law density distribution of the supernova ejecta, we can approximate the early expansion of the PWN by (Chevalier 2004):

$$R_{\rm PWN} = 1.87 \left(\frac{\dot{E}}{10^{38}\,{\rm erg/s}}\right)^{0.254} \left(\frac{E_0}{10^{51}\,{\rm erg}}\right)^{0.246} \left(\frac{M_0}{{\rm M}_\odot}\right)^{-0.5} \left(\frac{t}{10^3\,{\rm yr}}\right)^{1.254} \quad ({\rm pc}). \tag{1.8}$$

Here, $R_{\rm PWN}$ is the radius of the pulsar wind nebula, E_0 the explosion energy of the supernova, and M_0 the mass of the supernova ejecta. In this equation, the spin down luminosity $\dot{E}$ of the pulsar is assumed to be constant. At this early stage the peculiar velocity of the pulsar can be neglected and therefore the PWN is expected to be a spherical nebula at the centre of the expanding SN shockwave (see Fig. 1.1). In radio continuum observations we expect to see a freely expanding shell-type SNR with a symmetric PWN at its centre. Interestingly, in reality for young supernova remnants, we find either a bright pulsar wind nebula without detectable radio shell, such as the Crab Nebula or a bright radio shell without detectable PWN, such as Cas A. Both are supposed to be the result of a core-collapse supernova explosion. A radio image of another example for a young PWN without detected radio shell, 3C 58, is shown in Fig. 1.2.

3C 58 has a centrally peaked radio morphology, typical for a PWN. A bright network of filamentary structures lies on top of a broad envelope of diffuse emission that disappears into the noise at the edges. Most of the filamentary structures are proposed to be magnetic loops torn from the toroidal magnetic fields by kink instabilities (Slane et al. 2004). The diffuse emission is elongated east-west with an extent of about $9' \times 5'$, which translates to spatial dimensions of 5.5×3.4 pc at a distance of 2 kpc (Kothes 2013). This is comparable to the size of the Crab Nebula. The elongation in young PWNe is proposed to be the result of higher equatorial pressure associated with toroidal magnetic fields (Begelman and Li 1992; van der Swaluw 2003). Despite various deep radio continuum searches no evidence has been found for a shell-type supernova remnant that would mark the location of the expanding shockwave of the supernova explosion outside the PWN. There are many more examples for shell-less radio bright pulsar wind nebulae at various stages of evolution. There is still no satisfactory explanation for this phenomenon.

The only confirmed historical supernova explosion that shows both a radio PWN and a bright radio shell is the SNR G11.2–0.3 (Kothes and Reich 2001), which

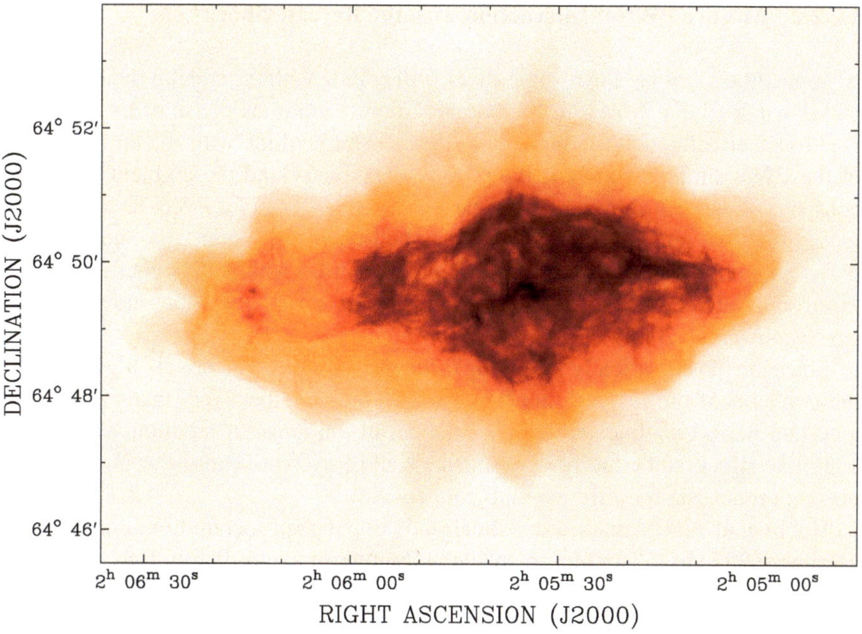

Fig. 1.2 Radio image of the PWN 3C 58 made with the data observed and published in Bietenholz (2006). These data were observed at 1420 MHz with the VLA in A, B, and C-configuration. The final image has a resolution of 1.4″

is linked to the historical supernova explosion of AD 385 (Stephenson and Green 2002). G11.2−0.3 contains the pulsar J 1811−1925, which produces with a current energy loss rate of $E_c = 9 \times 10^{36}$ erg s^{-1} (Torii et al. 1999) a faint pulsar wind nebula.

As the supernova shockwave sweeps up more and more circumstellar and interstellar matter a highly compressed shell of swept-up material forms on the inside of the expanding blast-wave and a reverse shock is created that eventually moves back into the interior of the remnant and shocks and heats up the supernova ejecta (see Fig. 1.1). The expanding shell-type SNR has entered the so-called energy-conserving Sedov-Taylor phase of its evolution (Sedov 1959). Early on, this has no immediate effect on the PWN, which continues to expand into the cool unshocked ejecta. According to the calculations by McKee and Truelove (1995) the SNR enters the Sedov-Taylor phase at the time:

$$t_{\rm ST} = 0.209 \left(\frac{E_0}{10^{51}\,\rm erg}\right)^{-0.5} \left(\frac{M_0}{\rm M_\odot}\right)^{\frac{5}{6}} \left(\frac{n_0}{\rm cm^{-3}}\right)^{-\frac{1}{3}} (10^3 \rm yr). \tag{1.9}$$

This is the time when the swept up material starts to dominate the hydrodynamics of the expanding SNR. Here, n_0 is the number density of the ambient medium assuming a hydrogen to helium ratio of 10:1. Typical times for $t_{\rm ST}$ are a few hundred up to maybe 2000 yr.

1.2.2.2 Mature PWNe: Interaction with the Reverse Shock

At some time $t_{col} > t_{ST}$ the reverse shock will collide with the edge of the expanding pulsar wind nebula. Now, the PWN enters a more complex phase of its evolution. Due to spherical geometry the reverse shock would collide with the entire surface of the PWN simultaneously if the pulsar is stationary and the ambient density is constant. Figure 1.3 shows an example for the expansion of a PWN inside a shell-type supernova remnant. In this case the shell-type SNR enters the Sedov-Taylor phase after about 800 yr and the reverse shock collides with the PWN after 3300 yr. From simulations we know that typical times for t_{col} are a few thousand years (van der Swaluw et al. 2004).

Since the pulsar typically is moving through the SNR at a relatively high velocity, the collision of the reveres shock with the surface of the PWN takes place over a certain period of time. A density gradient in the ambient medium would have a similar effect, since the reverse shock would travel back into the interior on a different timescale for different ambient densities.

Hobbs et al. (2005) made a statistical study of pulsar proper motions and found an average of 227 km s^{-1} for pulsars inside supernova remnants. In Fig. 1.4 a diagram is displayed showing another example for the expansion of a PWN inside a shell-type supernova remnant. In this case the pulsar is moving through the nebula at a velocity of $v_{PSR} = 227$ km s^{-1}. The other parameters are the same as in Fig. 1.3. Again the shell-type SNR enters the Sedov-Taylor phase after about 800 yr. However, the leading edge of the PWN collides with the reverse shock after about 3000 yr and the back edge of the PWN finally after about 3600 yr.

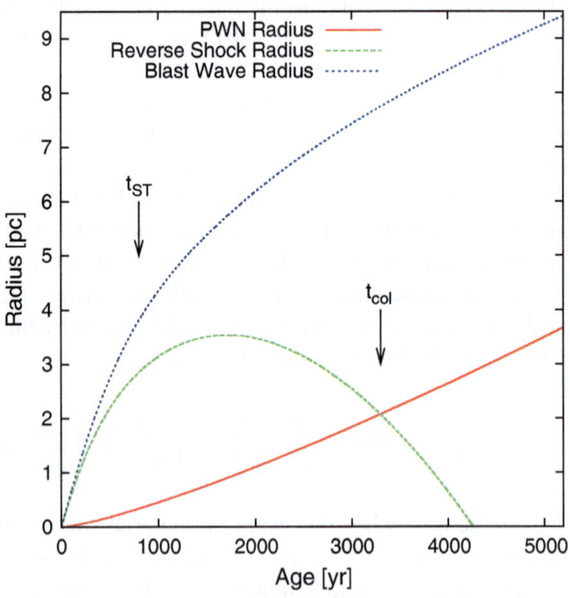

Fig. 1.3 Blast Wave radius, Reverse Shock radius, and PWN radius as a function of time. For this diagram the following parameters were assumed: $\dot{E} = 10^{37}$ erg s^{-1}, $E_0 = 10^{51}$ erg, $M_0 = 5\,M_\odot$, and $n_0 = 1$ cm^{-3}. The time when the SNR enters the Sedov-Taylor phase t_{ST} and the time when the reverse shock collides with the PWN t_{col} are indicated. The reverse shock and blast wave radii were estimated using the equation published in Table 2 of McKee and Truelove (1995). The PWN radius was calculated with Eq. (1.8)

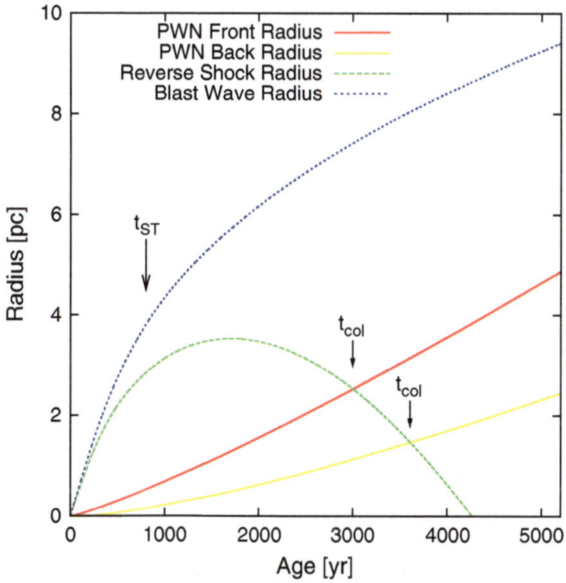

Fig. 1.4 Blast Wave radius, front and back Reverse Shock radius, and PWN radius as a function of time. For this diagram the following parameters were assumed: $\dot{E} = 10^{37}$ erg s^{-1}, $E_0 = 10^{51}$ erg, $M_0 = 5\ \mathrm{M}_\odot$, $n_0 = 1$ cm^{-3}, and $v_{\mathrm{PSR}} = 227$ km s^{-1}. The time when the SNR enters the Sedov-Taylor phase t_{ST} and the time when the reverse shock collides with the front and back of the PWN t_{col} are indicated. The reverse shock and blast wave radii were estimated using the equation published in Table 2 of McKee and Truelove (1995) using the modification by van der Swaluw et al. (2004) to take the proper motion of the pulsar into account. The PWN radius was calculated with Eq. (1.8)

After the collision of the SNR reverse shock with the edge of the pulsar wind bubble the passage of this shock through the PWN causes reverberations which produce oscillations in the pulsar wind bubble. At this point the evolution of the pulsar wind nebula becomes more complex and depends on many characteristics. The interaction timescale may last several thousand years (van der Swaluw et al. 2004).

One example of a composite supernova remnant, a SNR containing a shell-type remnant and a pulsar wind nebula, that currently shows interaction between the PWN and the reverse shock, is G327.1−1.1. A very thorough multi-waveband analysis of G327.1−1.1 is presented in Temim et al. (2015). This analysis reveals an SNR with a central pulsar moving at a velocity of $v_{\mathrm{PSR}} = 400$ km s^{-1}, expanding into a medium with a density gradient perpendicular to the pulsar trajectory. The reverse shock of the blast-wave hit the leading PWN edge after about 4000 yr and the back after 7000 yr. Now, at an age of 17,000 yr, the reverse shock in G327.1−1.1 almost dissipated, leaving a relic PWN behind.

1.2.2.3 Late PWN Evolution

After the reverse shock dissipates the oscillations fade away and the pulsar can once again power a small expanding pulsar wind nebula. At this point in its evolution the pulsar may have travelled a significant distance from its birthplace in the centre of the supernova remnant. There are now several possibilities mainly depending on the spin down luminosity of the pulsar and its velocity. It may continue to travel through the hot interior of the SNR with a small PWN around it, such as the pulsar B 1853+01 inside SNR W 44 with its small cometary-shaped radio nebula, discovered by Frail et al. (1996).

The pulsar may also have sufficient velocity to leave the host supernova remnant, so that its pulsar wind interacts with the interstellar medium outside the shell-type supernova remnant. A spectacular example is the SNR G5.4−1.2 with its pulsar B 1757−24. A radio image observed with the 100-m radiotelescope in Effelsberg at a frequency of 10.45 GHz (Kothes 1998) is shown in Fig. 1.5. The pulsar was

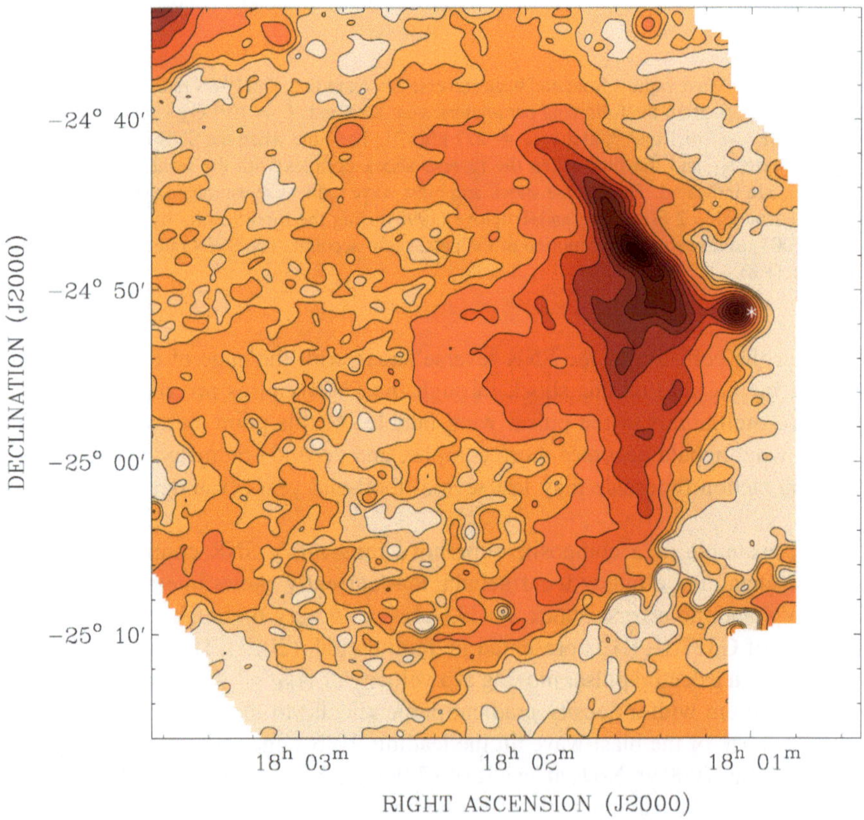

Fig. 1.5 10.45 GHz radio continuum image of the supernova remnant G5.4−1.2 observed with the 100-m Effelsberg radio telescope. The location of the pulsar B 1757−24 is indicated by the *white asterisk*

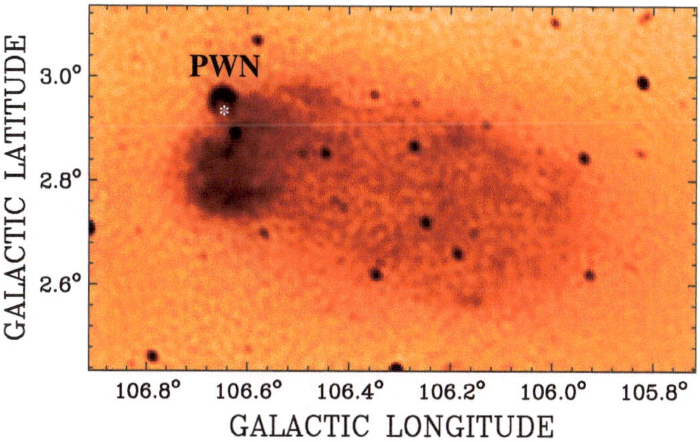

Fig. 1.6 1420 MHz radio continuum image of the supernova remnant G106.3+2.7 observed with the DRAO Synthesis Telescope as part of the Canadian Galactic Plane Survey (CGPS, Taylor et al. 2003). The location of the pulsar J 2229+6114 is marked by a *white asterisk*. The location of the pulsar wind nebula at the *top left* of the SNR is indicated

presumably born in the geometric centre of the shell-type SNR and travelled at high velocity in a westerly direction projected to the plane of the sky. Similar to PSR B 1853+01 inside SNR W 44 it creates a small cometary shaped radio pulsar wind nebula around it, which was discovered by Frail and Kulkarni (1991). However, in this case the pulsar wind interacts with the interstellar medium and not the hot and shocked ejecta inside the supernova remnant.

Another example for a PWN interacting with the ISM outside its host supernova remnant is the "Boomerang" pulsar wind nebula, which is part of SNR G106.3+2.7 (Fig. 1.6). In this case, however, the supernova exploded at the edge of a stellar wind bubble (SWB) (Kothes et al. 2001). This caused the reverse shock of the supernova blastwave, that expanded in the direction of the SWB shell, to collide very quickly with the PWN and then dissipate on a short time-scale. For the pulsar J 2229+6114 and its wind nebula this leads to a location at one edge of the SNR interacting with the ISM, while the SNR blastwave is still freely expanding into the interior of the stellar wind bubble on the other side.

Depending on the early energy input of the pulsar into the nebula and its velocity it may also leave a so-called relic PWN behind. This is a pulsar wind nebula that is no longer powered by the pulsar. The pulsar B 1757−24 related to SNR G5.4−1.2, again, is a very spectacular example, since it may have produced two relic pulsar wind nebulae. In Fig. 1.5 we can see the older possible relic PWN as a diffuse emission plateau in the interior of the shell-type SNR G5.4−1.2. This nebula was presumably produced in the early evolutionary stage of the SNR. While the reverse shock was interacting with the PWN the pulsar moved at high velocity in a westerly direction leaving a relic PWN behind. After the interaction with the supernova blastwave B 1757−24 left a second relic PWN behind, which is now seen in Fig. 1.5 as the bright compact source just outside the SNR shell. The pulsar seems to be located at the Western tip of this source, but high resolution radio continuum

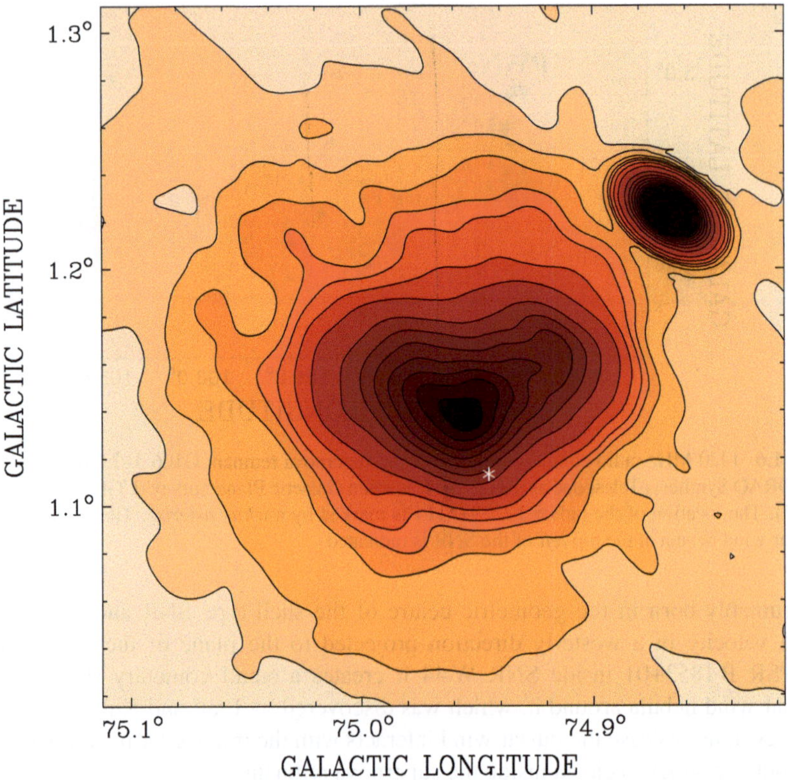

Fig. 1.7 1420 MHz radio continuum image of the supernova remnant CTB 87 observed with the DRAO synthesis telescope as part of the Canadian Galactic Plane Survey (Taylor et al. 2003). The location of the putative pulsar, that powers the PWN (Matheson et al. 2013), is indicated by a *white asterisk*

observations with the Jansky Very Large Array (JVLA) show, that this compact source actually consists of two nebulae, a diffuse nebula just outside the SNR shell, presumably the second relic PWN, and the pulsar wind nebula currently produced by the pulsar B 1757−24 (Frail and Kulkarni 1991).

Another example of a radio relic PWN is the SNR CTB 87 (see Fig. 1.7). CTB 87 belongs to the group of PWN, which show no evidence for a related shell-type SNR. Matheson et al. (2013) propose that CTB 87 is located off-centre inside a stellar wind cavity. The reverse shock has already collided with the PWN and dissipated, leaving CTB 87 as a relic PWN behind. The putative pulsar travelled at high velocity away from the origin of the supernova and is now located at the edge of the relic, projected to the plane of the sky (see Fig. 1.7).

Eventually the pulsar will leave its supernova remnant behind and travel supersonically through interstellar space by itself with a faint bow-shock pulsar wind nebula around it, leaving a tail of radio emission behind. One example for such a PWN may be G319.9−0.7 with its central engine the pulsar J 1509−5850 (Ng et al. 2010).

1.3 Radio Emission from Pulsar Wind Nebulae

1.3.1 *Synchrotron Emission from PWNe*

The pulsar wind creates a nebula, in which relativistic particles travel through the magnetic field producing highly linearly polarized synchrotron emission, which is observable from radio wavelength up to the γ-ray band. In the radio band the synchrotron emission of a pulsar wind nebula is typically characterized by a power-law:

$$S(\nu) \sim B_{\perp}^{\frac{1}{2}(\delta+1)} \nu^{-\frac{1}{2}(\delta-1)}, \tag{1.10}$$

where $S(\nu)$ is the flux density at frequency ν and $\alpha = -\frac{1}{2}(\delta - 1)$ represents the spectral index. $B_{\perp}$ is the magnetic field perpendicular to the line of sight. Radio PWNe typically have flat radio continuum spectra with spectral indices between -0.3 and 0.0 (see Fig. 1.8), which is comparable to optically thin free-free emission from HII regions. But in contrast to the thermal emission from an HII region the PWN's emission is highly linearly polarized, with the polarization vector perpendicular to the magnetic field at the place of origin. The synchrotron spectrum reflects the energy spectrum of the relativistic particles, which are accelerated at the

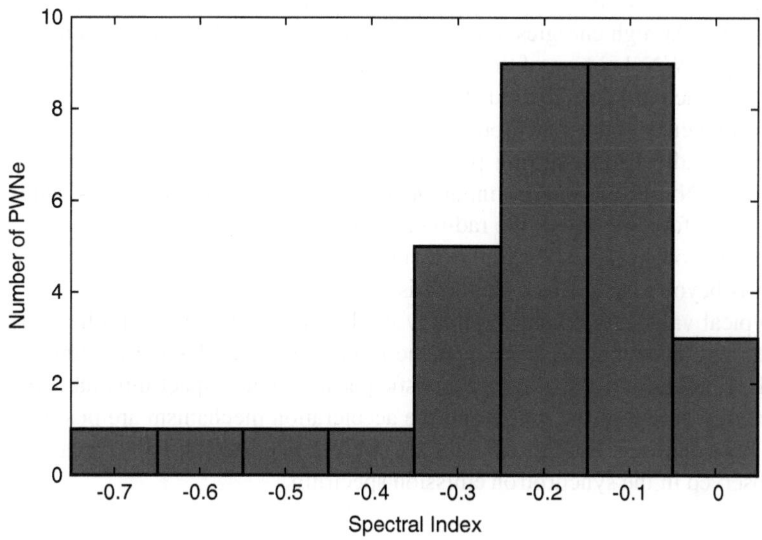

Fig. 1.8 Diagram of the number of pulsar wind nebulae as a function of radio spectral index α. The values were taken from Green (2014) and references therein

termination shock of the pulsar wind:

$$N(E)dE \sim E^{-\delta} dE, \tag{1.11}$$

where $N(E)$ is the number of relativistic particles with energy E.

In the X-ray and higher energy bands the synchrotron spectrum is typically characterized by a power-law of the photon distribution:

$$N_{\rm ph}(E) \sim E^{\Gamma}. \tag{1.12}$$

Here, $N_{\rm ph}(E)$ is the number of photons radiated at energy E and Γ is called the photon index. Γ is related to α by: $\Gamma = 1 - \alpha$. For the synchrotron emission from PWNe Γ is typically found to be around 2, increasing with distance from the pulsar due to "ageing" of the relativistic particles. The ageing is caused by the short synchrotron-emitting lifetime of X-ray emitting particles, which depends on the particle energy and the strength of the magnetic field. The strong dependence on particle energy is reflected in a cooling break in the synchrotron spectrum expressed by (Chevalier 2000):

$$\nu_c(\mathrm{GHz}) = 1.187\, B^{-3}(\mathrm{G})\, t^{-2}(\mathrm{yr}) \tag{1.13}$$

Here, ν_c is the cooling break frequency, B the magnetic field and t the age. The change of spectral index across the synchrotron cooling break is expected to be $\Delta\alpha \approx 0.5$. At high energies, the break frequency depends strongly on the location within the PWN, because of the short synchrotron lifetime and the continuous input of relativistic particles. Towards lower frequencies there will be one global cooling break frequency in the synchrotron spectrum.

In the radio band emitting particles have lifetimes much longer than the age of the PWN and the energy input of the pulsar is decreasing significantly with time. Therefore, the age of the radio-emitting particles can assumed to be constant. According to Chevalier (2000) this assumption is valid when the actual age t of the PWN is beyond the intrinsic characteristic age τ_0.

Typical values for α and Γ already imply a break in the synchrotron spectrum somewhere between the X-ray and the radio band. Besides the cooling breaks a break or breaks in the intrinsic relativistic particle energy spectrum that depends on the energy output of the pulsar and the acceleration mechanism are postulated and have been detected. But those breaks are not well understood. Those breaks can also be observed in the synchrotron emission spectrum.

1.3.2 *Radio Spectra of PWNe*

Most radio PWNe have flat radio continuum spectra with $-0.3 \leq \alpha \leq 0.0$. Figure 1.8 shows the distribution of spectral indices for all known radio pulsar wind nebulae. Included in this diagram are all radio pulsar wind nebulae that were included in Green (2014) plus the newly discovered PWN G141.2+5.0 (Kothes et al. 2014), which was the only discovery of a radio PWN after the latest update of the catalogue. The diagram contains PWNe with and without shells.

Clearly, the majority of PWNe have flat radio spectra with a peak of α around -0.2 to -0.1. There are four exceptions to that rule indicated in Fig. 1.8, which have very steep radio spectra more indicative of shell-type SRNs. Those are the PWNe G76.9+1.0 ($\alpha = -0.6$, Landecker et al. 1993; Kothes et al. 2006a), DA 495 ($\alpha = -0.45$, Kothes et al. 2008), CTB 87 ($\alpha = -0.41$, Kothes et al. 2017), and G141.2+5.0 ($\alpha = -0.69$, Kothes et al. 2014). As an example, a radio image and the radio continuum spectrum of the PWN G141.2+5.0 are shown in Fig. 1.9 (Kothes et al. 2014). G141.2+5.0 is a PWN in the latest phase of evolution after the passage of the reverse shock and shows the steepest radio continuum spectrum of all pulsar wind nebulae. Like the other three PWNe with steep radio spectra it also does not show a shell-type SNR. CTB 87 and DA 495 are also believed to be radio PWNe in a very late stage of evolution, which have already interacted with the reverse shock (Matheson et al. 2013; Kothes et al. 2008, respectively). For G76.9+1.0, however, the evolutionary stage is not clear. It certainly contains with PSR J 2022+3842 a very young and highly energetic pulsar (Arzoumanian et al. 2011).

While the acceleration at the termination shock of a pulsar wind produces particle energy spectra that translate to flat synchrotron spectra in the radio band, the steep spectrum PWNe likely have a different additional acceleration mechanism that re-accelerates particles, producing a steeper particle energy spectrum. A possible explanation for the much steeper radio spectral index would be that the reverse shock re-accelerated the relativistic particles with shock acceleration which might produce a steeper particle energy spectrum. But further study is required to find a satisfactory explanation for this phenomenon.

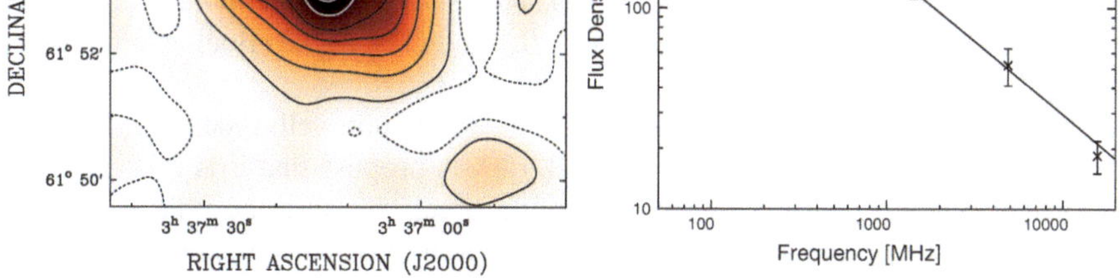

Fig. 1.9 *Left*: Radio continuum image of the PWN G141.1+5.0 at 1420 MHz observed with the DRAO ST (Kothes et al. 2014). *Right*: Radio continuum spectrum of G141.1+5.0. The radio flux density values were taken from Kothes et al. (2014) and Perrott et al. (2013)

1.3.3 Spectral Breaks in PWNe

In pulsar wind nebulae the synchrotron spectra show two different kinds of spectral breaks. First, there is the synchrotron cooling break, which marks the energy beyond which synchrotron radiation losses become significant for the relativistic particles in the lifetime of the PWN. This break is defined by Eq. (1.13). But other breaks have been found in synchrotron spectra of pulsar wind nebulae that cannot be explained by synchrotron cooling and are attributed to the acceleration mechanism at the termination shock which is not well understood.

The typical spectral indices observed in the radio and X-ray bands of $-0.3 \leq \alpha \leq 0.0$ and $\Gamma \approx 2.0$ ($\alpha = -\Gamma + 1.0 = -1.0$) already imply a steepening of the spectrum towards higher frequencies. This indicates at least one spectral break somewhere between the radio and the X-ray band. However, these breaks are usually found at higher frequencies, well beyond the radio band. Therefore the synchrotron lifetime of radio emitting particles is typically well beyond the age of the pulsar wind nebula. In this case the radio luminosity represents an integrated lifetime history of the PWN as was shown in Kothes (1998). A strong correlation was found between the radio surface brightness of the PWN at 1 GHz $\Sigma_{1\mathrm{GHZ}}$, the current spin down luminosity of the central pulsar $\dot{E}_c$, and the diameter of the PWN D:

$$\Sigma_{1\mathrm{GHZ}} = 10^{-20.57\pm0.04} \cdot \dot{E}_c^{0.99\pm0.02} \cdot D^{-0.99\pm0.03} \tag{1.14}$$

The index of 0.99 indicates negligible energy losses for radio emitting particles. On the other hand in the X-ray band a strong correlation was found between the current spin down luminosity and the X-ray luminosity by e.g. Possenti et al. (2002). Therefore, the X-ray synchrotron luminosity of such a nebula is created by young highly relativistic electrons which have a short lifetime and thus mainly depends on the current energy input of the pulsar. The radio emission, however, is created by electrons with lower energies which have a much longer lifetime and thus reflect the total energy content of the nebula.

There are several radio pulsar wind nebulae for which spectral breaks in the radio band have been predicted, but there are only five that have confirmed breaks in their radio synchrotron spectra: G21.5–0.9 ($\nu_{\mathrm{br}} \approx 40$ GHz, Salter et al. 1989a,b), DA 495 ($\nu_{\mathrm{br}} \approx 1.3$ GHz, Kothes et al. 2008), CTB 87 ($\nu_{\mathrm{br}} \approx 11$ GHz, Morsi and Reich 1987), the Boomerang PWN ($\nu_{\mathrm{br}} \approx 4.5$ GHz, Kothes et al. 2006b), and 3C 58 ($\nu_{\mathrm{br}} \approx 50$ GHz, Green and Scheuer 1992). There were possible breaks in the radio band found in other PWNe, e.g. in G16.7+0.1 and G29.7−0.3 (Bock and Gaensler 2005), but have not been confirmed yet.

The nature of the spectral break in G21.5−0.9 is not well understood and has not been studied in detail. But Bock et al. (2001) propose that it is related to the intrinsic particle spectrum and not a cooling break, because they did not detect any spectral steepening with radius in their observations at a frequency beyond the spectral break. In addition, the PWN would have to be very old to display a cooling break at such a low frequency.

Kothes et al. (2017) propose that the break in CTB 87 is not real, but the result of low surface brightness diffuse emission that was missed in many observations, in

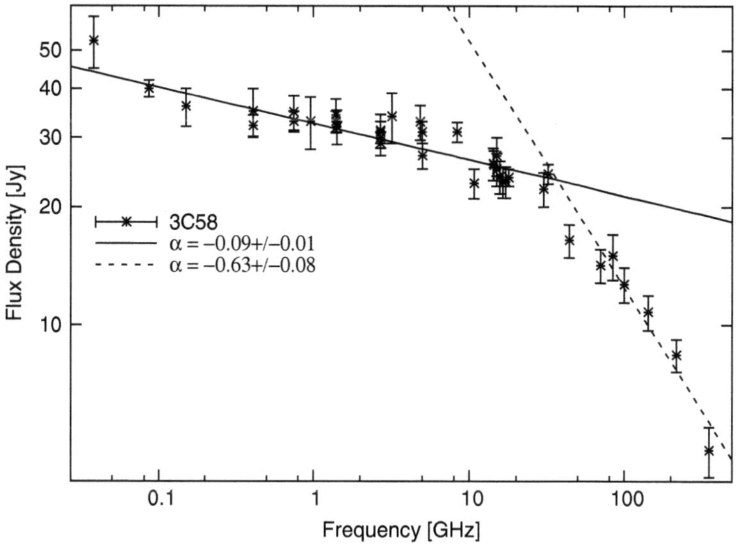

Fig. 1.10 Integrated radio continuum spectrum of the pulsar wind nebula 3C 58. The flux values were taken from Kothes (2016, and references therein)

particular at high radio frequencies. The break in the radio spectrum of 3C 58, shown in Fig. 1.10, is supposed to be related to the particle acceleration at the termination shock, because a cooling break would require unreasonably high magnetic fields, more than two orders of magnitude higher than determined from X-ray and TeV observations (Slane et al. 2008; Aleksić et al. 2014, respectively).

The breaks in the radio synchrotron spectra of DA 495 and the Boomerang pulsar wind nebulae are proposed to be synchrotron cooling breaks (Kothes et al. 2008, 2006b, respectively), which indicate mG strength magnetic fields inside the nebulae. A radio continuum image and the radio synchrotron spectrum of the Boomerang are shown in Fig. 1.11. The Boomerang PWN shows a radial steepening of the radio continuum spectrum (Kothes et al. 2006b), which is very unusual. In other pulsar wind nebulae this is only observed in the X-ray band. This implies rapid cooling of the radio emitting particles and confirms the nature of the spectral break as a synchrotron cooling break. However, a synchrotron cooling break at such a low frequency requires a very high magnetic field strength inside the nebula of 2.6 mG (Kothes et al. 2006b). According to Ng and Romani (2004) the radius of the termination shock inside the Boomerang PWN is $r_t = 9.2''$. This translates at a distance of 800 pc (Kothes et al. 2001) to a spatial radius of 0.036 pc, the smallest radius of a termination shock measured to date. Using Eq. (1.6) results in an internal magnetic field strength of 350 μG inside the nebula. This is almost a factor of 10 below the magnetic field required for the observed cooling break. However, the PWN could be somewhat older than deduced in Kothes et al. (2006b), which would decrease the necessary magnetic field.

Clearly, more detailed observations, in particular in the radio band, are necessary to study this PWN more closely to solve this mystery.

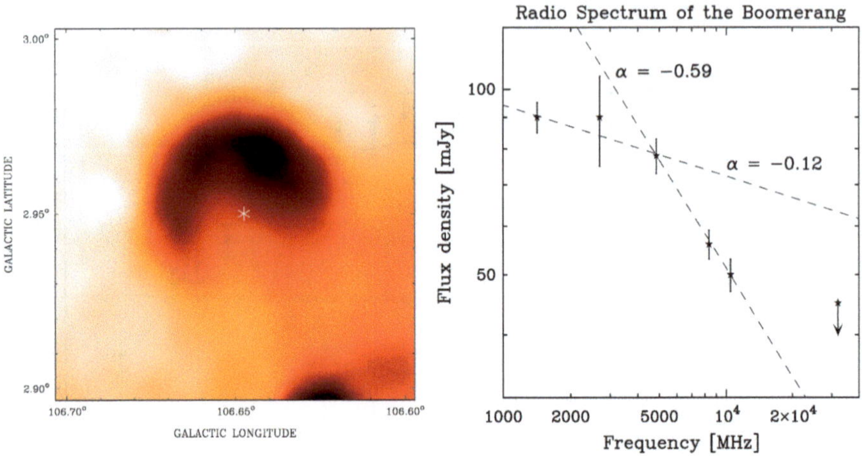

Fig. 1.11 *Left*: Radio continuum image at 1420 MHz of the Boomerang pulsar wind nebula inside the SNR G106.3+2.7 taken from Kothes et al. (2006b). *Right*: Radio continuum spectrum of the Boomerang PWN (Kothes et al. 2006b)

1.4 Magnetic Fields in PWNe

Radio polarimetry is an excellent tool to study the magneto-ionic medium and magnetic fields in particular. The synchrotron emission of pulsar wind nebulae is linearly polarized with the E-vector perpendicular to the magnetic field at the place of origin and synchrotron emission is most easily observed at radio frequencies. Therefore, from linear polarization measurements we can study the magnetic field inside pulsar wind nebulae perpendicular to the line of sight (see also Eq. (1.10)).

In addition, in the interstellar medium the polarization angle of linearly polarized emission is rotated in the presence of free thermal electrons and a magnetic field. This Faraday rotation is expressed as the rotation measure *RM*, which is defined by:

$$\phi_{\mathrm{obs}}(\lambda) = \phi_0 + RM\,\lambda^2 \ \ (\mathrm{rad}), \tag{1.15}$$

here, ϕ_0 is the intrinsic polarization angle of the linearly polarized emission and $\phi_{\mathrm{obs}}(\lambda)$ is the angle of polarization observed at wavelength λ. The Faraday rotation depends on the characteristics of the magneto-ionic media the synchrotron emission is travelling through and can be calculated via:

$$RM = \frac{e^3}{2\pi m_e^2 c^4} \int_s n_e \mathbf{B} d\mathbf{s} = 0.81 \int_s n_e B_{\parallel} d\dot{s} \ \ [\mathrm{rad\,m^{-2}}], \tag{1.16}$$

here e is the electron charge, m_e is the mass of an electron, c the vacuum speed of light, s the pathlength along the line of sight, n_e the electron density, and $B_{\parallel}$ is the magnetic field component parallel to the line of sight. The product of n_e and $B_{\|}$ is integrated from the source of the linearly polarized emission to the observer.

If we observe pulsar wind nebulae at various frequencies in the radio band we can deduce the Faraday rotation, which gives us information about the line of sight component of the magnetic field. We can also extrapolate the original polarisation angle from the observed angle and the rotation measure to study the magnetic field component perpendicular to the line of sight. In most pulsar wind nebulae the synchrotron emitting material is intermixed with the Faraday rotating plasma. Therefore we rely on models such as that published by Burn (1966) to investigate the magnetic field inside the PWNe, in particular to separate internal Faraday rotation from foreground effects.

Magnetic fields inside pulsar wind nebulae are typically assumed to be toroidal in MHD simulations, because of geometrical considerations (e.g. Begelman and Li 1992; van der Swaluw 2003). Radial magnetic fields should decrease with radius R by $B_{\rm rad} \sim R^{-2}$ and toroidal magnetic fields by $B_{\rm tor} \sim R^{-1}$. The pulsar wind emerges in a two-sided collimated outflow producing an elongated young synchrotron nebula, e.g. the Crab Nebula (Fig. 1.12) and 3C 58 (Fig. 1.13). This elongation is proposed to be the result of higher equatorial pressure associated with

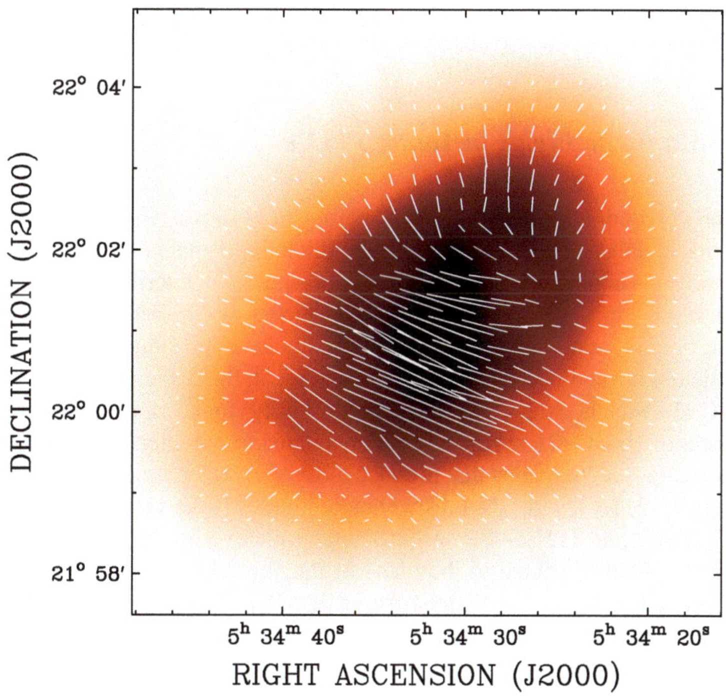

Fig. 1.12 32 GHz total power radio continuum image of the Crab Nebula observed with the 100-m Effelsberg radio telescope taken from Reich et al. (1998). Vectors are overlaid in B-field direction

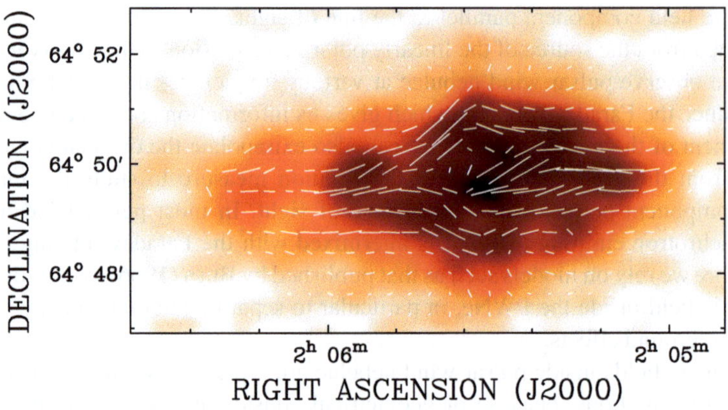

Fig. 1.13 32 GHz total power radio continuum image of the PWN 3C 58 observed with the 100-m Effelsberg radio telescope taken from Reich et al. (1998). Vectors are overlaid in B-field direction

the toroidal magnetic fields (e.g. Begelman and Li 1992; van der Swaluw 2003). Therefore toroidal magnetic fields should dominate inside the pulsar wind nebula at any significant distance from the central pulsar. However, this is not what we observe in radio polarimetric observations of PWNe.

1.4.1 Magnetic Fields in Young PWNe

In the early evolutionary stage of a PWN, before the collision with the reverse shock, the nebula mainly consists of relativistic particles and magnetic field. Therefore, very young pulsar wind nebulae do not produce significant internal Faraday rotation, because of the missing thermal electrons which are required to illuminate the magnetic field in rotation measure. The Crab Nebula is an example for such a young PWN. Bietenholz and Kronberg (1991) found that all the Faraday rotation observed in the Crab Nebula is produced externally. They found a smooth foreground component of about $-21\ \mathrm{rad\,m^{-2}}$, probably related to the interstellar medium, and small-scale thermal filaments that produce most of the depolarization.

In Figs. 1.12, 1.13, and 1.14 I display radio images of the young PWNe the Crab Nebula, 3C 58, and G21.5−0.9, respectively, with overlaid vectors, representing the magnetic field projected to the plane of the sky. These are observations from the 100-m radiotelescope in Effelsberg at 32 GHz (Reich et al. 1998). At this high radio frequency Faraday rotation is insignificant. Even a rotation measure of $\pm 1000\ \mathrm{rad\,m^{-2}}$ would produce a Faraday rotation of less than 5°. High rotation measures like that have only been observed in very dense environments like the Galactic centre.

The spin axis of the Crab pulsar B 0531+21 is parallel to the elongation of the synchrotron nebula (Ng and Romani 2004). Therefore, toroidal magnetic fields

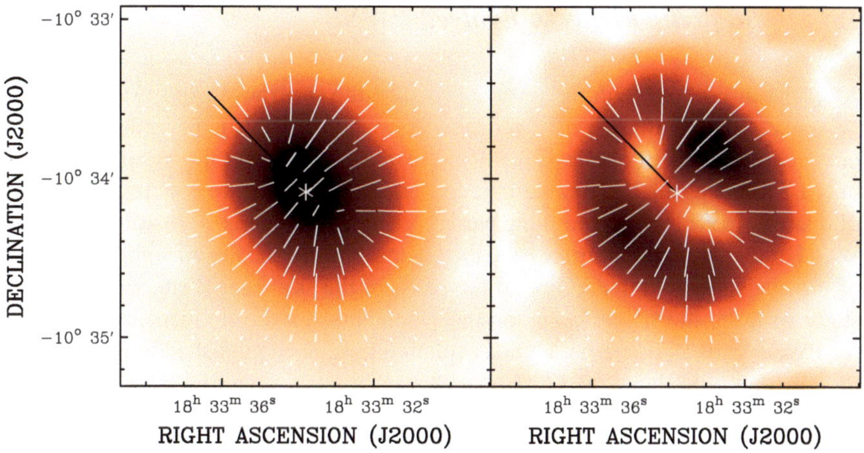

Fig. 1.14 *Left*: 32 GHz total power radio continuum image of the PWN G21.5−0.9 observed with the 100-m Effelsberg radio telescope taken from Reich et al. (1998). *Right*: Map of polarized intensity at 32 GHz of G21.5−0.9 from the same observations. Vectors are overlaid in B-field direction with the length proportional to the polarized intensity. The *white asterisk* indicates the location of pulsar J 1833−1034. The *black line* represents the direction of the pulsar's spin axis projected to the plane of the sky taken from Fürst et al. (1988)

projected to the plane of the sky should be observed perpendicular to it. In Fig. 1.12 the magnetic field in the Crab Nebula appears to be very complex in the Effelsberg observations. The Crab Nebula shows mostly toroidal fields, at least in the equatorial range, but the field is mostly radial at the edges. This is quite peculiar considering the 3-dimensional geometric expectations (see above).

In 3C 58 the spin axis of the pulsar is observed to be parallel to the elongation, similar to the Crab nebula (Slane et al. 2004). The magnetic field in 3C 58 seems to be mostly following its elongation, which is perpendicular to what is expected for a toroidal field (Fig. 1.13). In the centre part, close to the pulsar, a little bit of a tendency to a toroidal configuration can be seen. Again, at the edges of the nebula we find mostly radial magnetic fields. A magnetic field configuration similar to that in 3C 58 has also been found for the young PWN G54.1+0.3 by Lang et al. (2010).

G21.5−0.9 is another very young pulsar wind nebula, which has not interacted with the supernova reverse shock yet (Bietenholz and Bartel 2008). In the 32 GHz Effelsberg observations in Fig. 1.14 its magnetic field seems to be purely radial, projected to the plane of the sky. The spin axis of the pulsar J 1833−1034 inside G21.5−0.9 is proposed to be as indicated in Fig. 1.14 (Fürst et al. 1988). Although the magnetic field appears to be radial, in the equatorial plane of the pulsar it could be interpreted as toroidal. It may just look somewhat different than the other two elongated PWNe, because we look more along the elongation of the nebula.

If we take into account that these are 3-dimensional objects, the magnetic field for young pulsar wind nebulae derived from our radio observations seems to be quite consistent in all three examples shown here. We found an overall radial, maybe dipolar configuration with an additional toroidal component in the equatorial plane of the central pulsar.

1.4.2 Magnetic Fields in Evolved PWNe

After the interaction with the reverse shock of the supernova explosion, we find a greater variety of different emission structures and magnetic field configurations in pulsar wind nebulae. The reverse shock heated and shocked the expanding ejecta of the SN and pushed it back into the interior of the PWN. Now there is a mixture of relativistic synchrotron emitting particles and thermal Faraday rotating electrons. With Faraday rotation, we now have the means to probe the line of sight component of the magnetic field in addition to the plane of the sky component, which we can study with the linearly polarization measurements at radio frequencies. Examples for PWNe with high internal Faraday rotation have been published in Kothes et al. (2006b) and in Kothes et al. (2008).

A polarized intensity image with overlaid B-field vectors and a rotation measure map of the Boomerang PWN taken from Kothes et al. (2006b) are shown in Fig. 1.15. The magnetic field vectors clearly indicate a toroidal magnetic field for this PWN. However, according to the rotation measure map Faraday rotation must be dominated by a radial or dipolar magnetic field. A toroidal magnetic field would produce a RM gradient since it points towards us on one side and away from us on the other. A radial magnetic field has the smallest angle with the line of sight in the centre, symmetrically getting larger to either side. This produces a rotation measure peak in the centre, which is what we observe here. Evidently in this case the synchrotron emission is dominated by a toroidal magnetic field, while the line of

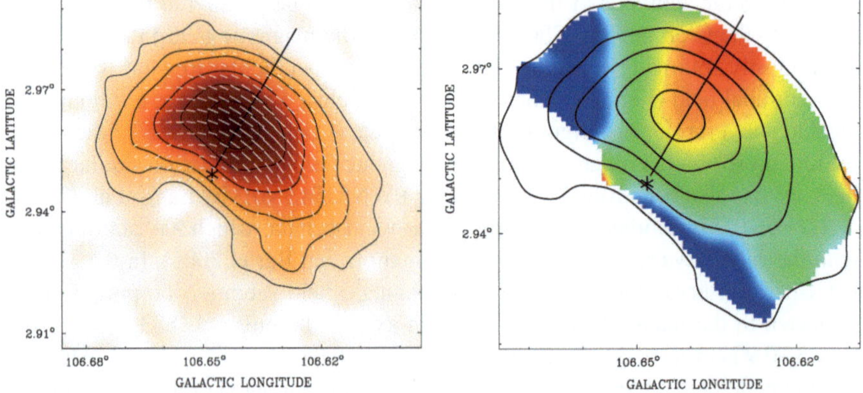

Fig. 1.15 *Left*: Radio polarized intensity image at 10.45 GHz of the Boomerang pulsar wind nebula inside the SNR G106.3+2.7 taken from Kothes et al. (2006b). The overlaid vectors are in B-field direction after the polarization angles were corrected for Faraday rotation. *Right*: Rotation measure map taken from Kothes et al. (2006b). The contours represent polarized intensity (see *left*). In both images the *black line* indicates the spin axis of the central pulsar projected to the plane of the sky as determined by Ng and Romani (2004). The location of pulsar J 0205+6449 is indicated by the *black asterisk*

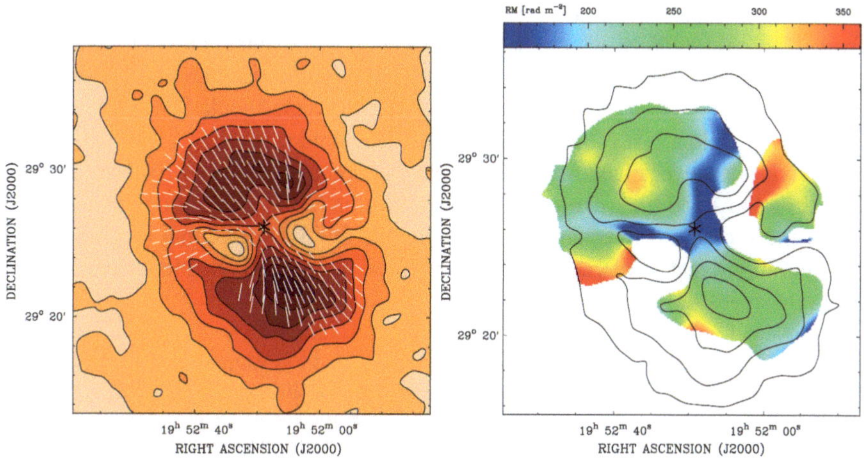

Fig. 1.16 *Left*: Radio polarized intensity image at 4850 MHz of the PWN DA 495 taken from Kothes et al. (2008). The overlaid vectors are in B-field direction after the polarization angles were corrected for Faraday rotation. *Right*: Rotation measure map taken from Kothes et al. (2008). The contours represent polarized intensity (see *left*). The location of the putative central neutron star (Arzoumanian et al. 2008) is indicated by the *black asterisk*

sight component is dominated by a radial or dipolar magnetic field. Interestingly, the Vela X pulsar wind nebula shows exactly the same structures in polarized intensity, the relative direction of the pulsar's spin axis and the observed magnetic field vectors, and the rotation measure (Dodson et al. 2003).

A polarized intensity image with overlaid B-field vectors and a rotation measure map of the PWN DA 495 taken from Kothes et al. (2008) are shown in Fig. 1.16. Here, the magnetic field vectors clearly indicate a dipolar magnetic field for this PWN. This finding is supported by the rotation measure map. Evidently in this case the dipolar magnetic field dominates the synchrotron emission and the line of sight component that produces the Faraday rotation. If we assume that the spin axis of the pulsar projected to the plane of the sky is parallel to the bridge in the centre of the polarized intensity emission (Kothes et al. 2008), there could be a faint equatorial toroidal component as well (Fig. 1.16).

Thorough radio polarisation measurements to determine 3-dimensional magnetic field structures inside PWNe are quite rare. There are a few more observations of highly evolved pulsars that exhibit bow-shock nebulae or a radio tail behind them. They also may have left a radio relic behind.

Ng et al. (2010) published a polarization study of the radio PWN G319.9−0.7, a bow-shock nebula with a long radio tail behind it. They found a magnetic field perpendicular to the elongation of the tail, which is interpreted as a helical magnetic field. Radio polarization studies of other pulsars with a radio tail reveal magnetic fields that follow the elongation of their tails, such as the mouse G359.23−0.82 (Yusef-Zadeh and Gaensler 2005), the frying pan G315.9−0.0 (Ng et al. 2012), and the snail inside SNR G327.1−1.1 (Ma et al. 2016). The snail is an extension from a radio relic PWN, which displays a rather complex magnetic field inside. Ma et al. (2016) argue that there could be differences in the magnetic field depending on whether it is driven by the pulsar wind or the reverse shock. The presence of a bow-shock could affect the shape of the magnetic field as well.

Clearly more work has to be done to find a coherent picture of PWNe magnetic field structures and their evolution. High resolution radio observations are required at a variety of radio frequencies to disentangle the 3-dimensional magnetic field configurations and construct a conclusive model of PWN magnetic fields at all different stages of their evolution.

Acknowledgements I would like to express my special thanks to Wolfgang Reich for providing his 32 GHz Effelsberg observations of the PWNe 3C 58, the Crab Nebula, and G21.5−0.9. I also would like to express my deepest appreciation to Michael Bietenholz for providing his beautiful radio map of 3C 58 for this publication. I would also like to thank Tom Landecker for careful reading of this manuscript. This research is partly based on observations with the 100-m radio telescope of the MPIfR (Max-Planck-Institut für Radioastronomie) at Effelsberg. The Dominion Radio Astrophysical Observatory is a national facility operated by the National Research Council Canada.

References

Aleksić, J., Ansoldi, S., Antonelli, L.A., Antoranz, P., Babic, A., Bangale, P., Barrio, J.A., Becerra González, J., Bednarek, W., Bernardini, E., Biasuzzi, B., Biland, A., Blanch, O., Bonnefoy, S., Bonnoli, G., Borracci, F., Bretz, T., Carmona, E., Carosi, A., Colin, P., Colombo, E., Contreras, J.L., Cortina, J., Covino, S., Da Vela, P., Dazzi, F., De Angelis, A., De Caneva, G., De Lotto, B., de Oña Wilhelmi, E., Delgado Mendez, C., Dominis Prester, D., Dorner, D., Doro, M., Einecke, S., Eisenacher, D., Elsaesser, D., Fonseca, M.V., Font, L., Frantzen, K., Fruck, C., Galindo, D., García López, R.J., Garczarczyk, M., Garrido Terrats, D., Gaug, M., Godinović, N., González Muñoz, A., Gozzini, S.R., Hadasch, D., Hanabata, Y., Hayashida, M., Herrera, J., Hildebrand, D., Hose, J., Hrupec, D., Idec, W., Kadenius, V., Kellermann, H., Kodani, K., Konno, Y., Krause, J., Kubo, H., Kushida, J., La Barbera, A., Lelas, D., Lewandowska, N., Lindfors, E., Lombardi, S., López, M., López-Coto, R., López-Oramas, A., Lorenz, E., Lozano, I., Makariev, M., Mallot, K., Maneva, G., Mankuzhiyil, N., Mannheim, K., Maraschi, L., Marcote, B., Mariotti, M., Martínez, M., Mazin, D., Menzel, U., Miranda, J.M., Mirzoyan, R., Moralejo, A., Munar-Adrover, P., Nakajima, D., Niedzwiecki, A., Nilsson, K., Nishijima, K., Noda, K., Orito, R., Overkemping, A., Paiano, S., Palatiello, M., Paneque, D., Paoletti, R., Paredes, J.M., Paredes-Fortuny, X., Persic, M., Prada Moroni, P.G., Prandini, E., Puljak, I., Reinthal, R., Rhode, W., Ribó, M., Rico, J., Rodriguez Garcia, J., Rügamer, S., Saito, T., Saito, K., Satalecka, K., Scalzotto, V., Scapin, V., Schultz, C., Schweizer, T., Shore, S.N., Sillanpää, A., Sitarek, J., Snidaric, I., Sobczynska, D., Spanier, F., Stamatescu, V., Stamerra, A., Steinbring, T., Storz, J., Strzys, M., Takalo, L., Takami, H., Tavecchio, F., Temnikov, P., Terzić, T., Tescaro, D., Teshima, M., Thaele, J., Tibolla, O., Torres, D.F., Toyama, T., Treves, A., Uellenbeck, M., Vogler, P., Zanin, R.: Discovery of TeV γ-ray emission from the pulsar wind nebula 3C 58 by MAGIC. Astron. Astrophys. **567**, L8 (2014)

Arzoumanian, Z., Safi-Harb, S., Landecker, T.L., Kothes, R., Camilo, F.: Chandra confirmation of a pulsar wind nebula in DA 495. Astrophys. J. **687**, 505–515 (2008)

Arzoumanian, Z., Gotthelf, E.V., Ransom, S.M., Safi-Harb, S., Kothes, R., Landecker, T.L.: Discovery of an energetic pulsar associated with SNR G76.9+1.0. Astrophys. J. **739**, 39 (2011)

Begelman, M.C., Li, Z.Y.: An axisymmetric magnetohydrodynamic model for the Crab pulsar wind bubble. Astrophys. J. **397**, 187–195 (1992)

Bietenholz, M.F.: Radio Images of 3C 58: expansion and motion of its wisp. Astrophys. J. **645**, 1180–1187 (2006)

Bietenholz, M.F., Kronberg. P.P.: Faraday rotation and physical conditions in the Crab Nebula. Astrophys. J. **368**, 231–240 (1991)

Bietenholz, M.F., Bartel, N.: The expansion and radio spectral index of G21.5–0.9: is PSR J1833–1034 the youngest pulsar? Astron. Astrophys. **386**, 1411–1416 (2008)

Bock, D.C.J., Gaensler, B.M.: Measurement of spectral breaks in pulsar wind nebulae with millimeter-wave interferometry. Mon. Not. R. Astron. Soc. **626**, 343–349 (2005)

Bock, D.C.J., Wright, M.C.H., Dickel, J.R.: The crab-like supernova remnant G21.5–0.9 at millimeter wavelengths. Astrophys. J. Lett. **561**, L203–L206 (2001)

Bolton, J.G., Stanley, G.J.: The position and probable identification of the source of the galactic radio-frequency radiation Taurus-A. Aust. J. Sci. Res. A Phys. Sci. **2**, 139 (1949)

Burn, B.J.: On the depolarization of discrete radio sources by Faraday dispersion. Mon. Not. R. Astron. Soc. **133**, 67 (1966)

Chevalier, R.A.: A model for the X-ray luminosity of pulsar nebulae. Astrophys. J. **539**, L45–L48 (2000)

Chevalier, R.A.: Pulsar wind nebulae: theoretical aspects and observational constraints. Adv. Space Res. **33**, 456–460 (2004)

Collins II, G.W., Claspy, W.P., Martin, J.C.: A reinterpretation of historical references to the supernova of A.D. 1054. Publ. Astron. Soc. Pac. **111**, 871–880 (1999)

Dodson, R., Lewis, D., McConnell, D., Deshpande, A.A.: The radio nebula surrounding the Vela pulsar. Mon. Not. R. Astron. Soc. **343**, 116–124 (2003)

Ferrand, G., Safi-Harb, S.: A census of high-energy observations of Galactic supernova remnants. Adv. Space Res. **49**, 1313–1319 (2012)

Frail, D.A., Kulkarni, S.R.: Unusual interaction of the high-velocity pulsar PSR1757-24 with the supernova remnant G5.4-1.2. Nature **352**, 785–787 (1991)

Frail, D.A., Giacani, E.B., Goss, W.M., Dubner, G.: The Pulsar wind nebula around PSR B1853+01 in the supernova remnant W44. Astrophys. J. Lett. **464**, L165 (1996)

Fürst, E., Handa, T., Morita, K., Reich, P., Reich, W., Sofue, Y.: Detection of axisymmetric filaments in the filled-center supernova remnant G21.5–0.9. Publ. Astron. Soc. Jpn. **40**, 347–356 (1988)

Gaensler, B.M., Slane, P.O.: The evolution and structure of pulsar wind nebulae. Annu. Rev. Astron. Astrophys. **44**, 17–47 (2006)

Gotthelf, E.V.: A spin-down power threshold for pulsar wind nebulax generation? In: Camilo, F., Gaensler, B.M. (eds.) Young Neutron Stars and Their Environments, IAU Symposium, vol. 218, p. 225 (2004)

Green, D.A.: A catalogue of 294 Galactic supernova remnants. Bull. Astron. Soc. India **42**, 47–58 (2014)

Green, D.A., Scheuer, P.A.G.: Upper limits on the infrared flux density of the 'filled-centre' supernova remnant 3C58. Mon. Not. R. Astron. Soc. **258**, 833–840 (1992)

Hewish, A., Bell, S.J., Pilkington, J.D.H., Scott, P.F., Collins, R.A.: Observation of a rapidly pulsating radio source. Nature **217**, 709–713 (1968)

Hobbs, G., Lorimer, D.R., Lyne, A.G., Kramer, M.: A statistical study of 233 pulsar proper motions. Mon. Not. R. Astron. Soc. **360**, 974–992 (2005)

Kothes, R.: A correlation between the rotational energy loss rate E of pulsars and the radio surface brightness Σ of their synchrotron nebulae. Memorie della Società Astronomica Italiana **69**, 971 (1998)

Kothes, R.: Distance and age of the pulsar wind nebula 3C 58. Astron. Astrophys. **560**, A18 (2013)

Kothes, R.: Supernova of AD 1181 and its Remnant: 3C 58. In: Alsabti, A.W., Murdin, P. (eds.) Handbook of Supernovae, pp 1–19. Springer, Cham (2016)

Kothes, R., Reich, W.: A high frequency radio study of G11.2-0.3, a historical supernova remnant with a flat spectrum core. Astron. Astrophys. **372**, 627–635 (2001)

Kothes, R., Uyaniker, B., Pineault, S.: The supernova remnant G106.3+2.7 and its pulsar-wind nebula: relics of triggered star formation in a complex environment. Astrophys. J **560**, 236–243 (2001)

Kothes, R., Fedotov, K., Foster, T.J., Uyanıker, B.: A catalogue of Galactic supernova remnants from the Canadian Galactic plane survey. I. Flux densities, spectra, and polarization characteristics. Astron. Astrophys. **457**, 1081–1093 (2006a)

Kothes, R., Reich, W., Uyanıker, B.: The Boomerang PWN G106.6+2.9 and the magnetic field structure in pulsar wind nebulae. Astrophys. J. **638**, 225–233 (2006b)

Kothes, R., Landecker, T.L., Reich, W., Safi-Harb, S., Arzoumanian, Z.: DA 495: an aging pulsar wind nebula. Astrophys. J. **687**, 516–531 (2008)

Kothes, R., Sun, X.H., Reich, W., Foster, T.J.: G141.2+5.0, a new pulsar wind nebula discovered in the cygnus arm of the milky way. Astrophys. J. Lett. **784**, L26 (2014)

Kothes, R., Reich, W., Safi-Harb, S., Matheson, H., Fürst, E.: A radio continuum and polarization study of PWN CTB 87. Astrophys. J. submitted (2017)

Landecker, T.L., Higgs, L.A., Wendker, H.J.: G76.9+1.0, a supernova remnant and unusual properties. Astron. Astrophys. **276**, 522 (1993)

Lang, C.C., Wang, Q.D., Lu, F., Clubb, K.I.: The radio properties and magnetic field configuration in the crab-like pulsar wind nebula G54.1+0.3. Astrophys. J. **709**, 1125–1137 (2010)

Ma, Y.K., Ng, C.Y., Bucciantini, N., Slane, P.O., Gaensler, B.M., Temim, T.: Radio polarization observations of the snail: a crushed pulsar wind nebula in G327.1-1.1 with a highly ordered magnetic field. Astrophys. J. **820**, 100 (2016)

Matheson, H., Safi-Harb, S., Kothes, R.: X-ray observations of the supernova remnant CTB 87 (G74.9+1.2): an evolved pulsar wind nebula. Astrophys. J. **774**, 33 (2013)

McKee, C.F., Truelove, J.K.: Explosions in the interstellar medium. Phys. Rep. **256**, 157–172 (1995)

Morsi, H.W., Reich, W.: 32 GHz radio continuum observations of four plerionic supernova remnants. Astron. Astrophys. Suppl. Ser. **69**, 533–540 (1987)

Ng, C.Y., Romani, R.W.: Fitting pulsar wind tori. Astrophys. J. **601**, 479–484 (2004)

Ng, C.Y., Gaensler, B.M., Chatterjee, S., Johnston, S.: Radio polarization observations of G319.9-0.7: a bow-shock nebula with an azimuthal magnetic field powered by pulsar J1509–5850. Astrophys. J. **712**, 596–603 (2010)

Ng, C.Y., Bucciantini, N., Gaensler, B.M., Camilo, F., Chatterjee, S., Bouchard, A.: An extreme pulsar tail protruding from the frying pan supernova remnant. Astrophys. J. **746**, 105 (2012)

Ostriker, J.P., Gunn, J.E.: On the nature of pulsars. I. Theory. Astrophys. J. **157**, 1395 (1969)

Pacini, F.: Energy emission from a neutron star. Nature **216**, 567–568 (1967)

Pacini, F., Salvati, M.: On the evolution of supernova remnants. Evolution of the magnetic field, particles, content, and luminosity. Astrophys. J. **186**, 249–266 (1973)

Perrott, Y.C., Scaife, A.M.M., Green, D.A., Davies, M.L., Franzen, T.M.O., Grainge, K.J.B., Hobson, M.P., Hurley-Walker, N., Lasenby, A.N., Olamaie, M., Pooley, G.G., Rodríguez-Gonzálvez, C., Rumsey, C., Saunders, R.D.E., Schammel, M.P., Scott, P.F., Shimwell, T.W., Titterington, D.J., Waldram, E.M., AMI Consortium: AMI Galactic Plane Survey at 16 GHz – I. Observing, mapping and source extraction. Mon. Not. R. Astron. Soc. **429**, 3330–3340 (2013)

Possenti, A., Cerutti, R., Colpi, M., Mereghetti, S.: Re-examining the X-ray versus spin-down luminosity correlation of rotation powered pulsars. Astron. Astrophys. **387**, 993–1002 (2002)

Reich, W., Fürst, E., Kothes, R.: High frequency observations of plerions and combined-type SNRs. Mem. Soc. Astron. Ital. **69**, 933 (1998)

Salter, C.J., Emerson, D.T., Steppe, H., Thum, C.: Observations at 90 and 142 GHz of nine extended galactic radio sources. Astron. Astrophys. **225**, 167–178 (1989a)

Salter, C.J., Reynolds, S.P., Hogg, D.E., Payne, J.M., Rhodes, P.J.: 84 gigahertz observations of five Crab-like supernova remnants. Astrophys. J. **338**, 171–177 (1989b)

Sedov, L.I.: Similarity and Dimensional Methods in Mechanics. Academic Press, New York (1959)

Slane, P., Helfand, D.J., van der Swaluw, E., Murray, S.S.: New constraints on the structure and evolution of the pulsar wind nebula 3C 58. Astrophys. J. **616**, 403–413 (2004)

Slane, P., Helfand, D.J., Reynolds, S.P., Gaensler, B.M., Lemiere, A., Wang, Z.: The infrared detection of the pulsar wind nebula in the galactic supernova remnant 3C 58. Astrophys. J. Lett. **676**, L33–L36 (2008)

Staelin, D.H., Reifenstein III, E.C.: Pulsating radio sources near the crab nebula. Science **162**, 1481–1483 (1968)

Stephenson, F.R., Green, D.A.: Historical supernovae and their remnants. Historical supernovae and their remnants. In: Stephenson, F.R., Green, D.A. (eds.) International Series in Astronomy and Astrophysics, vol. 5. Clarendon Press, Oxford (2002). ISBN 0198507666 5

Taylor, A.R., Gibson, S.J., Peracaula, M., Martin, P.G., Landecker, T.L., Brunt, C.M., Dewdney, P.E., Dougherty, S.M., Gray, A.D., Higgs, L.A., Kerton, C.R., Knee, L.B.G., Kothes, R., Purton, C.R., Uyaniker, B., Wallace, B.J., Willis, A.G., Durand, D.: The Canadian Galactic Plane Survey. Astron. J. **125**, 3145–3164 (2003)

Temim, T., Slane, P., Kolb, C., Blondin, J., Hughes, J.P., Bucciantini, N.: Late-time evolution of composite supernova remnants: deep chandra observations and hydrodynamical modeling of a crushed pulsar wind nebula in SNR G327.1-1.1. Astrophys. J. **808**, 100 (2015)

Torii, K., Tsunemi, H., Dotani, T., Mitsuda, K., Kawai, N., Kinugasa, K., Saito, Y., Shibata, S.: Spin-down of the 65 millisecond X-ray pulsar in the supernova remnant G11.2-0.3. Astrophys. J. Lett. **523**, L69–L72 (1999)

van der Swaluw, E.: Interaction of a magnetized pulsar wind with its surroundings. MHD simulations of pulsar wind nebulae. Astron. Astrophys. **404**, 939–947 (2003)

van der Swaluw, E., Downes, T.P., Keegan, R.: An evolutionary model for pulsar-driven supernova remnants. A hydrodynamical model. Astron. Astrophys. **420**, 937–944 (2004)

Yusef-Zadeh, F., Gaensler, B.M.: A radio study of the mouse, G359.23 - 0.82. Adv. Space Res. **35**, 1129–1136 (2005)

Chapter 2
Optical and Infrared Observations of Pulsar Wind Nebulae

Tea Temim and Patrick Slane

Abstract The extended nebulae formed as pulsar winds expand into their surroundings provide information about the composition of the winds, the injection history from the host pulsar, and the material into which the nebulae are expanding. Observations from across the electromagnetic spectrum provide constraints on the evolution of the nebulae, the density and composition of the surrounding ejecta, and the long-term fate of the energetic particles produced in these systems. Optical and infrared observations in particular reveal information that may not be observable at other wavelengths, including the presence of spectral features in the PWN particle spectrum, shocked supernova ejecta, and newly formed dust. Here we provide a summary of the growing number of optical and infrared observations of pulsar wind nebulae and what they tell us about the evolution of these systems and their host supernova remnants.

2.1 Introduction

The explosion of a supernova triggered by the collapse of a massive star produces several solar masses of stellar ejecta expanding at $\sim 10^4$ km s^{-1} into surrounding circumstellar (CSM) and interstellar (ISM) material. The resulting forward shock compresses and heats the ambient gas. As the shock sweeps up material, the deceleration drives a reverse shock back into the cold ejecta, heating the metal-enhanced gas. In many cases, though the actual fraction remains a currently-unsolved question, what remains of the collapsed core is a rapidly-spinning, highly magnetic neutron star that generates an energetic wind of particles and magnetic field confined by the surrounding ejecta. The evolution of this pulsar wind nebula

T. Temim (✉)
Space Telescope Science Institute, Baltimore, MD, USA
e-mail: ttemim@stsci.edu

P. Slane
Harvard-Smithsonian Center for Astrophysics, Cambridge, MA, USA
e-mail: slane@cfa.harvard.edu

D.F. Torres (ed.), *Modelling Pulsar Wind Nebulae*, Astrophysics and Space Science Library 446, DOI 10.1007/978-3-319-63031-1_2

(PWN) is determined by the properties of the central pulsar, its host supernova remnant (SNR), and the structure of the surrounding CSM/ISM.

Observations of PWNe from across the electromagnetic spectrum provide constraints on the evolution of the nebulae, the density and composition of the surrounding ejecta, the geometry of the central engines, and the long-term fate of the energetic particles. Optical and infrared (IR) observations, in particular, are extremely important for characterizing the PWN's evolutionary spectral breaks that are expected to occur at these wavelengths, as well as the emission from SN ejecta and dust produced when the PWNe expands into the surrounding ejecta material. In this paper, we will review the basic properties and evolution of PWNe (Sect. 2.2) and discuss what the optical and IR emission tells us about these systems (Sect. 2.3). A brief summary is presented in Sect. 2.4. We also refer the reader to the recent review by Slane (2017).

2.2 Properties and Evolution of PWNe

2.2.1 Basic Properties

At early stages of evolution, the pulsar wind is confined by the slow-moving ejecta in the host SNR, forming an expanding magnetic bubble of relativistic particles—the PWN. As the fast wind entering the nebula decelerates to meet the boundary condition imposed by the much slower expansion of the PWN, a wind termination shock (TS) is formed at a radius R_{TS} where the ram pressure of the wind is balanced by the pressure within the nebula:

$$R_{TS} = \sqrt{\dot{E}/(4\pi\omega c P_{PWN})}, \tag{2.1}$$

where ω is the equivalent filling factor for an isotropic wind and P_{PWN} is the total pressure in the nebula. The geometry of the pulsar system results in an axisymmetric wind (Lyubarsky 2002), forming a torus-like structure in the equatorial plane, along with collimated jets along the rotation axis. The higher magnetization at low latitudes confines the expansion here to a higher degree, resulting in an elongated shape along the pulsar spin axis for the large-scale nebula (Begelman and Li 1992; van der Swaluw 2003). This structure is evident in Fig. 2.1 (left), where observations of the Crab Nebula clearly reveal the jet/torus structure surrounded by the elongated wind nebula bounded by filaments of swept-up ejecta. MHD models of the jet/torus structure in pulsar winds reproduce many of the observed details of these systems (see Bucciantini 2011 for a review).

The relativistic particles in the PWN produce synchrotron radiation extending from the radio to the X-ray band, and upscatter ambient low-energy photons (from the cosmic microwave background, the stellar radiation field, and emission from ambient dust) producing inverse-Compton (IC) emission in the γ-ray band. As

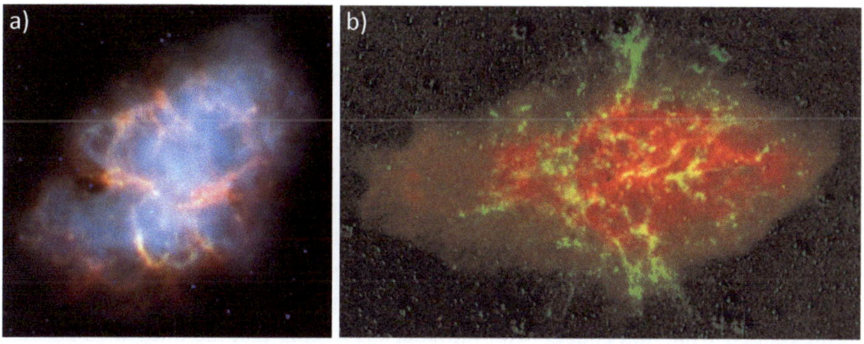

Fig. 2.1 (**a**) *Spitzer* Space Telescope composite image of Crab Nebula with the synchrotron-dominated near-IR images in *blue* and the line and dust continuum dominated 24 μm image in *red* and *yellow* (from Temim et al. 2006). (**b**) Composite image of 3C 58 with radio emission in *red* and Hα emission in *green* (from Fesen et al. 2008)

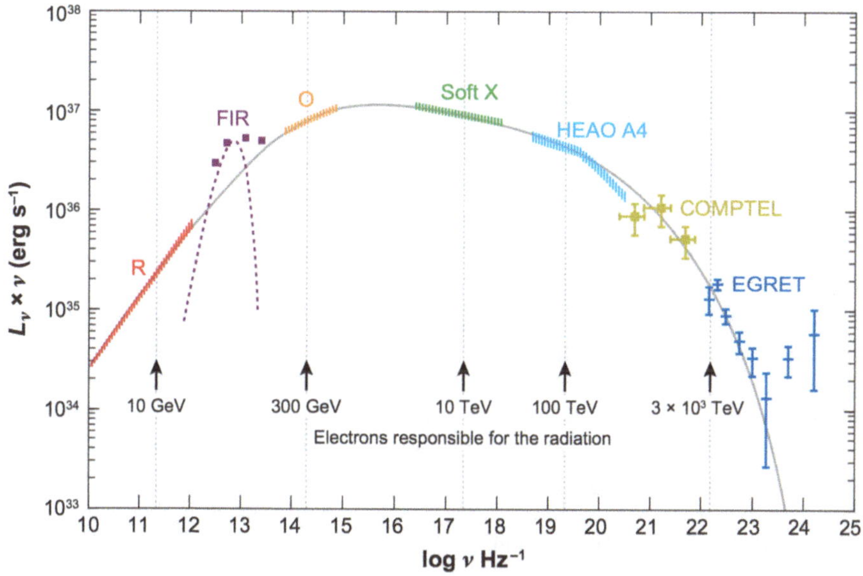

Fig. 2.2 The integrated broadband spectrum of the Crab Nebula from Atoyan and Aharonian (1996) and Hester (2008). The *black arrows* indicate the energies of the electron responsible for the emission at the particular frequencies for a magnetic field of 300 μG. The optical emission is shown in *orange* and indicates that the spectral index of the synchrotron emission breaks in this frequency range. The far-IR data in *purple* clearly show excess emission above the synchrotron continuum that is attributed to emission from dust

an example, the broadband spectrum of the Crab Nebula is shown in Fig. 2.2. Somewhere between the pulsar magnetosphere and the termination shock, the wind converts from being Poynting-dominated to being particle-dominated. Magnetic reconnection in the current sheet has been suggested as a mechanism for dissipating the magnetic field, transferring its energy into that of the particles (e.g., Lyubarsky 2003). Recent particle-in-cell simulations of relativistic shocks show that shock compression of the wind flow can drive regions of opposing magnetic fields together, causing the reconnection (Sironi and Spitkovsky 2011). This process can result in a broad particle spectrum, with a power-law-like shape $dN/dE \propto E^{-p}$ with $p \sim 1.5$. High energy particles in the equatorial regions can diffuse upstream of the shock, generating turbulence that supports acceleration of subsequent particles to high energies through a Fermi-like process, potentially creating a steeper high-energy tail with $p \sim 2.5$. The energy range spanned by the flat spectral region, and the maximum energy to which the steep spectrum extends, depend on properties of the striped wind that change with latitude, suggesting that the integrated particle injection spectrum may be quite complex (e.g., Slane et al. 2008). However, the maximum Lorentz factor that appears achievable is limited by the requirement that the diffusion length of the particles be smaller than termination shock radius; $\gamma_{max} \sim 8.3 \times 10^6 \dot{E}_{38}^{3/4} \dot{N}_{40}^{-1/2}$. This is insufficient to explain the observed X-ray synchrotron emission in PWNe, suggesting that an alternative picture for acceleration of the highest energy particles in PWNe is required (Sironi et al. 2013). As will be discussed in later sections, optical and IR observations are crucial for characterizing the particle spectra of PWNe, since spectral breaks are expected to occur in this wavelength range.

2.2.2 PWN Evolution

The evolution of a PWN within the confines of its host SNR is determined by both the rate at which energy is injected by the pulsar and by the density structure of the ejecta material into which the nebula expands. The location of the pulsar itself, relative to the SNR center, depends upon any motion given to the pulsar in the form of a kick velocity during the explosion, as well as on the density distribution of the ambient medium into which the SNR expands. At the earliest times, the SNR blast wave expands freely at a speed of $\sim$(5–10) $\times 10^3$ km s^{-1}, much higher than typical pulsar velocities of $\sim$200–1500 km s^{-1}. As a result, for young systems the pulsar will always be located near the SNR center.

The energetic pulsar wind is injected into the SNR interior, forming a high-pressure bubble that expands supersonically into the surrounding ejecta, forming a shock. The input luminosity is generally assumed to have the form (e.g. Pacini and Salvati 1973)

$$\dot{E} = \dot{E}_0 \left(1 + \frac{t}{\tau_0}\right)^{-\frac{(n+1)}{(n-1)}} \tag{2.2}$$

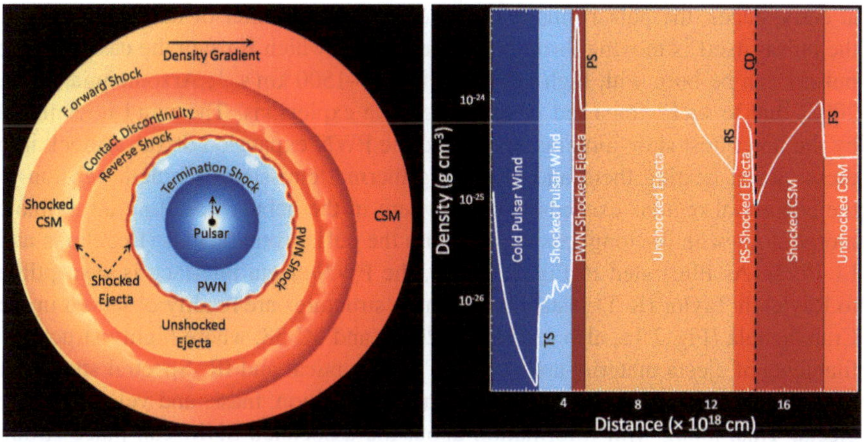

Fig. 2.3 *Panel on the left* shows the density image from a hydrodynamical simulation of a PWN expanding into an SNR. The nebula is displaced from the center of the SNR shell due to the pulsar's motion to the north and the expansion of the SNR into a CSM density gradient increasing to the right. The *right panel* shows the corresponding density profile for a radial slice through the simulated PWN/SNR system, with *colored regions* representing different physical regions identified in the SNR image. The optical and IR emission can either arise as non-thermal emission from the PWN itself (*blue* regions), the PWN-shocked ejecta region (*dark red*), or the unshocked ejecta region (*pale orange*) that might be photoionized by the PWN (from Slane 2017)

where

$$\tau_0 \equiv \frac{P_0}{(n-1)\dot{P}_0} \tag{2.3}$$

is the initial spin-down time scale of the pulsar. Here $\dot{E}_0$ is the initial spin-down power, P_0 and $\dot{P}_0$ are the initial spin period and its time derivative, and n is the so-called "braking index" of the pulsar ($n = 3$ for magnetic dipole spin-down). The pulsar has roughly constant energy output until a time τ_0, beyond which the output declines fairly rapidly with time.

Figure 2.3 (Slane 2017) illustrates the evolution of a PWN within its host SNR. The left panel shows a hydrodynamical simulation of an SNR evolving into a non-uniform medium, with a density gradient increasing from left to right. The pulsar is moving upward. The SNR forward shock, reverse shock and contact discontinuity separating the shocked CSM and shocked ejecta are identified, as is the PWN shock driven by expansion into the cold ejecta. The right panel illustrates the radial density distribution, highlighting the PWN TS as well as the SNR forward shock, contact discontinuity, and reverse shock.

The energetic pulsar wind injected into the SNR interior forms a high-pressure bubble that drives a shock into the surrounding ejecta. The sound speed in the relativistic fluid within the PWN is sufficiently high ($c_s = c/\sqrt{3}$) that any pressure variations experienced during the expansion are quickly balanced within the bubble;

at early stages, the pulsar thus remains located at the center of the PWN, even if the pulsar itself is moving through the inner SNR, which is often the case because pulsars can be born with high velocities ($\sim$200–1500 km s^{-1}; Arzoumanian et al. 2002) due to kicks imparted in the supernova explosions. The wind is confined by the innermost slow-moving ejecta, and the PWN expansion drives a shock into these ejecta, heating them and producing thermal emission. Magnetic tension in the equatorial regions exceeds that elsewhere in the nebula, resulting an a oblate morphology with the long axis aligned with the pulsar rotation axis (Begelman and Li 1992). As illustrated in Fig. 2.3 (left), the PWN/ejecta interface is susceptible to Rayleigh-Taylor (R-T) instabilities. These structures are readily observed in the Crab Nebula (Fig. 2.1a; also see Hester 2008) and 3C 58, where highly-structured filaments of ejecta material are observed in the optical and IR. Spectral studies of these filaments provide information on the composition, mass, and velocity of the ejecta. This, along with information about the associated SNR, can place strong constraints on the progenitor system.

2.3 Optical and Infrared Emission from PWNe

The emission from PWNe can be divided into two broad categories—that originating from the relativistic particles within the nebula and that produced by material that has been swept up by the nebula. Three main emission mechanisms are responsible for producing the optical and IR emission from PWNe; synchrotron emission from the PWN itself, line emission from the SN ejecta that is either shocked or photoionized by the PWN, and continuum emission from SN-formed dust that may surround the PWN.

The broadband spectrum of a PWN, along with the associated dynamical information provided by measurements of the pulsar spin properties, and the size of the PWN and its SNR, place very strong constraints on its evolution and on the spectrum of the particles injected from the pulsar. The estimates of the composition and mass of swept-up ejecta and dust can be used to probe the properties of the progenitor star. Combing the information about the PWN and its progenitor star can in turn be use to predict the long-term fate of the energetic particles in the nebula. Recent multiwavelength studies of PWNe, combined with modeling efforts of their evolution and spectra, have provided unique insights into several of these areas. Optical and IR observations of the thermal and non-thermal emission from PWNe are of particular importance, since spectral signatures that characterize the SN ejecta, dust, and the PWN's particle population lie at these wavelengths.

2.3.1 Non-thermal Emission from the PWN

The emission from relativistic particles in PWNe is a combination of synchrotron radiation and IC radiation associated with the upscattering of ambient photons. If we characterize the injected spectrum as a power law,

$$Q(E_e, t) = Q_0(t)(E_e/E_0)^{-\gamma} \tag{2.4}$$

the integrated particle energy is then

$$\int Q(E, t)EdE = (1 + \sigma)\dot{E}(t) \tag{2.5}$$

The resulting emission spectrum is found by integrating the electron spectrum over the emissivity function for synchrotron and IC radiation using, respectively, the nebular magnetic field and spectral density of the ambient photon field. The low energy particles in PWNe actually appear to have a flatter spectrum, leading to a flat radio spectrum ($\alpha \sim 0.0$–0.3 where $S_\nu \propto \nu^{-\alpha}$). Since the spectrum generally steepens between the mm and optical bands, observations of PWNe at optical and IR wavelengths are important for identifying the spectral breaks and characterizing the particle spectrum. However, detections of PWN non-thermal emission are scarce at these wavelengths, due to the effects of extinction in the optical and high-confusion at IR wavelengths. Nevertheless, PWNe of some of the brightest PWNe have been detected at optical and/or IR wavelengths, including the Crab Nebula, 3C 58, G21.5−0.9, and G292.0+1.8.

Studies of the non-thermal emission in the Crab Nebula at optical and IR wavelengths have focused on spatial variations in the spectral properties across the larger PWN (e.g., Veron-Cetty and Woltjer 1993; Temim et al. 2006) as well as the spectral properties and time variability of the small-scale knot and wisp structures (Tziamtzis et al. 2009). The global synchrotron spectral index of the Crab Nebula at optical and IR wavelengths is $\alpha = 0.8$ (Veron-Cetty and Woltjer 1993) and $\alpha = 0.5$ (Douvion et al. 2001; Temim et al. 2006), respectively. These global values represent the average indices for the entire nebula. There are however significant spatial variations, with the index generally steepening with distance from the pulsar as a result of synchrotron losses. The observed range various from 0.6 to 1.0 in the optical (Veron-Cetty and Woltjer 1993) and 0.3 to 0.8 in the IR (Temim et al. 2012a).

Detailed analyses of the spatial variations in the PWN spectrum show evidence for a superposition of multiple synchrotron components along the line of sight that may indicate the presence of distinct populations of particles. Bandiera et al. (2002) show that inhomogeneities in α between radio and mm wavelengths are best explained by a separate synchrotron component with a morphology resembling the Crab Nebula's torus in X-rays. The morphological changes of the nebula with increasing frequency would then be explained by the changes in the relative contribution of these component, with the more extended nebula dominating at

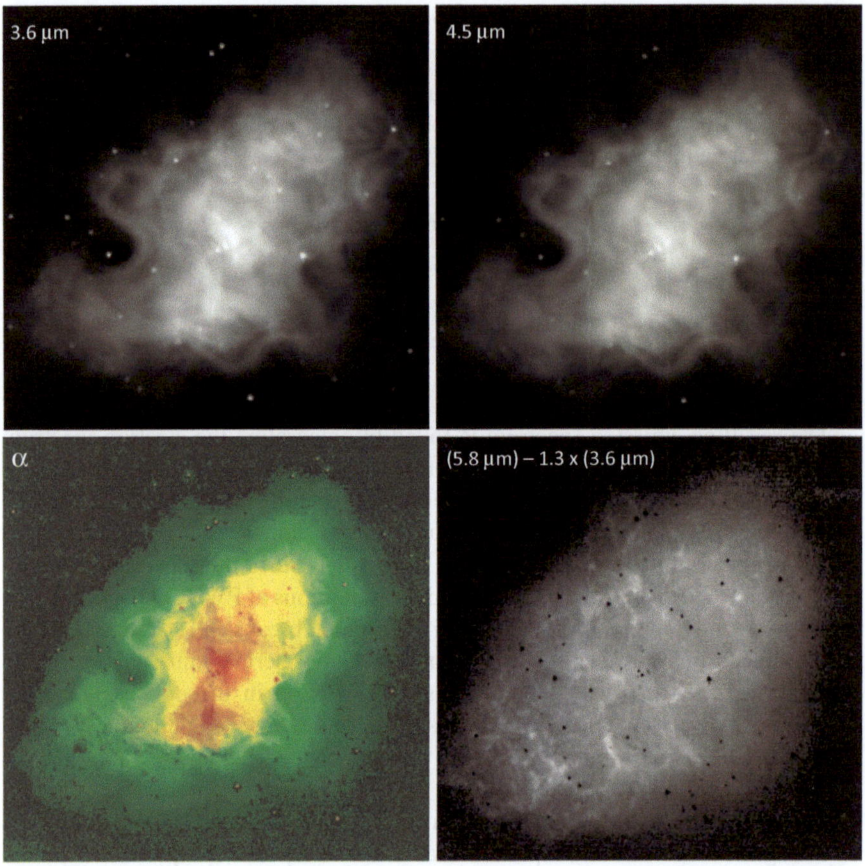

Fig. 2.4 *Spitzer* Space Telescope 3.6 and 4.5 μm images of the Crab Nebula are shown in the *top panels*. In both images, the emission almost entirely originates from synchrotron emission from the PWN. The *bottom left panel* shows the spectral index (α) map calculated from the two images, with the scale ranging from $\alpha = 0.3$ in *red* to $\alpha = 0.8$ in *dark green*. The spectral map shows evidence for two distinct spectral components, a flat torus superposed on a smooth extended nebula whose index steepens with distance from the pulsar. A flat torus spectrum is further supported by the residual image in the *bottom right panel*, showing the difference between the 5.8 μm image and the 3.6 μm multiplied by a constant of 1.3. The torus and jet structures are seen to disappear entirely from the residual image (from Temim et al. 2006)

radio and the inner torus component at X-rays wavelengths. The evidence for these components was also seen in the analysis of *Spitzer* IR observations (Temim et al. 2006). The top panels of Fig. 2.4 show the 3.6 and 4.5 μm images of the Crab Nebula that are dominated by synchrotron emission. The bottom left panel shows the spectral index map derived from these images that suggest that two distinct spectral components are responsible for the emission, a flat torus superposed on a smooth extended nebula whose index steepens with distance from the pulsar. This is more easily seen in the residual image in the bottom right panel of Fig. 2.4. When

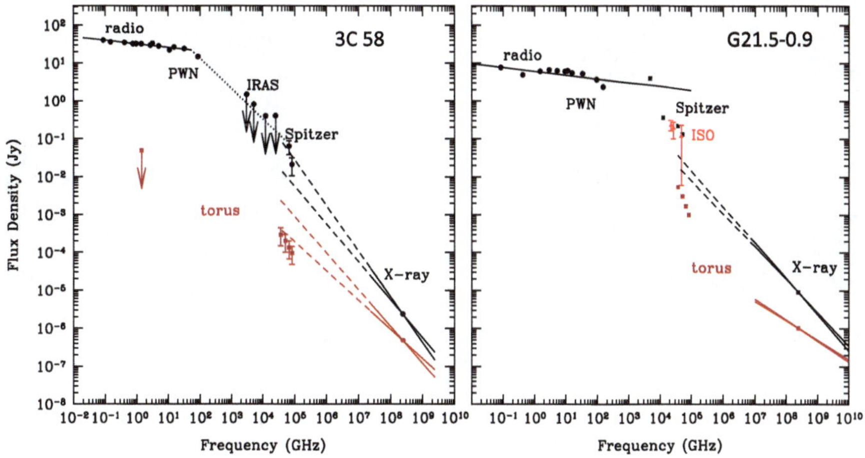

Fig. 2.5 The integrated broadband spectrum for the PWN in 3C 58 is shown in the *left panel* (Slane et al. 2008). We have compiled similar data for the PWN in G21.5−0.9, shown in the *right panel*. The measurement for the extended PWN are shown in *black* and the torus region in *red*. The *dashed lines* are extrapolations from the measured X-ray spectral index. The addition of the optical and IR data shows that the two component have different spectral properties

the scaled 3.6 μm image is subtracted from the 5.8 μm image of the nebula, the torus and jet structures disappear entirely, leaving behind a smooth extended nebula whose index steepens with distance from the pulsar.

Other PWNe also show evidence for different spectral properties between the torus and the surrounding extended nebula, including 3C 58 (Slane et al. 2008), G21.5−0.9 (Fig. 2.5), and G292.0+1.8 (Zyuzin et al. 2009; Zharikov et al. 2013). Figure 2.5 shows the integrated broadband spectra for the tori (in red) and the extended PWNe (in black) in 3C 58 and G21.5−0.9. The the two components are seen to have different spectral indices and locations of the spectral breaks. While it seems clear that the injection particle spectrum in these PWNe is a not a single power law, the exact nature of these distinct components remains an open question.

2.3.2 Emission from Shocked SN Ejecta

As the PWN expands into the surrounding SN ejecta, as described in Sect. 2.2.2, it heats the material and makes it observable at a range of wavelengths, including the optical and IR. The resulting emission, often confined to filaments, is a combination of radiation from shocked gas and continuum emission from dust condensed from the cold ejecta in the early adiabatic expansion of the SNR. The thermal emission depends on the velocity of the PWN shock driven into the ejecta which, in turn, depends on the spin-down power of the central pulsar and the density and velocity profile of the ejecta. For slow shocks, line emission may be observed in the IR and

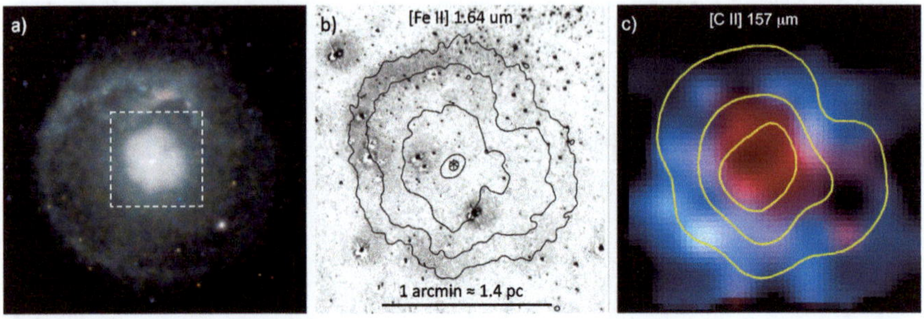

Fig. 2.6 (**a**) *Chandra* image of G21.5−.9. The pulsar is located at the center and is surrounded by a PWN. The faint outer shell is the SNR, and a portion of the emission between the PWN and the outer shell is scattered flux from the PWN. (**b**) Infrared image at 1.64 μm showing a shell of [Fe II] emission from ejecta that has been swept up and shocked by the expanding PWN (From Zajczyk et al. 2012) (**c**) *Herschel* Space Observatory image of the [C II] emission surrounding the X-ray PWN outlined in *yellow contours*. The red and blue shifted [C II] line emission is shown in *blue*, while the emission at rest is shown in *red*. The morphology is consistent with a shell of [C II] that is either shocked or radiatively heated by the PWN

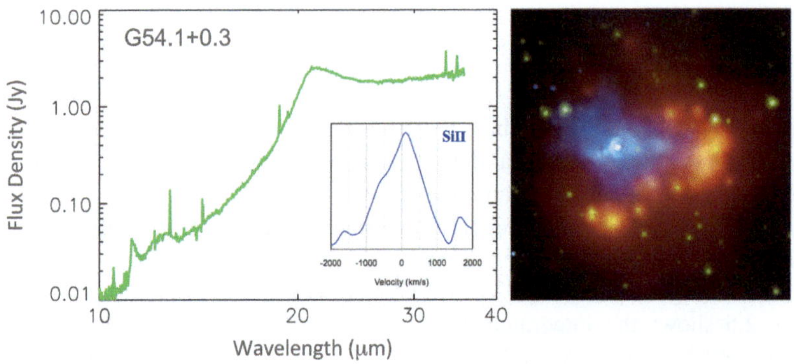

Fig. 2.7 The *Spitzer* mid-IR spectrum of the shell surrounding the pulsar wind nebula in SNR G54.1+0.3 is shown on the *left*, along with a velocity profile of the [Si II] 34.8 μm line in *blue*. The highly broadened emission line profiles indicate a shell expansion velocity of 1000 km s^{-1} and confirm that the emission arises from rapidly expanding SN ejecta. The dust continuum spectrum in *green* shows a distinct feature peaking at 21 μm, arising from the same dust species as the ejecta dust in Cas A (Temim et al. 2017). A three-color composite image of SNR G54.1+0.3 is shown on the *right*, with the *Chandra* X-ray emission in *blue*, and *Spitzer* 8 and 24 μm emission in *green* and *red*, respectively. The PWN emitting in the X-rays is surrounded by IR emission from SN ejecta and dust. The dust is primarily heated by the stars that are members of the he cluster in which the progenitor exploded. The IR point sources that stand out in *yellow* arise from the hotter dust in the immediate vicinity of the stars (Image credit: X-ray: NASA/CXC/SAO/Temim et al. (2010); IR: NASA/JPL-Caltech, also appears in Williams and Temim 2016)

optical bands, such as that observed from the Crab Nebula (Fig. 2.1), G21.5−0.9 (Fig. 2.6), and G54.1+0.3 (Fig. 2.7), while for faster shocks the emission may appear in the X-ray band, as observed in 3C 58 (Slane et al. 2004). This line emission can provide important information on the ejecta properties, such as the mass, composition, and expansion velocity.

In the Crab Nebula, for example, the total mass of the ejecta swept up by the PWN is $\sim 5\ M_{\odot}$ (Fesen et al. 1997), and the expansion velocity is ~ 1300 km s^{-1} (Temim et al. 2006). The Crab is one of a small set of young PWNe for which there is no evidence of the surrounding SNR, other than the swept-up ejecta. Other examples include 3C 58 and perhaps G54.1+0.3, although there is some evidence for radio and X-ray emission that might be associated with an SNR shell in the latter (Bocchino et al. 2010). The lack of bright (or any) SNR emission in these systems is generally assumed to result from some combination of low explosion energy, as might result from low-mass progenitors that produce electron-capture SNe, and a very low surrounding density, as could result from mass loss through stellar winds in the late phase of massive star evolution.

For the Crab Nebula, the available evidence appears to be consistent with a low-mass progenitor (Yang and Chevalier 2015). For G54.1+0.3, on the other hand, an IR shell surrounding the X-ray PWN is observed to encompass a collection of what appear to be O-type stars that presumably formed in the same stellar cluster as the PWN progenitor, indicating that this system resulted from a high mass star (Temim et al. 2010). The IR emission appears to arise from a combination of slow shocks driven into the surrounding ejecta and unshocked supernova dust that is being radiatively heated by emission from the embedded stars. *Spitzer* IR observations show highly broadened emission lines from elements such as Si, S, Cl, and Fe, that indicate expansion velocities as high as 500 km s^{-1} (see Fig. 2.7). Gelfand et al. (2015) used a one-zone model for the evolution of a PWN inside an SNR to show that this expansion velocity at the radius of the PWN in G54.1+0.3 is consistent with an ejecta mass of $10 - 15\ \mathrm{M}_{\odot}$, and a sub-energetic SN explosion with an energy of $\mathrm{E}_{51} = 0.1 - 0.2$. The line ratios in select regions of the IR shell are consistent with a PWN driving a slow 25 km s^{-1} shock into dense ejecta clumps.

The optical emission from 3C 58 shows evidence for R-T structures, as can be seen from the morphology of the Hα emission in Fig. 2.1b. Optical imaging and spectroscopy of 3C 58 indicate the presence of two distinct kinematic emission components, a ~ 250 km s^{-1} component attributed to CSM from the pre-SN mass loss, and a high velocity 770–1100 km s^{-1} ejecta component (Fesen et al. 2008). A skin of thermal X-ray emission is observed to surround the PWN (the PWN is seen in red in Fig. 2.1b), with indications of enhanced Ne and Mg abundances as would be expected from shocked ejecta (Bocchino et al. 2001; Slane et al. 2004). Ejecta abundance and mass measurements (0.5–1.0 $\mathrm{M}_{\odot}$) from the X-rays, combined with expansion measurements from the optical and the presence of slow H-rich material, indicate that 3C 58 resulted from a Type IIP SN explosion (Chevalier 2005).

For more typical systems, the ambient density (and/or SN explosion energy) is sufficiently high to form a distinct SNR shell of swept-up CSM/ISM material, accompanied by reverse-shocked ejecta, as illustrated in Fig. 2.3. An exceptional example is G21.5–0.9. X-ray observations (Fig. 2.6a) show a bright central nebula that coincides with a flat-spectrum radio nebula. The nebula is accompanied by a faint SNR shell (Slane et al. 2000; Matheson and Safi-Harb 2005), and radio timing measurements with the Parkes telescope reveal the 62 ms pulsar J1833-1034 in the center of the nebula (Camilo et al. 2006). Ground-based IR observations (Zajczyk et al. 2012) reveal a ring of [Fe II] emission associated with ejecta that has been swept up by the expanding PWN (Fig. 2.6b; contours are X-ray emission from the dashed square region from Fig. 2.6a). *Herschel* spectroscopy of the PWN region in G21.5−0.9 also shows evidence for ejecta emission. Figure 2.6c shows the [C II] 157 μm line emission map with the X-ray PWN contours overlaid in yellow. The blue- and red-shifted lines indicating expansion velocities of ∼400 km s^{-1} are shown in red, while the lines centered on the rest wavelength are shown in blue. The map is consistent with a scenario in which a shell of ejecta material surrounds the PWN. We are then seeing the high-velocity components when looking through the center of the PWN, but lower (or zero) velocities along the limb-brightened edges of the PWN. The material is either shocked and/or radiatively-heated by the PWN.

Herschel spectroscopy of the PWN in Kes 75 also shows highly broadened emission lines, the first clear evidence for SN ejecta in this SNR. Figure 2.8 shows the profile of the oxygen line at 63.2 μm, extracted from the entire PWN region. The line broadening corresponds to an ejecta expansion velocity of ∼500 km s^{-1}. Another PWN with evidence for SN-ejecta at optical and IR wavelengths is B0540−69.3 in the Large Magellanic Cloud (Morse et al. 2006; Williams et al. 2008). *Spitzer* spectroscopy indicates expansion velocities as high as 650 km s^{-1}, while the line intensities and ratios support a scenario in which the PWN drives a slow 20 km s^{-1} shock into dense ejecta clumps and produces a photoionized halo of O III emission (Williams et al. 2008). The ejecta abundance favor a 20–25 M$_{\odot}$ progenitor, consistent with a Type IIP explosion.

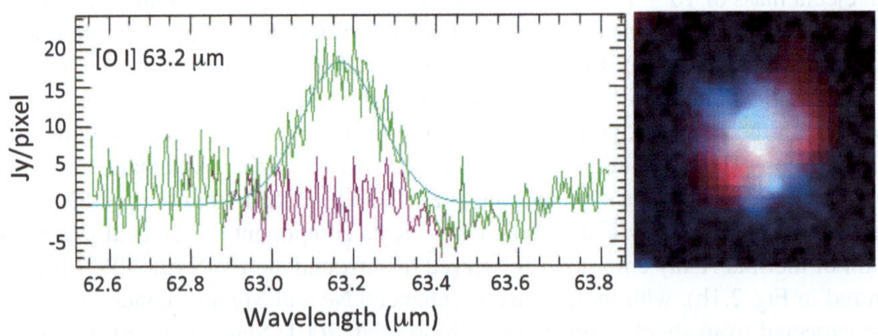

Fig. 2.8 The *left panel* shows the average *Herschel* spectrum centered on the [O I] line, covering the Kes 75 PWN region displayed in the *right panel*. The line originates from the shell of ejecta emission around the PWN and has a FWHM of 1000 km s^{-1}. The *right panel* image of the PWN in Kes 75 shows the *Chandra* X-ray emission in *blue* and *Herschel* 70 μm emission in *red*. The IR emission that surrounds the PWN is likely dominated by continuum emission from dust that condensed from the SN ejecta

2.3.3 *Emission from SN-Formed Dust*

Young PWNe are also ideal laboratories for studying the properties of SN-condensed dust grains. Despite the fact that interstellar dust plays a crucial role in the astrophysical processes governing galaxy evolution, the origin and nature of dust in the ISM is still not well understood. Theoretical dust condensation models suggest that SNe are capable of producing large quantities of dust, up to 1 $M_{\odot}$. However, observational evidence for significant quantities of dust has so far only been found in a couple of systems. What fraction of the dust formed in a SN explosion will eventually survive the SN reverse shock to be injected into the ISM is also an open question (for a review of formation and destruction of dust in SNRs, see Williams and Temim 2016).

Dust grains are expected to form in the SN ejecta on timescales of a few hundred days after the explosion and then rapidly cool to far-IR and submillimeter-emitting temperatures. As young PWNe expand into the surrounding SN ejecta, they also heat the dust grains to higher temperatures, making them observable at mid-IR wavelengths where spectral signatures are found that may reveal the composition of the grains. Measurements of emission from ejecta dust prior to interaction with the SN reverse shock are of particular importance in estimating dust formation rates in SNe. Properties of the pristine SN-formed dust that has not yet been processed by shocks or mixed with the CSM may also provide important information about the SN progenitor.

Dust that surrounds a PWN is either collisionally heated by the PWN-shocked gas or radiatively heated by the synchrotron emission. The emission is in the form of a blackbody-like spectrum whose properties depend on the temperature, composition, and grain-size distribution of the dust grains. There are now a handful of PWNe that show clear evidence for the presence of ejecta dust in their immediate surroundings. The best known example is the Crab Nebula, in which the dust grains reside in the cores of the ejecta filaments seen in Fig. 2.1a. The presence of dust was originally inferred from an IR excess above the synchrotron emission (e.g. Glaccum et al. 1982; Marsden et al. 1984) and absorption features across he optical ejecta filaments (e.g. Woltjer and Veron-Cetty 1987; Fesen and Blair 1990). Observations with *Spitzer* confirmed that the dust continuum emission spatially correlates with the ejecta filaments, and revealed a featureless spectrum that could be described by 10^{-3} – 10^{-2} $M_{\odot}$ of warm (50–60 K) carbon or silicate dust (Temim et al. 2012b). A much larger quantity of colder dust was discovered by *Herschel* far-IR observations of the Crab Nebula, covering wavelengths up to 500 μm (Gomez et al. 2012). The total mass was derived from a two-temperature fit to the spectral energy distribution (SED) of the dust and found to be 0.1–0.25 $M_{\odot}$. A PWN-dust heating model that invoked a continuous distribution of grain sizes and temperatures, shown in Fig. 2.9, resulted in at least a factor of two reduction in dust mass, with an estimated value of 0.02–0.13 $M_{\odot}$ (Temim and Dwek 2013). Since the Crab Nebula is thought to have resulted from a low mass progenitor (see Sect. 2.3.2), this lower

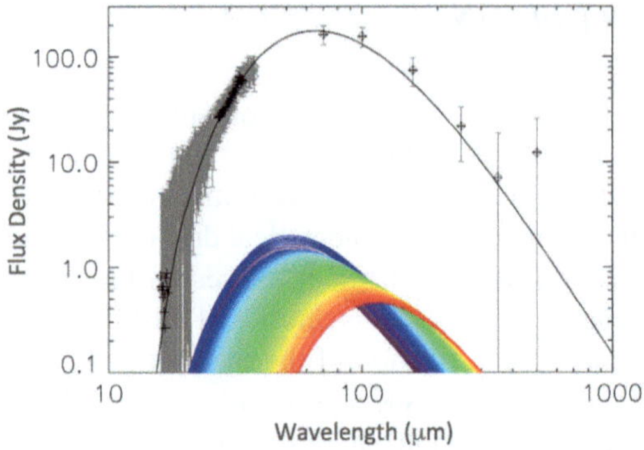

Fig. 2.9 IR spectrum and flux densities of the synchrotron- and line-subtracted dust continuum emission from the Crab Nebula's filaments, as observed with *Spitzer* and *Herschel*. The emission has been modeled with a continuous size-distribution of dust grains that are radiatively heated by the synchrotron emission from the PWN. The *rainbow-colored curves* represent the emission from the individual dust grains, while the *black solid line* that fits the data represents the sum of these individual components. The total dust mass found by this method is a factor of ~2 lower than for a two-temperature model using the same composition and is an equally good fit to the data (from Temim and Dwek 2013)

dust mass may be expected. The heating model also implies that the grains are relatively large, which would be expected for a Type IIP progenitor.

The IR shell surrounding the PWN in G54.1+0.3 (see Fig. 2.7) also contains SN-formed dust. *Spitzer* spectroscopy of the shell shows that the continuum is characterized by a distinctive dust feature that peaks at 21 μm and is attributed to a magnesium silicate grain species that is also found in the ejecta of Cas A. If this grain species is responsible for producing the spectral feature, it would dominate the dust mass in the shell. The total dust mass estimated from the fitting of the mid- to far-IR SED of the shell is at least $M_\odot$, comparable to the dust mass in SN 1987A (Temim et al. 2017). However, accounting for a distribution of grain sizes and properties with a self-consistent grain heating model may lower this estimate. Dust properties and masses found by this particular model suggest a progenitor mass of 16–27 $M_\odot$. Figure 2.10 shows the dust temperature map for the G54.1+0.3 shell produced by spatially resolved fitting of the IR SED. The locations of the stars are marked by white "x" symbols. The dust temperature clearly peaks at the locations of the stars, confirming that the dust is primarily heated by the surrounding O and B stars, as discussed in Sect. 2.3.2.

Other PWNe that show surrounding IR emission from dust include B0540−69.3, Kes 75, and G21.5−0.9. They all show a dust continuum excess above the power-law arising from synchrotron emission. Williams et al. (2008) analyzed the *Spitzer* observations of B0540−69.3 and found a fairly small amount of SN-ejecta dust (10^{-3} $M_\odot$) that is shock-heated by the PWN. The distribution of dust in Kes 75 is seen in the right panel of Fig. 2.8, where the X-ray emission from the PWN is shown in blue and IR emission from dust in red. The IR emission forms a shell around the

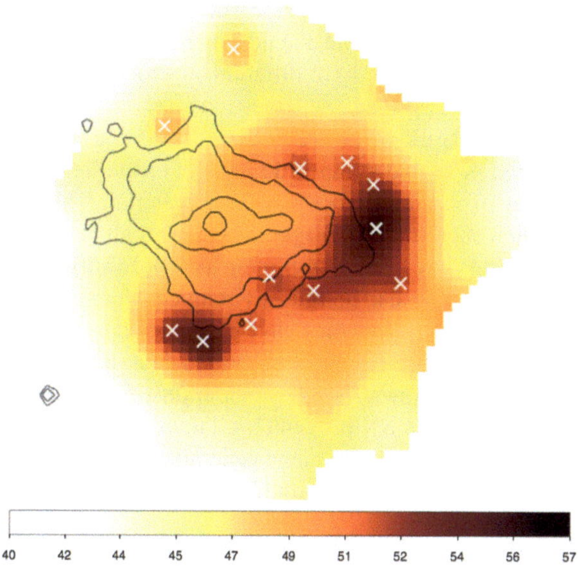

Fig. 2.10 Temperature map in units of degrees Kelvin for the dust surrounding the PWN in G54.1+0.3. The "*times*" symbols represent the location of the O and B stars, while the *black contours* represent the X-ray PWN. The temperature peaks at the location of the stars, implying that these O and B stars are the primary heating sources for the dust (from Temim et al. 2017)

PWN and likely originates from SN-formed dust. These systems have also been detected by far-IR *Herschel* observations that may uncover larger masses of colder dust. Future studies of these and other systems will improve our understanding of the dust properties that form in SN ejecta and help answer the question of what fraction of SN-formed dust will survive the eventual passage of the SN reverse shock and be injected into the ISM.

2.4 Summary

The evolution and structure of a PWN are determined by both the properties of the host pulsar and the surrounding SNR environment into which the nebula expands. Observations across the electromagnetic spectrum allow us to constrain the nature and evolution of PWNe, and the properties of the ejecta of the SNRs in which they were born. These observations continue to inform theoretical models of relativistic shocks which, in turn, have broad importance across the realm of high-energy astrophysics. Optical and IR observations are of particular importance, since they reveal properties that may not be observable at other wavelengths. Measurements of the non-thermal emission from PWNe at these wavelengths provide information on the evolutionary spectral breaks and the particle injection spectrum. These observations show that the particle injection spectrum is not a simple power-law and that the well-defined tori and more extended nebula appear to arise from distinct particle populations and generally have different spectral properties. The nature of these distinct components is still an open question.

Optical and IR observations of the thermal emission around PWNe yield information about the mass and composition of SN ejecta into which the nebulae expand, and on the expansion velocity of that material. This in turn leads to important information about the SN progenitors. So far, all but one (Kes 75) of the PWNe with surrounding optical and IR emission have been attributed to Type IIP SN explosions. Young PWNe are also unique environments for studying the properties of SN-condensed dust grains before they have been altered by the SN reverse shock. A number of PWNe show evidence for SN-formed dust in the surrounding ejecta, whose properties not only allow us to learn about dust formation in SNe, but also provide information about the SN progenitors. Future studies of PWNe at optical and IR wavelengths, particularly using new generations of instruments with higher sensitivity and spatial resolution (such as the JWST), will lead to significant progress in all of these areas.

References

Arzoumanian, Z., Chernoff, D.F., Cordes, J.M.: The velocity distribution of isolated radio pulsars. Astrophys. J. **568**, 289–301 (2002)

Atoyan, A.M., Aharonian, F.A.: On the mechanisms of gamma radiation in the Crab Nebula. Mon. Not. R. Astron. Soc. **278**, 525–541 (1996)

Bandiera, R., Neri, R., Cesaroni, R.: Astron. Astrophys. **386**, 1044 (2002)

Begelman, M.C., Li, Z.-Y.: An axisymmetric magnetohydrodynamic model for the Crab pulsar wind bubble. Astrophys. J. **397**, 187–195 (1992)

Bocchino, F., Warwick, R.S., Marty, P., Lumb, D., Becker, W., Pigot, C.: The X-ray nebula of the filled center supernova remnant 3C 58 and its interaction with the environment. Astron. Astrophys. **369**, 1078 (2001)

Bocchino, F., Bandiera, R., Gelfand, J.D.: XMM-Newton and SUZAKU detection of an X-ray emitting shell around the pulsar, wind nebula G54.1+0.3. Astron. Astrophys. **520**, A71 (2010)

Bucciantini, N.: MHD models of pulsar wind nebulae. Astrophys. Space Sci. Proc. **21**, 473 (2011)

Camilo, F., Ransom, S.M., Gaensler, B.M., Slane, P., Lorimer, D.R., Reynolds, J., et al.: PSR J1833–1034: discovery of the central young pulsar in the supernova remnant G21.5–0.9. Astrophys. J. **637**, 456–465 (2006)

Chevalier, R.: Young core-collapse supernova remnants and their supernovae. Astrophys. J. **619**, 839 (2005)

Douvion, T., Lagage, P.O., Cesarsky, C.J., Dwek, E.: Dust in the Tycho, Kepler and Crab supernova remnants. Astron. Astrophys. **373**, 281 (2001)

Fesen, R., Blair, W.P.: Optical identification of dust within the Crab Nebula's filaments. Astrophys. J. Lett. **351**, L45 (1990)

Fesen, R.A., Shull, J.M., Hurford, A.P.: An optical study of the circumstellar environment around the crab nebula. Astron. J. **113**, 354–363 (1997)

Fesen, R., Rudie, G., Hurford, A., Soto, A.: Optical imaging and spesctoscopy of the galactic supernova remnant 3C 58 (G130.7+3.1). Astrophys. J. Suppl. Ser. **174**, 379 (2008)

Gelfand, J.D., Slane, P.O., Temim, T.: The properties of the progenitor supernova, pulsar wind, and neutron star, inside PWN G54.1+0.3. Astrophys. J. **807**, 30 (2015)

Glaccum, W., Harper, D.A., Loewestein, R.F., Pernic, R., Low, F.J.: Far infrared measurements of the crab nebula. Bull. Am. Astron. Soc. **14**, 612 (1982)

Gomez, H.L., et al.: A cool dust factory in the crab nebula: a Herschel study of the filaments. Astrophys. J. **760**, 96 (2012)

Hester, J.J.: The crab nebula: an astrophysical chimera. Annu. Rev. Astron. Astrophys. **46**, 127–155 (2008)

Lyubarsky, Y.E.: On the structure of the inner Crab Nebula. Mon. Not. R. Astron. Soc. **329**, L34–L36 (2002)

Lyubarsky, Y.E.: The termination shock in a striped pulsar wind. Mon. Not. R. Astron. Soc. **345**, 153 (2003)

Marsden, P.L., Gillett, F.C., Jennings, R.E., Emerson, J.P., de Jong, T., Olnon, F.M.: Far-infrared observations of the Crab nebula. Astrophys. J. **278**, 29 (1984)

Matheson, H., Safi-Harb, S.: The plerionic supernova remnant G21.5–0.9: in and out. Adv. Space Res. **35**, 1099–1105 (2005)

Morse, J.A., Smith, N., Blair, W.P., Kirshner, R.P., Winkler, P.F., Hughes, J.P.: Hubble space telescope observations of oxygen-rich supernova remnants in the magellanic clouds. III. WFPC2 imaging of the young, crab-like supernova remnant SNR 0540–69.3. Astrophys. J. **644**, 188 (2006)

Pacini, F., Salvati, M.: On the evolution of supernova remnants. Evolution of the magnetic field, particles, content, and luminosity. Astrophys. J. **186**, 249–266 (1973)

Sironi, L., Spitkovsky, A.: Acceleration of particles at the termination shock of a relativistic striped wind. Astrophys. J. **741**, 39 (2011)

Sironi, L., Spitkovsky, A., Arons, J.: The maximum energy of accelerated particles in relativistic collisionless shocks. Astrophys. J. **771**, 54 (2013)

Slane, P.: Pulsar Wind Nebulae. In: Handbook of Supernovae. Springer, Switzerland (2017)

Slane, P., Chen, Y., Schulz, N.S., Seward, F.D., Hughes, J.P., Gaensler, B.M.: Chandra observations of the crab-like supernova remnant G21.5–0.9. Astrophys. J. **533**, L29–L32 (2000)

Slane, P., Helfand, D.J., van der Swaluw, E., Murray, S.S.: New constraints on the structure and evolution of the pulsar wind nebula 3C 58. Astrophys. J. **616**, 403–413 (2004)

Slane, P., Helfand, D.J., Reynolds, S.P., Gaensler, B.M., Lemiere, A., Wang, Z.: The infrared detection of the pulsar wind nebula in the galactic supernova remnant 3C 58. Astrophys. J. **676**, L33 (2008)

Temim, T., Dwek, E.: The importance of physical model for deriving dust masses and grain size distributions in supernova eject: radiatively heated dust in the crab nebula. Astrophys. J. **774**, 8 (2013)

Temim, T., Gehrz, R.D., Woodward, C.E., Roellig, T.L., Smith, N., Rudnick, L., et al. Spitzer space telescope infrared imaging and spectroscopy of the crab nebula. Astron. J. **132**, 1610–1623 (2006)

Temim, T., Slane, P., Reynolds, S.P., Raymond, J.C., Borkowski, K.J.: Deep chandra observations of the crab-like pulsar wind nebula G54.1+0.3 and spitzer spectroscopy of the associated infrared shell. Astrophys. J. **710**, 309–324 (2010)

Temim, T., Slane, P., Arendt, R., Dwek, E.: Infrared and X-ray spectroscopy of the Kes 75 supernova remnant shell: characterizing the dust and gas properties. Astrophys. J. **753**, 72 (2012a)

Temim, T., Sonneborn, G., Dwek, E., Arendt, R.G., Gehrz, R.D., Slane, P., Roellig, T.L.: Properties and spatial distribution of dust emission in the crab nebula. Astrophys. J. **745**, 46 (2012b)

Temim, T., Dwek, E., Arendt, R.G., Borkowski, K.J., Reynolds, S.P., Slane, P., Geldand, J.D., Raymond, J.C.: A massive shell of supernova-formed dust in SNR G54.1+0.3. Astrophys. J. **836**, 129 (2017)

Tziamtzis, A., Lundqvist, P., Djupvik, A.A.: The Crab pulsar and its pulsar-wind nebula in the optical and infrared. Astron. Astrophys. **508**, 221 (2009) van der Swaluw, E.: Interaction of a magnetized pulsar wind with its surroundings. MHD simulations of pulsar wind nebulae. Astron. Astrophys. **404**, 939–947 (2003)

Veron-Cetty, M.P., Woltjer, L.: Spectrophotometry of the continuum in the crab nebula. Astron. Astrophys. **270**, 370 (1993)

Williams, B.J., Temim, T.: Infrared emission from supernova remnants: formation and destruction of dust. In: Handbook of Supernovae. Springer, Switzerland (2016). doi:10.1007/978-3-319-20794-0_94-1

Williams, B.J., et al.: Ejecta, dust, and synchrotron radiation in SNR B0540-69.3: a more crab-like remnant than the crab. Astrophys. J. **687**, 1054 (2008)

Woltjer, L., Veron-Cetty, M.P.: Discovery of continuum emission in the jet and of absorption in the filaments of the Crab Nebula. Astron. Astrophys. **172**, 7 (1987)

Yang, H., Chevalier, R.C.: Evolution of the crab nebula in a low energy supernova. Astrophys. J. **806**, 153 (2015)

Zajczyk, A., Gallant, Y.A., Slane, P., Reynolds, S.P., Bandiera, R., Gouiffs, C.: Infrared imaging and polarimetric observations of the pulsar wind nebula in SNR G21.5-0.9. Astron. Astrophys. **542**, A12 (2012)

Zharikov, S.V., Zyuzin, D.A., Shibanov, Y.A., Mennickent, R.E.: The PSR J1124-5916 wind nebula in the near-infrared. Astron. Astrophys. **554**, A120 (2013)

Zyuzin, D.A., Danilenko, A.A., Zharikov, S.V., Shibanov, Y.A.: Astron. Astrophys. **508**, 855 (2009)

Chapter 3
Pulsar Wind Nebulae Viewed Through the Gamma-Ray Eye

Fabio Acero

Abstract This chapter reviews the recent advancements in the observations of pulsar wind nebulae using gamma-ray instruments and put them in context of all classes of objects observed in gamma-rays.

3.1 Introduction

In the last decade, our view of the gamma-ray sky has dramatically improved in terms of sensitivity and angular resolution both at high (HE: 100 MeV–100 GeV) and very-high energies (VHE: 100 GeV–100 TeV). Progress made on the ground with stereoscopic Cherenkov telescopes (e.g. MAGIC, HESS, VERITAS) observing at VHE and the launch in 2008 of the Large Area Telescope (LAT) operating at HE onboard the *Fermi* satellite have led to the current golden age of gamma-ray astronomy allowing to study gamma-ray sources over six orders of magnitude in energy when combining ground and space telescopes! Those new gamma-ray eyes have provided us a much deeper view of the most energetic objects in our Galaxy and in our Universe. Together with gamma-ray bursts and accreting compact sources, stellar explosions and their remnants are amongst the most energetic and violent phenomena known to date. In the case of core-collapse supernova, a rapidly spinning, highly magnetized neutron star (a pulsar) can be formed as a leftover of the stellar explosion. The pulsar and it's environment are the site of efficient particle acceleration and can result in intense gamma-ray emission.

Pulsed non-thermal emission from particles accelerated in the magnetosphere of the pulsar can be seen radiating from radio to gamma-ray energies. In addition a fraction of the rotational energy of the pulsar is transferred to a relativistic wind that will interact with the termination shock where particles will be accelerated. The role of diffuse shock acceleration and/or magnetic field reconnection in the acceleration mechanism is still unclear and is discussed in more details in this book.

F. Acero (✉)
Laboratoire AIM, IRFU/SAp - CEA/DRF - CNRS, Université Paris Diderot, Bat. 709, CEA-Saclay, Gif-sur-Yvette Cedex, France
e-mail: fabio.acero@cea.fr

D.F. Torres (ed.), *Modelling Pulsar Wind Nebulae*, Astrophysics and Space Science Library 446, DOI 10.1007/978-3-319-63031-1_3

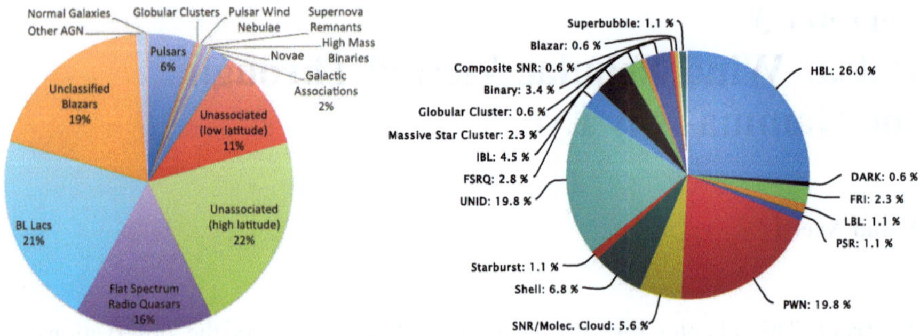

Fig. 3.1 Fraction of different source classes at HE (*left panel*) from the *Fermi*-LAT 3FGL catalog (reproduced from Bühler 2015) and from the TeVCat* online catalog (*right panel*). *http://tevcat.uchicago.edu

The shocked pulsar wind is confined by the pressure of the surrounding supernova remnant and creates a bubble of highly energetic particles: the pulsar wind nebula (PWN).

In young PWNe, the synchrotron emission from the accelerated particles peaks in the X-ray band and extends up to several MeV (a few 100 MeV in the case of the Crab nebula) while the inverse Compton (IC) scattering off ambient photon field can produce photons reaching TeV energies. While the X- and gamma-ray emission from PWNe both reveal highly energetic particles, the electrons responsible of the IC emission at 1 TeV are typically less energetic ($\sim$10 TeV for CMB target photons) than the electrons emitting synchrotron radiation at 1 keV ($\sim$50 TeV for a typical magnetic field of 10 μG). Due to this difference in energy, the lifetime of gamma-ray emitting electron is larger than that of X-ray emitting electrons. This makes the gamma-ray energy range perfectly suited to search for PWNe as these nebulae act like calorimeter in this energy range and the old (relic) nebulae can still be bright gamma-ray emitters while being very faint or extinct in X-rays.

The HESS galactic plane survey has revealed a high number of PWNe and they represent the most numerous class of VHE galactic source ($\sim$20% as shown in Fig. 3.1, right panel). While the lifetime of HE emitting electrons is larger than the lifetime of VHE emitting electrons, the source population in the galactic plane as revealed by *Fermi*-LAT is not dominated by PWNe where they represent only about 1% of the sources (see Fig. 3.1, left panel). One of the main reason for this striking difference is that the magnetospheric emission of the pulsar peaks at GeV energies (e.g. the *Fermi*-LAT PSR catalog: Abdo et al. 2013) and outshines the nebula in this energy range. With 205 pulsars reported as of February 2016,[1] they represent the most numerous class of galactic source at high energies.

[1]https://confluence.slac.stanford.edu/display/GLAMCOG/Public+List+of+LAT-Detected+Gamma-Ray+Pulsars.

In addition to the pulsar, the host supernova remnant (SNR) surrounding the nebula can emit bright HE emission peaking around a few GeV for middle aged SNRs (e.g. the *Fermi*-LAT SNR catalog: Acero et al. 2016). Due to the limited angular resolution of the *Fermi*-LAT instrument[2] (PSF $\sim 1^\circ$ at 1 GeV), disentangling spatially the PSR, PWN, and SNR is not a simple task.

A large fraction of the Galactic sources at HE and VHE are unknown in nature and are labeled as unidentified. A description of this population and the link with PWNe is described in Sect. 3.2 and different approach to tackle this issue are discussed in Sect. 3.3. A demographic study of the gamma-ray emitting PWNe is presented in Sect. 3.4, a brief introduction to the unexpected flares of the Crab nebula in Sect. 3.5, and a description of a possible new type of nebulae around magnetars in Sect. 3.6.

3.2 Unidentified Sources and PWNe

Unidentified sources—sources that are currently not firmly identified with plausible counterparts at other wavelengths—represent about a third of the sources at HE in the *Fermi*-LAT 3FGL catalog (Acero et al. 2015) and about 20% at VHE as shown in Fig. 3.1. With a rate of about one pulsar per century (Faucher-Giguère and Kaspi 2006) in our Galaxy, and a lifetime of the PWN at VHE of the order of 50–100 kyrs, the expected number of gamma-ray emitting PWNe is of the order of 500–1000. Taking into account that relic nebula are no longer confined by the host SNR and are usually extended up to degrees scale, they densely populate the Galactic plane. In addition, besides a generous amount of gamma-ray emission, the associated pulsar is faint and can be well offset from the center of the nebula (see Sect. 3.4.2).

Concerning the multi-wavelength counterparts of the relic nebula, they can be hard to detect in X-rays as the electron population has cooled down (due to radiative losses) and the peak of synchrotron emission has shifted to the optical/UV regime where the nebula faint and diffuse emission is difficult to detect. For the aforementioned reasons, PWNe are natural candidates to explain at least a fraction of the VHE unidentified sources (as in the case of HESS J1303−631 discussed in Sect. 3.4.2).

The next generation of VHE observatory, the Cherenkov Telescope Array (CTA) will provide a gain of sensitivity of about an order of magnitude while improving the angular resolution by about a factor 2 from slightly better than 0.1° for current generation telescope to $\sim 0.05^\circ$ [3] (somewhat better resolution could be achieved with PSF optimized analysis at the expense of some effective area). The much increased

[2]Sixty-eight percentage of containment radius. Reference: http://www.slac.stanford.edu/exp/glast/groups/canda/lat_Performance.htm.

[3]Quoting the 68% containment radius at 1 TeV from: https://portal.cta-observatory.org/Pages/CTA-Performance.aspx.

sensitivity should reveal a large number of new sources in the Galactic plane survey but due to the limited increase in angular resolution, the source confusion in the plane will probably increase and the accumulation of unresolved VHE relic nebula might contribute to create a Galactic gamma-ray background (similar to what is seen in X-rays). In order to help disentangling the PWNe in the plane, the new generation of radio telescopes (LOFAR, SKA) can help to search for the associated pulsar and the current *XMM-Newton* and *Chandra* X-ray telescopes and their 16 years of data archive (and the eRositta X-ray survey telescope with a target launch in 2017) to identify the faint and compact relic nebula. Both the radio and the X-rays will provide key multi-wavelength observations to search for plausible counterparts and resolve this diffusion emission. Another powerful tool to identify relic nebulae amongst unassociated sources in the plane is to take advantage of the three decades in energy (~50 GeV–50 TeV) covered by CTA and test the energy-dependent morphology of the relic nebula that shrinks with energy closer to the pulsar powering the system.

In conclusion, CTA will be a PWN discovering machine with the capacities to detect several hundreds of new nebulae (Renaud and CTA Consortium 2011). This will allow us to perform deep population studies of PWNe at gamma-ray energies but they will potentially create a source of diffuse emission in the Galactic plane that will bring the issue of source confusion that can be tackled with the solutions described above.

3.3 The HE to VHE Connection

When the pulsar timing solution is known, it is possible to reduce it's contamination by searching for additional sources in the off-pulse time interval at the cost of significantly reducing the exposure time. A different approach is to look only at data above 10 GeV where the pulsed emission contribution is drastically reduced. This method was applied in Acero et al. (2013) and revealed 30 HE sources out of 58 VHE sources investigated. Among those 30 sources, the study confirmed or discovered five firm identification as PWN (the Crab nebula, Vela X, HESS J1825−137, HESS J1514−591 and HESS J1356−645) and 11 promising PWN candidates. One of those candidates, HESS J1420−607 in the Kookaburra region, perfectly illustrates the power of the HE to VHE connection (and in general of having a large leverage arm in energy). The spectral energy distribution (SED) represented in Fig. 3.2 shows that if we limit to the HE range ($E \ll 100$ GeV), only the pulsar component of this source is visible. While if we leverage on the 100 MeV–1 TeV energy range, both the pulsar and the PWN component, two facets of the same system, are revealed. Those different spectral features (peaking at different energies) and the timing aspect evoked earlier are key assets to disentangle the different components in the PSR + PWN systems.

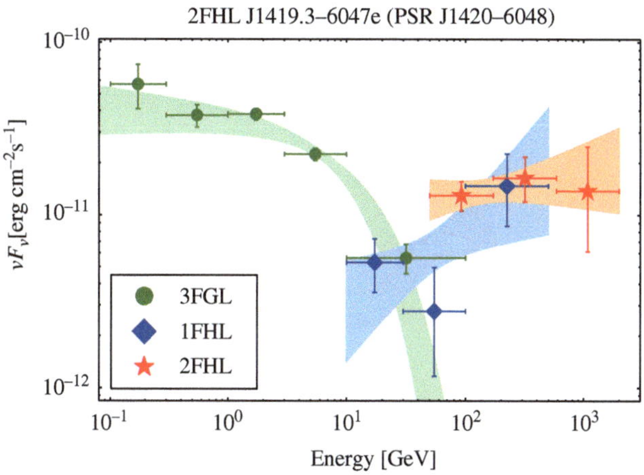

Fig. 3.2 The two facets of a same system dominated by the point-like emission of the pulsar PSR J1420−6048 at energies below 10 GeV and the extended source (2FHL J1419.3−6047e) associated with the surrounding PWN. The *data points* are extracted from different *Fermi*-LAT catalogs which use different energy thresholds in their analysis. Reproduced from Ackermann et al. (2016)

3.4 Demography of the Gamma-Ray PWNe

In the population of PWNe detected in gamma-rays, two main class of systems have been observed: (1) the young nebulae powered by an energetic fast spinning pulsar where the bulk of the X- and gamma-ray emitting electrons have been recently accelerated and (2) old relic nebulae still very bright in gamma-rays but where most of the highly energetic electrons (emitting X-rays) have been burned off. The young systems exhibit bright synchrotron an IC emission across the spectrum. In the first few kilo-years of the nebula, the highly energetic electrons have not yet suffered heavy synchrotron losses and the SED peaks at X-ray energies. The flux ratio $F_{1\ \mathrm{keV}}/F_{1\ \mathrm{TeV}}$ is of the order of 50 for a 1 kyr system in the simulation setup described in Mayer et al. (2012) whose SED is reproduced in Fig. 3.3. With time, the most energetic electrons will cool down and there are not enough newly accelerated electrons to compensate the losses as the rotational energy loss rate of pulsar ($\dot{E}$) decreases with time. Therefore the X-ray flux rapidly decreases with time while the gamma-ray emission, less sensible to losses, remains stable leading to a flux ratio of $F_{1\ \mathrm{keV}}/F_{1\ \mathrm{TeV}} \sim 0.1$ at 200 kyrs. This explains the correlations observed between the flux ratio $F_{1\ \mathrm{keV}}/F_{1\ \mathrm{TeV}}$ versus the pulsar characteristic age and flux ratio versus $\dot{E}$. Detailed studies of this correlation are presented in e.g. Mattana et al. (2009) and in a forthcoming population study of TeV PWN in the H.E.S.S. Galactic Plane Survey.

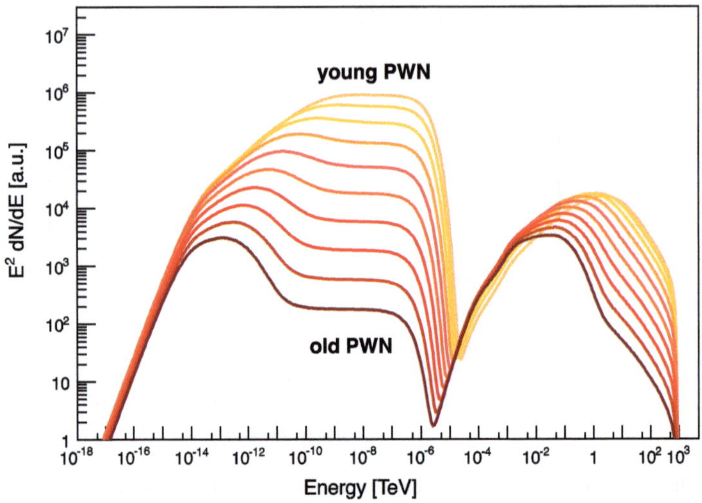

Fig. 3.3 Time dependent SED ranging from radio to VHE of a PWN. The *color scale* represents the age of the PWN, starting with a young system (500 years, *yellow*) and proceeding in equidistant steps on a logarithmic time scale to an old system (200 kyrs, *dark red*). Reproduced from Mayer et al. (2012)

3.4.1 Few Young and Energetic Nebulae

The prototype of a young and energetic PWN is the Crab nebula which has been considered, because of it's high brightness in a large fraction of the spectra, as a calibration source for many instruments. The detection of extreme flares with fluxes increasing up to 30 times the steady flux in the 100 MeV range have shattered this idea of a standard candle (see Sect. 3.5). For a change, I will focus here on another young, energetic and fast spinning pulsar (τ_c = 4 kyrs, $\dot{E} = 4.9 \times 10^{38}$ erg s^{-1}, P=16 ms): PSR J0537−6910. While this pulsar has a spin down luminosity slightly higher than the Crab, it does not outshine the Crab nebula as PSR J0537−6910 is an extragalactic source located in the nearby large Magellanic cloud at a 50 kpc distance. However in terms of luminosity, N157B (the nebula surrounding the pulsar) is the most luminous gamma-ray PWN known to date. It's luminosity at 1 TeV is nearly 20 times that of the Crab nebula while it's luminosity at 1 keV is ∼4 lower than the Crab (see Fig. 3.4). As only a handful of young and energetic pulsars are known, the comparison of those two extreme objects play a key role to separate the role of intrinsic PWN properties from the impact of environmental properties.

N157B, while being a young PWN, has a SED where the X- and gamma-ray energy flux are comparable which is expected only for middle-aged systems as shown in Fig. 3.3. The two main reasons presented by H.E.S.S. Collaboration et al. (2015b) to explain this difference are (1) a low magnetic field (∼45 μG) compared to the Crab (∼127 μG) which comparatively reduces the synchrotron emission from N157B and (2) the strong infrared radiation field of the LH 99 star cluster where the

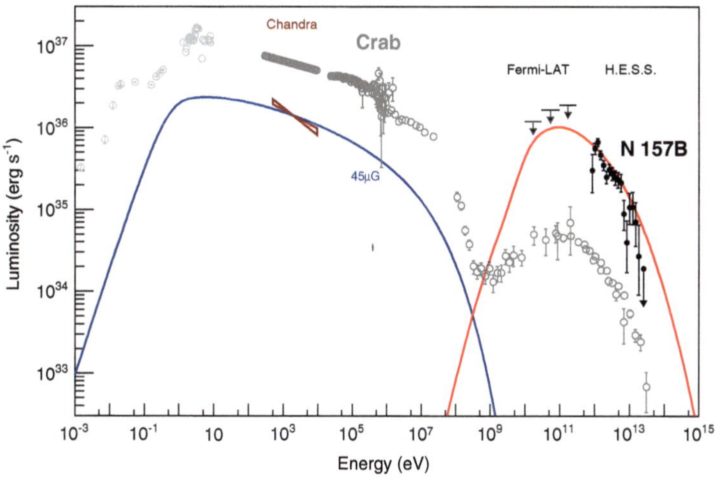

Fig. 3.4 Spectral energy distribution showing the intrinsic luminosity of both PWNe N157B and the Crab, both powered by young and energetic pulsars. Note the large difference for both objects in the L_X/L_γ ratio. Reproduced from H.E.S.S. Collaboration et al. (2015b)

nebula is likely embedded which considerably boosts the IC emission. Interestingly, while N157B exhibits a high pulsar spin-down luminosity, the fraction of this $\dot{E}$ that is transferred to the nebula (i.e. the efficiency of the accelerator) is of 11% compared to 50% for the Crab (see H.E.S.S. Collaboration et al. 2015b, and references therein). Therefore the phenomenal gamma-ray luminosity seems to be driven more by environmental factors (the photon field density) than from the nebula intrinsic particle acceleration efficiency.

3.4.2 A Large Population of Old and (Re)Tired Nebula

The gamma-ray emitting electrons suffer little from the radiative losses and trace the entire history of energetic particles deposited in the nebula. As discussed in Sect. 3.1, the gamma-ray energy range is well-suited to study PWN at a later stage of evolution. At an age of a few 10 kyrs, the reverse shock of the SNR makes it way to the center of the remnant and compresses the nebula (Blondin et al. 2001). This reverse shock can be asymmetric if the density of the surrounding medium is not homogenous which will cause a displacement of the nebula with regards to its particle injection source (the pulsar). In addition, the pulsar itself is born with a kick velocity due to asymmetries in the supernova explosion. For a projected kick velocity of the order of ~200 km/s (Hobbs et al. 2005), the pulsar can travel a distance of 2 pc over 10 kyrs. Combining both effects, the morphology of the PWNe viewed in gamma-rays is quite diverse and generally offset from the pulsar. A few examples are the PWN inside the SNR G327.1−1.1, the source HESS J1825−139,

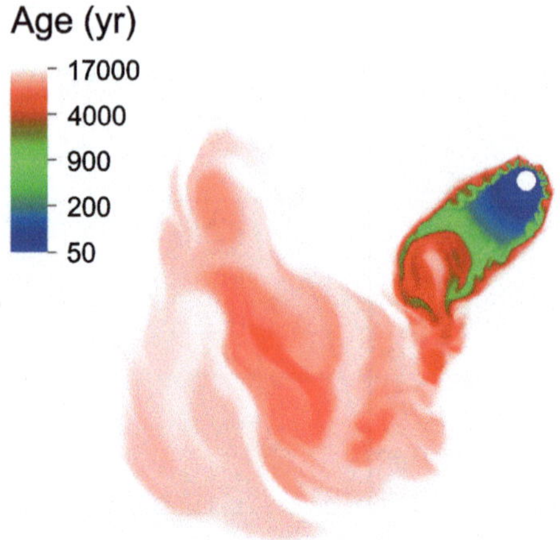

Fig. 3.5 Map of the age of the particles injected by the pulsar in the nebula from the hydrodynamical simulation of the composite SNR G327.1−1.1 presented in Temim et al. (2015). The population of young and energetic particles gives rise to X-ray emission while the older population (crushed by the asymmetric reverse shock) emits at radio wavelength and gamma-ray energies

and HESS J1303−631. In the first case, the system is about 20 kyrs old and both the SNR and the PWN are seen in radio and in X-rays. At radio wavelength, the bulk of the nebula consists of a spherical feature plus a narrow finger-like feature pointing to a compact X-ray source considered as the putative pulsar powering the nebula. As the X-rays trace only the most recently accelerated particles (see Fig. 3.5), the putative pulsar is surrounded by an X-ray nebula significantly smaller and offset to the radio peak (Temim et al. 2009, 2015). In the VHE regime, the spatial extension is difficult to measure due to the limited angular resolution of the H.E.S.S. telescope but the centroid of the emission is located in an intermediate position between the pulsar candidate and the bulk of the radio emission (Acero et al. 2012). The different location of the radio/X/gamma-ray emission is a text book example of an offset nebula due to an asymmetric reverse shock and the pulsar proper motion as shown in the simulation in Fig. 3.5.

The older nebula HESS J1825−139, and HESS J1303−631 could be what the PWN inside SNR G327.1−1.1 would evolve into. They are visible mostly in VHE gamma-rays, are very faint in radio and X-rays and are referred as relic nebula. While the population of the radio electrons act as a calorimeter, the nebula is no longer compressed by the surrounding SNR and the magnetic field within the expanding nebula decreases. The fainter magnetic field reduces the synchrotron radio surface brightness of the nebula. This combined with the difficulty of radio interferometers to detect faint large scale structures in the Galactic plane partly explain the lack of bright radio counterparts. In X-rays, the continuous decrease of the spin down power since birth results in a reduced population of young particles injected by the pulsar. In the study of HESS J1303−631 presented in H.E.S.S. Collaboration et al. (2012), an X-ray counterpart is observed, compatible with the radio position of PSR J1301−6305 ($\dot{E} = 1.7 \times 10^{36}$ erg s^{-1}, P = 184 ms). This pulsar is the most energetic within 6° of the HESS source and is located ∼0.15° away from the peak of the VHE emission. A small extension, roughly in the direction of the center of the VHE peak, is measured (elongation of a few arcmins) around the compact X-ray source and is interpreted as the modest population of recently accelerated particles in the ageing pulsar. Since HESS J1303−631 is a bright VHE source (a flux of 17% of the Crab flux above 380 GeV) observed for a total of 108 h, the statistics is sufficient to carry out a morphological analysis in several energy bands ($E_1 = 0.84 - 2$ TeV, $E_2 = 2 - 10$ TeV and $E_3 > 10$ TeV) and a Gaussian morphology was fit to each excess image. The resulting parameters show both a displacement of the peak towards the pulsar and a shrinking of the extension with increasing energies. The first energy band act as a calorimeter where most of the electrons produced in the pulsar history are still visible whereas the last energy band only shows a younger population of particles closer to the pulsar. This energy-dependent morphology implies a spectral steepening away from the pulsar, a property observed in X-rays and visible in this case at the high end of the VHE range. Note that the VHE extension in the last energy band (gaussian width of ∼5 arcmin) is still larger than the X-ray extension (gaussian width of ∼1 arcmin).

HESS J1303−631 is one of the best examples of unassociated (or "dark" sources) which are extremely bright at VHE and faint in all other wavelengths. Deep multi-wavelength studies were necessary to identify the associated pulsar but one of the key ingredients in the understanding of this source was the VHE gamma-ray energy dependent morphology pointing towards the region of high energy particles. This test can be performed with gamma-ray data alone and can provide a first guess for the numerous unassociated sources that are expected in the CTA era (see Sect. 3.2). The main reason of the power of VHE observations to detect relic nebula lies in that fact that the CMB energy density (the main seed photons for IC scattering) is constant while the magnetic field in the relic nebula decreases with time and can reach very low values (∼2 μG in HESS J1303−631).

3.5 Pulsar Wind Nebulae Flares

Since the huge supernova "flare" in July 1054 AD, observed on Earth by astronomers in the Asian and Arab world, the Crab pulsar left after the supernova explosion and in particular its surrounding nebula have been considered as a rather steady source in recent astronomy history and played an important role in high-energy astronomy. The source was used as a calibration source and its flux is even used as a reference unit (*e.g. A flux of 2 crabs*) in hard X-rays and VHE gamma-rays.

The discovery in April 2011 of a huge flare (an increase by a factor of 30 for $E > 100$ MeV) observed by the *Fermi*-LAT and AGILE satellites (Abdo et al. 2011; Tavani et al. 2011) came as a surprise. The rise time was a fraction of a day and the flux remained larger than the monthly averaged for about a week. The short timescales (<1 day) of the flare suggests a small emitting region ($< 10^{-3}$ pc, ∼0.1 arcsec at a 2.2 kpc distance). Since this event, several flares have been observed although the 2011 event remains the brightest to date. As of January 2017 eight

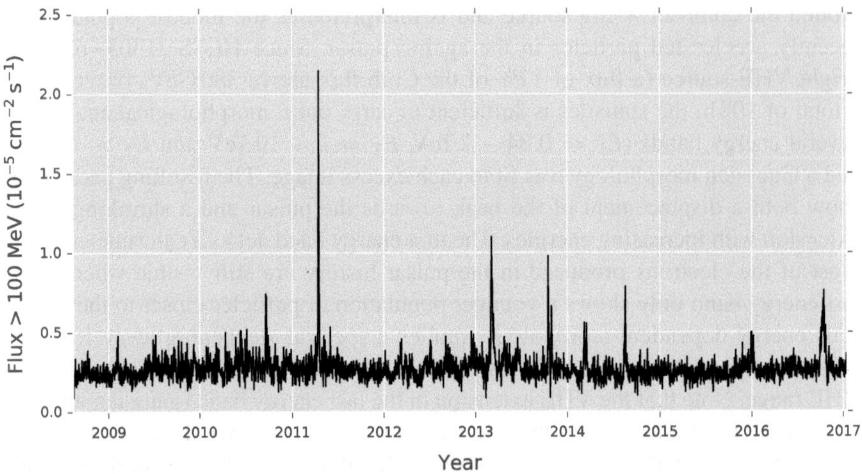

Fig. 3.6 Light curve of the Crab nebula as observed with the *Fermi*-LAT instrument. This daily light curve is the product of an automated analysis at the LAT Instrument Science Operations center performed on a sample of monitored sources

flares exceeding a flux of $F > 0.5 \times 10^{-5}\,\mathrm{cm^{-2}\,s^{-1}}$ (for E > 100 MeV) have been observed as shown in Fig. 3.6.

The HE spectrum of the Crab nebula is the sum of the high end of the synchrotron component dominating for $E < 1$ GeV and the IC component dominating the higher energies. The spectral analysis of the flares at the maximum levels revealed that the large increase in flux is due to large variations in the synchrotron component while the IC remains stable with time. The SED shown in Fig. 9 of Bühler and Blandford (2014) demonstrates that the spectral behaviour notably varies between flares. While the 2011 flares exhibited a new rising spectral component with a peak at ~400 MeV, the 2009 and 2010 flares showed an increased flux of the synchrotron component without significant changes in the spectral shape compared to the quiescent spectrum. Following the flare discovery, observation campaigns have been carried out in radio, IR-optical, X-rays and none of them have reported large flux variations (Weisskopf et al. 2013). In the VHE domain, no significant increase has been observed with the H.E.S.S. telescope (H.E.S.S. Collaboration et al. 2014), the VERITAS telescope (Aliu et al. 2014) or the MAGIC telescope (ATel #2967). Since the flaring feature is seen mostly at the tail of the synchrotron component emitted by highly energetic particles, those flux variations will affect the high end of the IC component (>50 TeV). However due to the reduced cross section in the Klein-Nishina regime at those energies, the IC counterpart of the synchrotron flares will be much attenuated. Follow-up observations of future Crab flares with the increased sensitivity at $E > 50$ TeV of CTA will be able to probe the IC counterpart of the HE Crab flares.

For a more detailed discussion about the observational aspects and the underlying mechanisms powering those flares see the chapter dedicated to the Crab in this book.

3.6 Magnetar Wind Nebulae: MWNe

Magnetars are a sub-class of neutron stars with a surface dipole magnetic fields much greater than the typical 10^{10}–10^{12} G measured in rotation powered pulsars. The McGill magnetar catalog (Olausen and Kaspi 2014) currently lists 23 confirmed magnetars with most sources having magnetic fields values greater than 10^{14} G, long spin periods ($P \sim$ 2–12 s) and low values of the spin down energy loss rate ($\dot{E} \sim 10^{32} - 10^{35}\,\mathrm{erg\,s^{-1}}$). In X-rays, those sources are bright ($L_X \sim 10^{32}$–$10^{36}\,\mathrm{erg\,s^{-1}}$) and most of them have been observed to exhibit short (~0.1 s) X-ray to soft gamma-ray burst features. Due to their burst activity with sometimes giant flares, a collection of sources that are now classified as magnetars were thought to be a variety of gamma-ray burst labeled as soft gamma-ray repeaters (SGRs). We note that in the last decade, observations have revealed magnetar-like burst activity in typical neutron star with 10^{12} G magnetic field (e.g. PSR J1846−0258) blurring the line between the characteristics of magnetars and rotation powered pulsars.

The source Swift J1834.9−0846 is located at the geometrical center of the SNR W41 and has been discovered during a bright hard X-ray burst in 2011 where the flux level increased by three orders of magnitude. This neutron star is a magnetar with a period of 2.48 s, a surface dipole magnetic field of 1.4×10^{14} G and a spin down power of $2.1 \times 10^{34}\,\mathrm{erg\,s^{-1}}$. The deep X-ray observations following-up the 2011 flare presented in Younes et al. (2012) revealed an extended non-thermal emission around the magnetar, a putative nebula, which is a property so far only observed in rotation powered pulsars and not around magnetars. This extended emission was argued by Esposito et al. (2013) to be a dust scattering halo associated with previous flares. A prediction of this scenario is that the dust scattered flux should decay with time. Based on several *XMM-Newton* observations, Younes et al. (2016) showed that the flux and spectral shape of this extended emission was stable in two observations taken 2.5 and 3.1 years after the source outburst in 2011 and compatible in flux using two short observations pre-burst. This provides a strong evidence for Swift J1834.9−0846 being the first magnetar showing a surrounding nebula, a characteristics previously reserved to rotation powered pulsars, therefore narrowing further the bridge between magnetars and rotation powered pulsars. Concerning the nebula's brightness in X-rays, it is not yet clear whether it is enhanced by the compression from the reverse shock (Younes et al. 2016; Torres 2017) or by an extra source of power like the decay of the magnetar's ultra-strong magnetic field (Younes et al. 2016).

This X-ray introduction to magnetars now leads me to the gamma-ray point of this section: an extended VHE source, HESS J1834−087, is observed coincident with the SNR W41 and a *Fermi*-LAT extended emission is also detected morphologically matching the VHE emission (H.E.S.S. Collaboration et al. 2015a). The VHE extended emission is best modelled with a sum of two components: one point-like and one significantly extended. The point-like component is compatible with the position of the magnetar and an association with a nearby OH maser (tracing the interaction of the SNR shock with a molecular cloud) is ruled out. Note that the extension measured in X-rays ($\sim$50 arcsec $\simeq$0.01°) would appear as point-like for the VHE instruments. In the study carried out by H.E.S.S. Collaboration et al. (2015a) the PWN scenario for the compact VHE source is discussed and is considered as possible. The large extended emission seen both at HE and VHE could be associated with the SNR shock interacting with its surroundings as seen for several middle-aged SNRs. While the PWN origin of the point-like VHE source remains to be confirmed, this would potentially be the first Magnetar Wind Nebula (MWN) observed at gamma-ray energies adding a sub-class in the growing population of nebulae in this high-energy range.

3.7 Conclusion

The breakthrough developments in terms of sensitivity and angular resolution at HE and VHE energies have provided a new view of the gamma-ray sky. In this new story, PWNe play a major role in the Galactic plane especially at VHE (E > 100 GeV). While X-ray observations reveal the most energetic an most recently injected particles in the nebula, the HE and VHE band (where the the electrons lifetime is much longer) is ideal to study the time integrated evolution of the nebula and disentangle the role of intrinsic pulsar properties with regards to environmental factors (pressure from the surrounding medium, photon radiation fields, etc). Effects such as the pulsar proper motion or the reverse shock of the host SNR in which the nebula is expanding tend to shift the nebula with regards to the pulsar position. Those offset/relic nebula are faint sources in every wavelength but gamma-rays, are difficult objects to identify as PWNe and are often classified as unassociated sources. Combining HE and VHE observations, providing a leverage on nearly six orders of magnitude in energy, to measure the energy dependent morphology of the nebula (shrinking towards the pulsar) has allowed to unveil the nature of several unassociated sources.

In the near future, the Cherenkov Telescope Array will be a PWN discovery machine with several hundreds of new PWNe expected. On one hand this will be a breakthrough for population studies but on the other hand, this large population of nebula will account for a large fraction of CTA's unassociated sources and their collection might create a gamma-ray diffuse emission in the Galactic plane. Extensive studies using multi-wavelength observations with existing and upcoming radio and X-ray observatories will be key assets to tackle this issue.

The survey capacity of HE instruments (e.g. *Fermi*-LAT covering the whole sky every 3 h) has allowed to discover unexpected flares from the Crab nebula shattering this idea of a standard candle for high-energy astronomy. Future observations with CTA will test the high-energy end of the IC counterpart in the Crab flares providing a better insight of the mechanism at stake.

Acknowledgements F. Acero would like to thank M. Lemoine-Goumard and the MODE SNR/PWN workshops for valuable discussions.

References

Abdo, A.A., Ackermann, M., Ajello, M., et al.: Astrophys. J. **734**, 28 (2011)
Abdo, A.A., Ajello, M., Allafort, A., et al.: Astrophys. J. Suppl. Ser. **208**, 17 (2013)
Acero, F., Djannati-Ataï, A., Förster, A., et al.: ArXiv e-prints [arXiv 1201.0481] (2012)
Acero, F., Ackermann, M., Ajello, M., et al.: Astrophys. J. **773**, 77 (2013)
Acero, F., Ackermann, M., Ajello, M., et al.: Astrophys. J. Suppl. Ser. **218**, 23 (2015)
Acero, F., Ackermann, M., Ajello, M., et al.: Astrophys. J. Suppl. Ser. **224**, 8 (2016)
Ackermann, M., Ajello, M., Atwood, W.B., et al.: Astrophys. J. Suppl. Ser. **222**, 5 (2016)
Aliu, E., Archambault, S., Aune, T., et al.: Astrophys. J. **781**, L11 (2014)
Blondin, J.M., Chevalier, R.A., Frierson, D.M.: Astrophys. J. **563**, 806 (2001)
Bühler, R.: ArXiv e-prints [arXiv 1509.00012] (2015)
Bühler, R., Blandford, R.: Rep. Prog. Phys. **77** 066901 (2014)
Esposito, P., Tiengo, A., Rea, N., et al.: Mon. Not. R. Astron. Soc. **429**, 3123 (2013)
Faucher-Giguère, C.-A., Kaspi, V.M.: Astrophys. J. **643**, 332 (2006)
H.E.S.S. Collaboration, Abramowski, A., Acero, F., Aharonian, F., Akhperjanian: Astron. Astrophys. **548**, A46 (2012)
H.E.S.S. Collaboration, Abramowski, A., Aharonian, F., et al.: Astron. Astrophys. **562**, L4 (2014)
H.E.S.S. Collaboration, Abramowski, A., Aharonian, F., et al.: Astron. Astrophys. **574**, A27 (2015a)
H.E.S.S. Collaboration, Abramowski, A., Aharonian, F., et al.: Science **347**, 406 (2015b)
Hobbs, G., Lorimer, D.R., Lyne, A.G., Kramer, M.: Mon. Not. R. Astron. Soc. **360**, 974 (2005)
Mattana, F., Falanga, M., Götz, D., et al.: Astrophys. J. **694**, 12 (2009)
Mayer, M.J., Brucker, J., Jung, I., Valerius, K., Stegmann, C.: ArXiv e-prints [arXiv 1202.1455] (2012)
Olausen, S.A., Kaspi, V.M.: Astrophys. J. Suppl. Ser. **212**, 6 (2014)
Renaud, M., CTA Consortium: Mem. Soc. Astron. Ital. **82**, 726 (2011)
Tavani, M., Bulgarelli, A., Vittorini, V., et al.: Science **331**, 736 (2011)
Temim, T., Slane, P., Gaensler, B.M., Hughes, J.P., Van Der Swaluw, E.: Astrophys. J. **691**, 895 (2009)
Temim, T., Slane, P., Kolb, C., et al.: Astrophys. J. **808**, 100 (2015)
Torres, D.F.: Astrophys. J. **835**, 54 (2017)
Weisskopf, M.C., Tennant, A.F., Arons, J., et al.: Astrophys. J. **765**, 56 (2013)
Younes, G., Kouveliotou, C., Kargaltsev, O., et al.: Astrophys. J. **757**, 39 (2012)
Younes, G., Kouveliotou, C., Kargaltsev, O., et al.: Astrophys. J. **824**, 138 (2016)

Chapter 4
Observational Links Between Fermi-LAT Pulsars and Their Nebulae

Emma de Oña Wilhelmi

Abstract GeV pulsar and TeV pulsar wind nebulae represent the largest Galactic population in the GeV and TeV regime respectively. Even when sharing the same central engine, that is, an energetic and young pulsar, the processes ruling the two regimes differ greatly. At GeV energies, the radiation is pulsed and the environment conditions on the pulsar magnetosphere and pulsar wind prevail. At TeV energies, the steady, integrated over time, nebula outshines. Nevertheless those processes should be ultimately connected being different aspects from a continuous pulsar/PWNe system. We review the state-of-the-art knowledge on these objects and describe in a quantitative manner the general trends observed experimentally, which will be further investigated.

4.1 Introduction

Pulsars or rotating neutron star are highly magnetised, fasting-rotating (ms to s) compact objects which emit radiation in all wavelength, even up to several tens of TeVs, offering an opportunity to study physic processes under extreme conditions. Along with electromagnetic emission, pulsars dissipate their rotational energy via relativistic winds of particles. Since the relativistic bulk velocity of the wind is supersonic with respect the ambient medium, such a wind produce a termination shock. The wind particles, moving through the magnetic field and the ambient photons, produce radiation that we observe as pulsar wind nebulae (PWNe). As the pulsars themselves, the PWN emits at all wavelengths from radio to TeV. The complex involving pulsars, their winds and, eventually, a surrounding supernova blast (see Fig. 4.1) covers a wide range on dimension scales, from a few kilometres to more than 20 pc in some cases. The small region comprised by the light cylinder (with typical size of $\sim 10^9$ cm) that surrounds the rotating $\sim 10^6$ cm-radius star, confines an exceptionally energetic magnetised plasma, with magnetic field strengths as high as $B \sim 10^{9-12}$ G. This small volume, however, drives the particle

E. de Oña Wilhelmi (✉)
Institute of Space Sciences, (IEEC-CSIC) Campus UAB, Carrer de Can Magrans, s/n, Catalonia, Spain
e-mail: wilhelmi@ice.csic.es

D.F. Torres (ed.), *Modelling Pulsar Wind Nebulae*, Astrophysics and Space Science Library 446, DOI 10.1007/978-3-319-63031-1_4

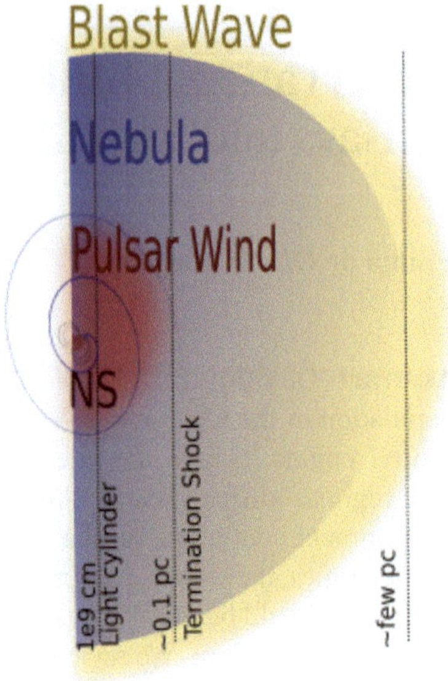

Fig. 4.1 Schematic view of a pulsar, its pulsar wind nebulae and hosting supernova remnant

flow that surrounds it, forming a relativist winds that extends up to ~0.1 pc. At even larger scales, reaching tens of parsecs in evolved systems or PWN embedded in regions of low density, the wind inflates a large bubble of particles and magnetic field, forming the magnetised nebula, which is at the same time strongly affected by the shock blast left behind by the supernova explosion. The understanding of pulsars and their nebula requires thus not only a multi-wavelength approach but also the study of transport mechanisms and cooling processes at very different length scales.

Since 2005, pulsar and PWN astrophysical research at high energies has experienced a veritable revolution, mainly driven by the increase of the sample of those sources at GeV and TeV energies, from a handful to a few tens or hundreds (in case of PWNe and pulsars respectively). The first five high energy pulsars were detected by the *Energetic Gamma-Ray Experiment Telescope* (EGRET) at energies $E > 100$ MeV. Crab, Vela, Geminga, PSR B1055−52, PSR B1706−4 and PSR B1951+32 (Nolan et al. 1996) were clearly identified through their timing signature, whereas several of the 271 EGRET unidentified sources were also believed to be related to pulsars and millisecond pulsars (see Thompson et al. 2005 for a review). Regarding their nebulae, only one source was established before 2005: the Crab nebula discovered in 1989 (Weekes et al. 1989). Interestingly, the flux measured from the Crab nebula has been used as the standard candle in very-high energy astronomy ever since, nonetheless recent observations with high sensitivity instruments have revealed a rich and complex phenomenology on its (previously believed to be) steady emission (see Zanin and Acero contributions in this book). Concerning pulsars, this scenario dramatically changed with the advent of more sensible detectors in satellites at GeV such *AGILE* (Astro-rivelatore Gamma a Immagini LEggero) (Tavani 2011) in 2007 and specially with the Fermi Gamma-ray Space Telescope (*Fermi*) (Atwood et al. 2009) with the LAT (Large Area

Telescope) instrument on board, launched in 2008. The number of pulsars detected by summer 2016 amounts 161 (Johnson et al. 2014), allowing the study of both individual and collective properties of the pulsar population. Several competing gamma-ray pulsars models (see next section for a short review based on Takata et al. (2006); Muslimov and Harding (2004); Romani and Yadigaroglu (1995); Grenier and Harding (2015); Kalapotharakos et al. (2016) and references therein) have been contrasted with the high-energy pulsar population, predicting their spectral features and light curves according to their birth properties, beaming and evolution (Gonthier et al. 2007; Watters and Romani 2011; Pierbattista et al. 2016). Those comparisons favor models in which the gamma-ray emission is originated in the outer part of the magnetosphere and they also require a large amount of pairs (electrons and positron) to sustain the radiation level. Even when they can reproduce many of the average radio and high-energy characteristics, they also highlight significant differences, as for instance the need to call of different regions/mechanisms to explain different light-curve structures. Alongside to those discoveries and theoretical development in the GeV regime, a number of high-sensitive, good-imaging Cherenkov telescope arrays have been exploring the TeV sky, resulting on a dramatically increase of the number of TeV-emitting sources on the Galactic plane. Observations with MAGIC, Veritas and H.E.S.S. telescope arrays and their upgrades have revealed 72 new TeV sources lying on the Galactic plane.[1] Out of them, 35 are classified as PWNe, although only 19 are firmly identified (based mainly on their multi-wavelength properties). In particular, the Galactic plane survey performed by the H.E.S.S. collaboration from 2005 until 2015 has allowed a complete population study on nebulae associated with bright pulsars, in a region defined by a range of longitude between 250° and 65° and narrow latitude strip of −3.5° and 3.5°.

Fed by the large amount of information provided by satellites and ground-based telescopes, theoretical and phenomenological models have dramatically evolved. However, the interplay and relations connecting the pulsed emission, which dominates in the MeV and GeV regime and the non-pulsed emission of the nebula, which dominates at TeV energies, are still far from being solved. The first seems to be regulated by the freshly injected electron and positron wind in the pulsar magnetosphere, and its propagation towards the termination shock. In the second, the gamma-ray emission detected relates to the integrated wind along the pulsar lifetime and it is subject to changes in the environment around the pulsar (i.e. the evolution within the hosting SNR or the region in which the explosion occurred). Several excellent summaries on the status of GeV pulsar observations can be found in i.e. Ray and Parkinson (2011); Caraveo (2014); Grenier and Harding (2015); Breed et al. (2016). Likewise, a throughout review on PWN was provided by several authors (i.e. Kargaltsev et al. 2013; de Oña Wilhelmi 2011; Gallant et al. 2008). Here, we intend to emphasise the link between the two populations, focusing in the shared energetic and environmental aspects that connect the different spectral features they display.

[1]http://tevcat2.uchicago.edu/.

In Sect. 4.2 we will outline the state-of-the-art of pulsar modelling and latest simulations. Likewise, we will summarise the current knowledge of PWNe, emphasising the main problems we face on the PWNe field. In Sect. 4.3 we will discuss the general properties of pulsars and PWNe populations and their interplay. In Sect. 4.4 we report on new types of pulsars and PWNe and how do they tally with the known populations. Finally the last Section will be devoted to the usual general discussion.

4.2 High-Energy Pulsars and Their Relativistic Winds

The standard pulsar paradigm to explain pulsed emission from neutron stars relies in acceleration of particles in low-density plasma regions or gaps within the pulsar magnetosphere. Those particles loose energy via synchrotron-curvature processes when they travel along the strong magnetic field lines. Different altitudes and regions in the magnetosphere were proposed to explain the origin of the multi-wavelength pulsed emission observed: at low altitude above the star, usually attribute to radio pulses, or further out near the light cylinder. This cylinder is defined by the radius $R_{LC} = c/\Omega$ from the spin axis, beyond which the field lines cannot co-rotate with the neutron star without violating the speed of light, but instead they spiral outwards forming a cold, relativistic wind (see Fig. 4.2). The resolution of

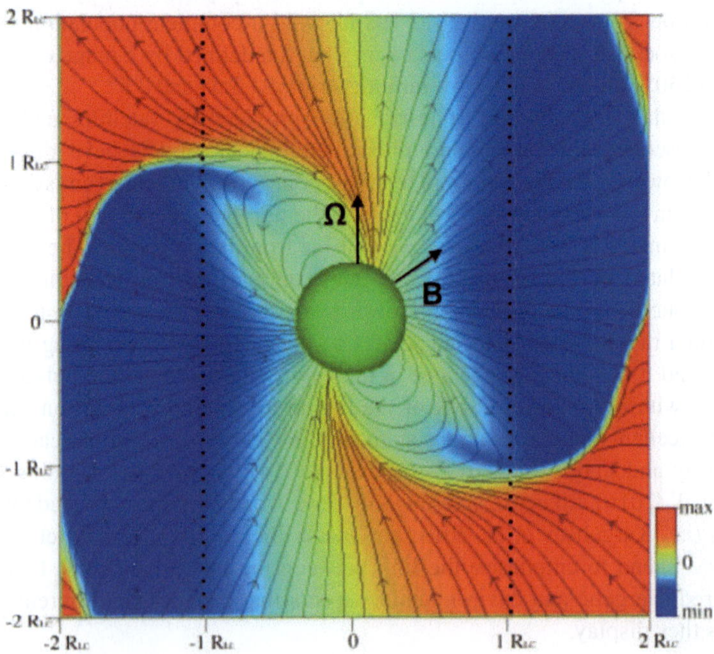

Fig. 4.2 *Magnetic field lines* obtained from MHD simulations of an oblique pulsar magnetosphere with magnetic inclination of 60° in the corotating frame plane (Spitkovsky 2006). *Color* represents the magnetic field perpendicular to the plane

force-free magnetosphere plasma models (Tchekhovskoy et al. 2016; Contopoulos 2016; Cerutti et al. 2016) can trace how currents are located in the surrounding of pulsars. Earlier models treated the magnetosphere as a vacuum, locating the acceleration and high-energy emission either above the pulsar caps (polar cap models), near the light cylinder (outer gap models) or along the last open field lines from the neutron star surface to near the light cylinder (slot gap). They appeal to synchrotron-curvature (SR, CR) mechanisms from primary and secondary electrons and positrons, as well as self-synchrotron Compton (SSC) in some particular cases, to explain the broad-band emission of pulsars.

The field is quickly evolving, thanks in part to the better computational power but mainly driven by the large date set of experimental results supplied in last years. New particle-in-cell (PIC) and magneto-hydro-dynamic (MHD) simulations are being currently developed, showing a quite different picture. This two approaches are based on either solving self-consistently electromagnetic fields, current and charge densities of a rotating star with a dipole moment through MHD simulations, or computing the individual particle response to the electromagnetic fields, through N-body simulations, or through PIC simulations that also compute the self-consistent change in the fields due to the particles. In recent MHD simulations of dissipative pulsar magnetospheres (Cerutti et al. 2016) (that is, including a parallel electric field, away from ideal MHD simulations), the interplay between the magnetosphere and the relativist pulsar wind, which blows away from the pulsar, plays an important role. By solving the electromagnetic structure of an oblique magnetosphere, including particle acceleration, and the emission of radiation self-consistently, in a 3D spherical particle-in-cell simulations, current sheets appear either along the last closed magnetosphere magnetic lines, and/or on the equatorial plane, extending beyond the magnetosphere. These current sheets are optimal places for magnetic reconnection to happen (Cerutti et al. 2016; Mochol and Pétri 2015) and have been suggested as origin of the gamma-ray emission at GeV and TeV, even beyond the magnetosphere. According to those models, young pulsars with high magnetic field and strong pair production are weakly dissipative whereas in old ones, the current sheets are dominated by the equatorial ones on the wind zone. When the rotator is oblique, naturally these sheets corrugate producing a striped wind in which particles can also undergo relativistic magnetic reconnection and emit gamma-rays via self-synchrotron-Compton (Pétri 2015; Mochol and Pétri 2015). Those stripped winds are also believed to depend on the pulsar age, becoming radiative-dominant (young ones, such as the Crab pulsar), or escaping-dominant (older pulsars such as Vela pulsar). A different mechanism for pulsed emission at high and very-high energies was proposed by Aharonian et al. (2012), in which the relativistic winds becomes kinetic dominant, transferring energy for the strong magnetic field to particles, and off-scattering pulsed optical and X-ray photon fields. As a results, a new Compton component will mimic the low energy timing signature (if close enough to the light cylinder) emerging above a few tens of GeV. Although these models in general satisfactorily explain light curves and phase-resolved spectra at GeV energies, they also suffer from some drawbacks, such the maximum Lorenz factor achieved, the lack of unscreened electric field in an ideal MHD or the origin of the supply charges.

4.3 Beyond the Wind: The Pulsar Wind Nebulae

PWNe are bubbles of relativistic particles and magnetic field formed by the interaction of the relativistic pulsar wind with the surrounding Supernova Remnant (SNR) or the interstellar medium (ISM). Models connecting the pulsed emission with the pulsar wind correlate with the structure of the surrounding nebula for young pulsars. Instead, in the case of evolved ones, the morphology and energetics of the PWN is strongly modified by first the interaction with the hosting SNR and then by the expanding within the surrounding medium of the nebula along the years. Upon passing through the termination shock, the wind kinetic energy is randomised, and the purely toroidal magnetic field is carried out and accumulates within the volume of the nebula. Since Kennel and Coroniti (1984) published the first MHD description of the plasma flow in the Crab Nebula, there has always been an unexplained ingredient in the pulsar to nebula modelling: the low level of magnetisation of the flow immediately upstream of the pulsar wind termination shock. This is what is commonly referred to as the σ-problem: the pulsar wind, carrying most of the rotational energy lost by the pulsar, is Poynting-dominated (highly magnetised) at its base, while MHD modelling of PWNe requires it to be particle dominated at the shock. 2D MHD simulations (Komissarov and Lyubarsky 2004; Del Zanna et al. 2006) somewhat alleviated the strong requirements of the (Kennel and Coroniti 1984) model, and more recently 3D MHD simulations (Porth et al. 2014) have suggested that actually an equal amount of energy in particles and field at the shock might be required to reproduce the Crab Nebula morphology. This latter conclusion, however, comes from the idea that, thanks to kink instabilities, the magnetic field becomes progressively more tangled and dissipates in the bulk of the nebula, i.e. in the radiation region, rather than in the cold pulsar wind. Unfortunately, observations of the pulsar wind are hampered by its own nature, being cold and ultra-relativistic. However, the recent developments on pulsar physics related to the location of the pulsed emission regions, extending beyond the magnetosphere, have opened a new line of research, in which high energy observations, and in particular observations of pulsed emission above 100 GeV, might unveil a new inverse Compton component, and therefore pinpoint a region where the wind evolved from a magnetic-dominant to a kinetic-dominant wind, transferring the energy injected by the pulsar to particles which can be accelerated and up-scatter soft photon fields.

Detailed morphological studies of PWNe with high-resolution X-ray satellites such as Chandra have revealed complex structures in these nebulae. In general, young pulsars show an axial symmetry around what is believe to be the pulsar rotation axis, interpreted as a result of the latitude dependence of the pulsar wind energy flux (Lyubarsky 2002; Komissarov and Lyubarsky 2003; Del Zanna et al. 2004; Bogovalov and Tsinganos 2005), whereas some of the more evolved systems show a crushed morphology, resulting from the interaction with the host SNR. Prominent examples such Crab and Vela can be described satisfactorily using 2D and 3D simulations, but they fail to reconstruct the multi-wavelength spectrum (Aleksić et al. 2015; Abramowski et al. 2012). On the other hand, spatially independent models in which the time evolution of the electron spectrum is taking into account (Martín et al. 2012; Tanaka and Takahara 2013; Fang and Zhang 2010; Gelfand et al. 2009) can fit the global spectrum of PWNe, by assuming different injection electron spectral shape, contribution of photon field (namely far infra-red, near infra-red and CMB background) and magnetic field strengths. Those models can unveil general trends, but they fail to explain satisfactorily the particular details of the observed emission, in particular in X-rays and TeV energies. This is the case even for the Crab nebula, which exhibits the strongest flux, for which dedicated models applied to the latest results obtained with the MAGIC telescopes (Aleksić et al. 2015) cannot reproduced the broadband shape. In general, the assumption of a constant magnetic field dominating the inverse Compton emission is not realistic. Instead a turbulent medium in the nebula magnetic field could explain some of the features observed in TeV PWNe, including the, in some objects, extremely low magnetic field required to adjust the X-ray and TeV flux level (see i.e. H.E.S.S. Collaboration et al. 2012). A two-folded approach, in which the proper time-dependent radiative mechanisms are combined with MHD realistic simulations seems to be providing answers on the correct direction, although the complexity of the morphology in evolved PWNe and computational expensive modelling is still a challenge.

4.4 Connecting Pulsars and PWNe

Observations of gamma-ray pulsars have been strongly boosted after the launching of the AGILE satellite and, specially, the *Fermi* satellite in 2007–2008. The gamma-ray telescopes on board of these satellites are sensitive to energies above 60 MeV and have allowed the discovery of more than 160 pulsars (Abdo et al. 2013) (see Fig. 4.3). From those, 61 were known prior through radio or X-ray observations, and typified as young radio loud pulsars (YRL). Thanks to both, the good sensitivity of the instrument and the strong flux of these sources, many of the, in principle, unidentified sources were revealed to be pulsars via the discovery of periodic signals using blind searches techniques (and dubbed young radio quiet, YRQ, pulsars) (Abdo et al. 2009). The gamma-ray emission is believed to originated in beam-like regions that sweep the sky co-rotating with the neutron star. In general, these beams seem to be wider than the radio emission beams, increasing thus the probability to intersecting with the line of sight and resulting on a *loud* gamma-ray emitter but *quiet* radio pulsar. Finally, a large fraction of the pulsars reported on the catalog (39) are classified as millisecond pulsars (MSPs) (Grégoire and Knödlseder 2013), a hitherto unknown class of strong gamma-ray emitters. Those MSP are characterised by very short periods and period derivatives (after being spun-up by a companion) and are located on the lower-left corner in Fig. 4.3. Beside these three groups, MS pulsars, YRL, and YRQ, the increase of sensitivity in longer data sets has revealed gamma-ray emission from new systems, such black widows and red back systems (see Bogdanov in this book). At TeV energies, the dominant gamma-ray

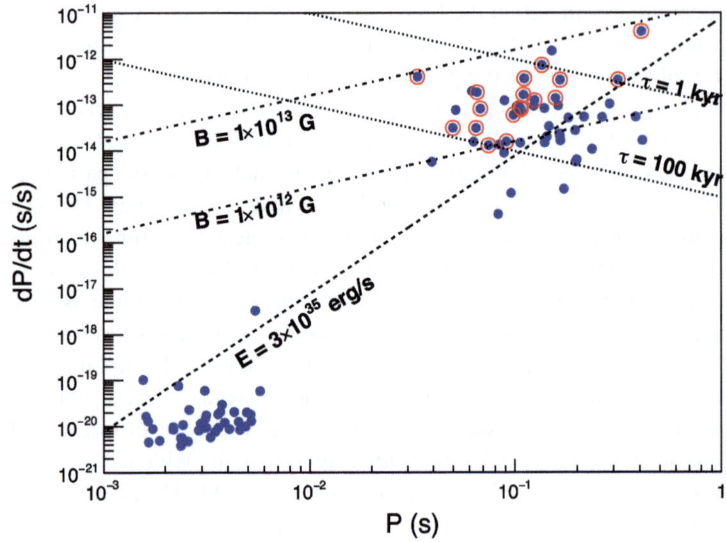

Fig. 4.3 P-Ṗ diagram corresponding to pulsars listed on the 2nd LAT pulsar catalog. LAT pulsars associated with TeV nebula are surrounded by a *red circle*. The estimate pulsar age τ, magnetic field strength B, and spin-down power E, obtained from the primary observable P and Ṗ are marked with *dashed-lines*

emission comes from the pulsar nebula. Indeed, PWNe are the most effective Galactic objects in producing very-high energy emission. In general, these very-high PWNe are related with young and energetic pulsars which power a magnetised nebula. In this scenario, particles are accelerated to very-high energies along their expansion into the pulsar surroundings and the interactions of relativistic leptons with the magnetic field and low energy radiation (of synchrotron origin, thermal, or microwave background), non-thermal radiation is produced from the lowest possible energies up to ~100 TeV. On the other hand, for a few μG magnetic fields, young electrons create a small synchrotron nebula around the pulsar which should be visible in X-rays, in contrast of a often much larger TeV nebula, generated by inverse Compton processes. Typically only young pulsars (τ < 100 kyrs) with large spin-down energy ($\dot{E}$ >10^{33} erg s^{-1}) produce prominent PWNe. Concerning these TeV nebula, two kind of objects are usually considered according to their observational features: young systems such as the Crab nebula (Aleksić et al. 2015; Aharonian et al. 2006a; Archer and VERITAS Collaboration 2015), G0.9+0.1 (Aharonian et al. 2005a) or MSH 15–52 (Aharonian et al. 2005b), which often still lie within their composite supernova shell, and evolved (extended and resolved) systems, such as Vela X (Abramowski et al. 2012), HESS J1825−137 (Aharonian et al. 2006b) or HESS J1809−193 (Aharonian et al. 2007). The young systems show a good match with the morphology seen in X-rays, while in the latter group, very often the pulsar powering the TeV PWN is found offset with respect to the center of the TeV emission, with large size ratios between the X-ray and VHE gamma-ray emission regions. The evolution of the SNR blast wave into an inhomogeneous ISM and/or the high velocity of the pulsar, together with a low magnetic field value (∼5 μG), may explain these large offset as being the relic nebulae from the past history of the pulsar wind inside its host SNR.

The second *Fermi* pulsar catalog provides very valuable information when comparing the population of LAT pulsars and the TeV PWNe. The catalog registers the positions and characteristics of 117 pulsars (see Fig. 4.3), from which the distance, pulsar parameters, energetics and spectral energy distribution, can be extracted. To investigate possible links between these high-energy pulsars and their TeV nebulae, we used the population of PWNe presented by H.E.S.S., which includes PWNe lying on the inner part of the Galaxy but also objects on the outer regions reported by other experiments, which mainly lie on the galactic anti-center and the Cygnus region. From the 61 PWNe/Pulsars investigated, 34 are firmly identified as PWNe whereas for 27 sources, upper limits on the integral flux in the 1–10 TeV energy range of the order of $\sim 10^{-12}$ cm/s were derived. Out of the 34 TeV sources, 14 have been associated with bright, young, LAT pulsars. Table 4.1 shows the LAT pulsars for which a TeV nebula has been associated, together with their location in the Galaxy and their classification as radio loud or quiet.

In Fig. 4.3 we show the P-$\dot{P}$ diagram for the LAT pulsars in the catalog (in blue), where pulsars with an associated TeV nebula are marked with a red circle. These pulsars have rotational energies $\dot{E} > 3 \times 10^{35}$ erg/s as expected for typical pulsars associated to TeV nebulae. The dashed lines in Fig. 4.3 define the region of parameters derived from the period and period derivative corresponding to the pulsars associated to TeV nebulae. They are typically highly magnetised, with magnetic fields above 1×10^{12} G and relatively young, in a range of ages between ∼1

Table 4.1 LAT pulsars associated with TeV nebulae

LAT PSR	TeV PWN	l (°)	b (°)	Dist[a] (kpc)	Type
J0007+7303	J0007+7303	119.65	10.60	1.40	YRQ
J0205+6449	J0205+6449	130.72	3.08	1.96	YRL
J0534+2200	B0531+21	184.56	−5.78	2.0	YRL
J1016−5857	J1016−5857	284.08	−1.88	2.90	YRL
J1023−5746	J1023−5746	284.17	−0.41	8.00	YRQ
J1028−5819	J1028−5819	285.07	−0.46	2.33	YRL
J1119−6127	J1119−614	292.15	−0.54	8.40	YRL
J1357−6429	J1356−645	309.92	−2.51	2.50	YRL
J1420−6048	J1420−607	313.54	0.23	5.61	YRL
J1418−6058	J1418−609	313.31	0.11	1.60	YRL
J1718−3825	J1718−3825	348.95	−0.43	3.60	YRL
J1747−2958	J1747−2809	0.86	0.09	4.75	YRL
J1801−2451	B1800−21	8.40	0.15	5.22	YRL
J1907+0602	J1907+0602	40.08	−0.88	3.21	YRQ

[a]The distance corresponding to the LAT pulsars is assumed here

and 100 kyrs. No TeV PWN has been found around a millisecond pulsar, however, in the presence of a companion wind, TeV emission could be in principle expected (Bednarek et al. 2016).

4.4.1 Distribution of LAT Pulsars and TeV PWNe in the Galaxy

The distributions in the Galaxy of the LAT pulsars and TeV PWNe are shown in Fig. 4.4, in a two-dimensional projection. The figure also shows an schematic representation of the spiral arm of the Milky Way according to the parametrisation of Vallée (2008). The Galactic center position and the Sun position are marked with stars. On the left plot (Fig. 4.4a), the YRQ and YRL LAT pulsars are marked with grey crosses and the MSP ones with yellow crosses. The TeV PWNe positions are marked with blue circle and red inverted triangles, for the detected and non-detected ones, respectively. The ones with an associated LAT PSRs are marked with a light blue circle. For clarity, the right plot (Fig. 4.4b) shows only the LAT YRL and YRQ and the TeV sources with LAT counterpart. From the galactocentric distribution of LAT pulsars, it is clear that there is a observational bias towards a region between -10 and -5 kpc in the X-galactocentric coordinate and -5 kpc and 5 kpc in the Y-galactocentric coordinates. The TeV PWNe instead seem to be located in a wider region, clustering around the position of the Carina-Sagittarius arm. The Sagittarius arm is considered one of the minor spiral arm in our Galaxy, but hosts a large number of HII regions, young stars and giant molecular clouds. In fact, the TeV emission depends not only on the pulsar energetics but also on the surrounding medium, that provides the up-scattered photon field in the inverse Compton emission. It is reasonable then to believe that pulsars located in regions with large stellar activity are more likely to be bright in gamma-rays. This correlations was investigated by Pedaletti et al. (2015) and also pointed out by Torres et al. (2014). On the other hand, the LAT telescope suffers greatly from the large diffusion emission produced in the inner part of the Galaxy. That might explain the reduced number of LAT PSR close to the direction of the Galactic center, for X-galactocentric distance larger than -5 kpc. Note that the strong gamma-ray emission from pulsars allow the detection of very distant ones, whereas in the case of TeV PWN, a compromise between the size of the nebula (close nebulae might be too large to be detected with the current instruments FoV with enough sensitivity) and the intensity of the emission (not too far to be too faint) might also resulting on an observational biased towards distances between $\sim$2 and $\sim$5 kpc.

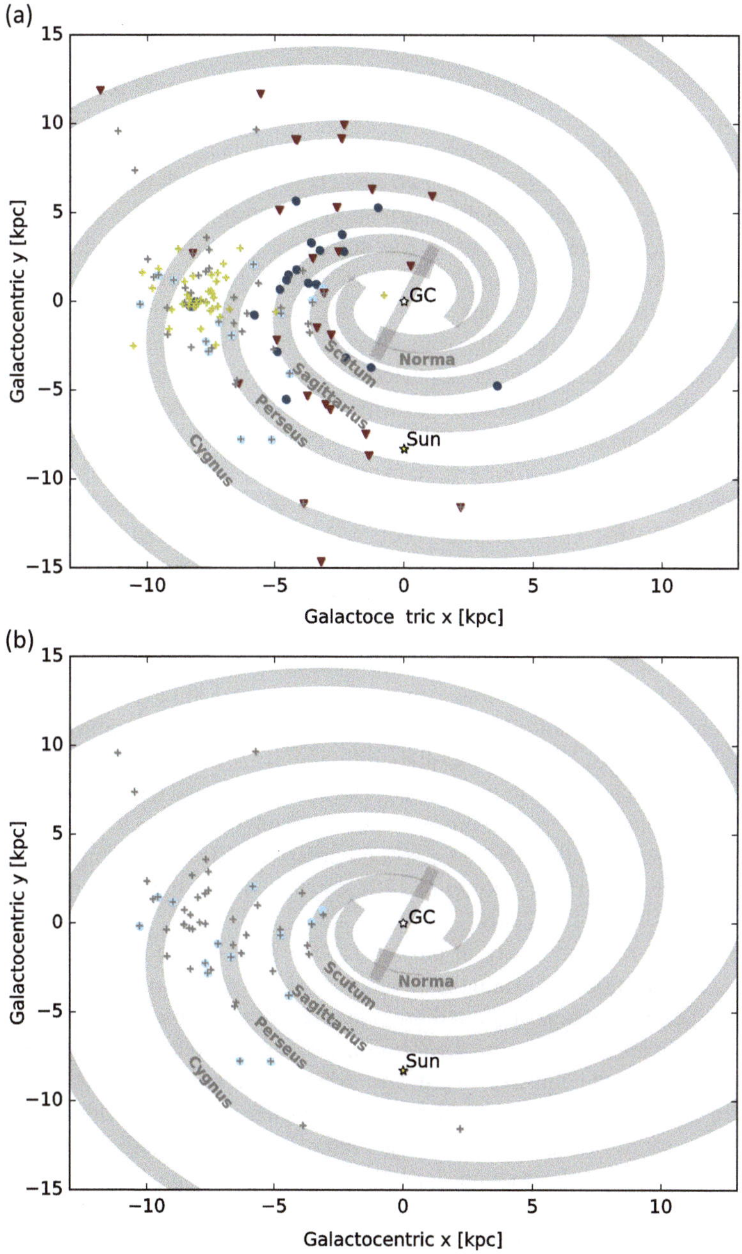

Fig. 4.4 At the *top* (**a**) two-dimensional projection of the positions of the LAT pulsars contained the 2nd *Fermi* Pulsar Catalog and the TeV PWNe (see text). The *grey crosses* mark the position of LAT YRQ and YRL pulsars whereas the *yellow* ones refer to MSP. The *blue dots* sign the TeV PWNe whereas we used *red inverted triangles* to mark the position of energetic pulsars for which an upper limit on steady TeV emission was provided. Finally the *light blue dots* mark TeV PWNe with an associated LAT pulsars. At the *bottom* (**b**) Only the *grey crosses* (LAT YRQ and YRL) and *blue circles* (TeV PWN with associated LAT PSR) are indicated for clarity

4.4.2 Energetics

The study of the spectral energy distribution should in principle serves as a bridge between pulsars and PWNe. The pulsar energy reservoir ($\dot{E}$) is shared between the pulsed emission and the radiation emitted by the nebula. However, the different dimensions and time scales of the two objects together with the lack of accurate information about magnetic field distribution and strength in the small and large scale, and photon fields present in the surrounding medium, render a complex picture with non-trivial solution. The spectral energy distribution of pulsed gamma-ray emission from most pulsars peak around a few GeV, above which the flux quickly decreases exponentially with energy. This exponential cutoff is expected when synchrotron-curvature is appealed to work in the pulsar magnetosphere, where particles reach a maximum energy limited by either magnetic absorption or radiation losses. Nevertheless, longer observations of bright GeV pulsars show an unexpected deviation from a pure exponential cutoff, questioning the previous established mechanisms (Li et al. 2016; Ahnen et al. 2016; Ansoldi et al. 2016; Breed et al. 2016). This deviation observed in phase-resolved spectra can arise from the caustic emission (Dyks and Rudak 2003; Viganò et al. 2015; Prosekin et al. 2013), i.e, overlapping of photons emitted at different heights and along different magnetic field lines or from a different interplay between acceleration and energy lose mechanisms. In particular, two pulsars have been recently detected by Cherenkov telescopes: deep observations of the Crab pulsar have revealed a second component of pulsed radiation that extends from GeV energies to above a few TeV (Ansoldi et al. 2016; VERITAS Collaboration et al. 2011). This unexpected result has drawn a large attention in the astrophysics community and it has triggered a large number of research theoretical projects aiming to unveil the mechanisms powering GeV/TeV pulsars and their winds; a second pulsar was detected using the H.E.S.S. II telescope above 20 GeV and up to 120 GeV, also departing from an exponential cutoff shape (Breed et al. 2016).

The gamma-ray luminosity in pulsars L_γ can be calculated from the observed flux G_{100} above 100 MeV, when the distance d is known, assuming a given beaming factor f_Ω, i.e. the fraction of the radiation that reaches the observer. The luminosity thus can be expressed as $L_\gamma = 4\pi d^2 f_\Omega G_{100}$, which introduces an unknown factor corresponding to the fan-like beam size when calculating the efficiency of the radiation transferred from the pulsar $\dot{E}$ to the pulsed emission does. A general trend close to $L_\gamma \propto \dot{E}$ has been observed, although Pierbattista et al. (2016) show that the range of variation depends on the pulsar type and magnetospheric model assumed. In the case of PWNe, the emitting electrons responsible for the TeV emission do not suffer from very severe radiative losses as in the case of i.e. X-ray nebulae, and the majority of them may survive from (and hence probe) early epochs of the PWNe evolution. Therefore the TeV emission observed corresponds to the accumulation of high-energy electrons from different evolution times. This interpretation has been further supported by the discovery of the spectral softening of the very-high energy nebulae HESS J1825−137 and HESS J1303−631 as a function of the distance from the pulsar (Aharonian et al. 2006b; Abramowski et al. 2012). The spectrum of PWNe is mostly determined by the accumulation of the injection electron spectrum but also by the nature of the target photon field for the inverse Compton scattering. Torres et al. (2014) studied the spectral energy distribution of 10 PWNe, modelling their emission using a time-dependent description of the nebulae's electron population. The injection parameters derived from the TeV modelling do not appear to be particularly correlated with the pulsar properties. On the other hand, Abeysekara and Linnemann (2015) studied the correlation between the LAT pulsars and their corresponding TeV PWNe in terms of luminosity. Abeysekara and Linnemann (2015) found a linear correlation (R=0.82) between the PSR luminosity (assuming a beaming factor of f_{Ω}) and TeV PWNe luminosity, which contrary to the pulsar one, comprised a 4π area.

We investigated the relation ship between the TeV PWN luminosity and the various parameters than defined the pulsed emission at GeV, namely, the spin-down energy of the pulsar $\dot{E}$, the GeV luminosity L_{γ}, the spectral index and the pulsar efficiency, defined as the fraction of rotational energy transferred to the GeV pulsed emission. The TeV PWNe luminosity has been compared with the pulsar rotational energy, resulting on a lack of dependency between the two parameters (Mattana et al. 2009). However, the larger sample studied by Klepser et al. (2016) resulted on a clear correlation of luminosity with pulsar spin-down, suggesting a dependency of $L_{1-10\mathrm{TeV}} = \dot{E}^{0.6\pm0.2}$. This correlation is somehow dimmed when focusing only in the LAT pulsars sample. The dashed-black line on Fig. 4.5 top shows such relation. The black-dashed line results from a linear fitting which clearly does not represent well the population. However, it is interesting to note that the fit is mainly affected by two young pulsars, the Crab and 3C58. The dashed-dotted line indicates a certain trend resulting from fitting the population excluding the two later, although the large errors prevent a claim in this direction. Likewise, a trend seems to be present when comparing the PWNe luminosity with the pulsar one, somehow expected from the correlation between the pulsar luminosity and the spin-down power. Note again, that we do not claim at this stage any quantitative correlation at this stage. When comparing the luminosity with the pulsar efficiency, no tendency or correlation is observed, likewise the spectral index. This is expected since the two parameters rather depend on the propagation and geometry of particles within the magnetosphere and not on the injection spectrum or total pulsar energetics.

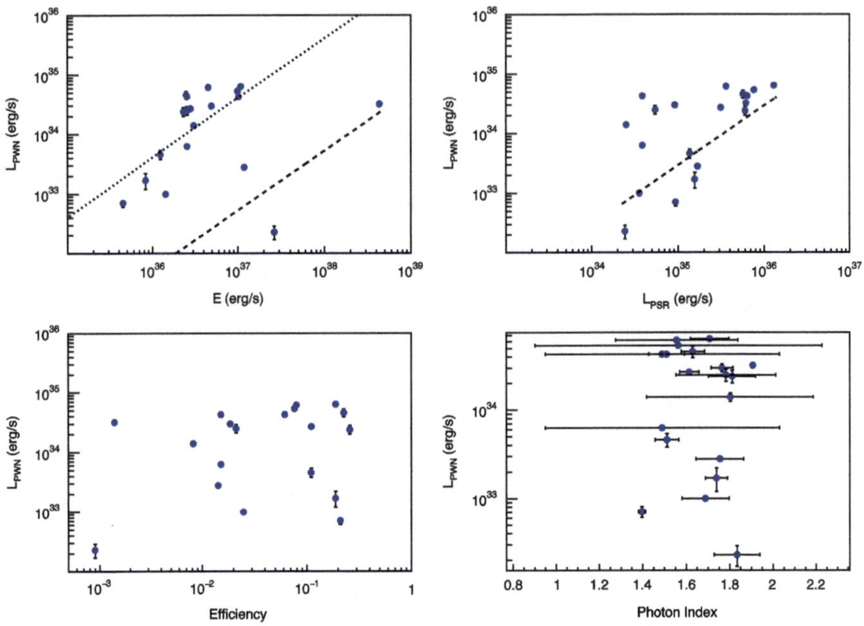

Fig. 4.5 Correlation between the TeV PWN Luminosity (see Table 4.1 and different parameters that define pulsars spectral energy distribution. The *top panels* show the correlation with the spin-down energy $\dot{E}$ and the pulsar luminosity (*left* and *right* respectively). The *bottom panels* show the TeV luminosity as compare with the pulsar efficiency and the pulsed spectral photon index (*left* and *right*)

4.5 New Type of Pulsars and Pulsar Wind Nebula

Beside MSP pulsars, the LAT has revealed new type of compact accelerators to high energies previously unknown. Among them are the transitional millisecond pulsars (such i.e. PSR J1023+0038 (Tam et al. 2010), see Fig. 4.6, or XSS J12270−4859 (Johnson et al. 2010)), black window systems (such 2FGL J1311.7−3429 (Romani 2012)) or, less certain, accreting millisecond pulsar such SAX J1808.4−3658 (de Oña Wilhelmi et al. 2016). The discovery of millisecond pulsars switching between states powered either by the rotation of their magnetic field or by the accretion of matter has recently proved the tight link shared by millisecond radio pulsars and isolated neutron stars in low-mass X-ray binaries. Those transitional millisecond pulsars also show an intermediate state in which the neutron star is likely surrounded by an accretion disk and emits coherent X-ray pulsations. In that intermediate state, the X-ray emission does not reach the level observed on accreting neutron stars, but at GeV energies it emits a large luminosity compared to the previous state. Several possibilities have been suggested to explain the complex phenomenology observed on those systems, implying acceleration of electrons in the pulsar wind or/and the intra-shock regime produced by the disk (Stappers et al. 2014; Takata et al. 2014;

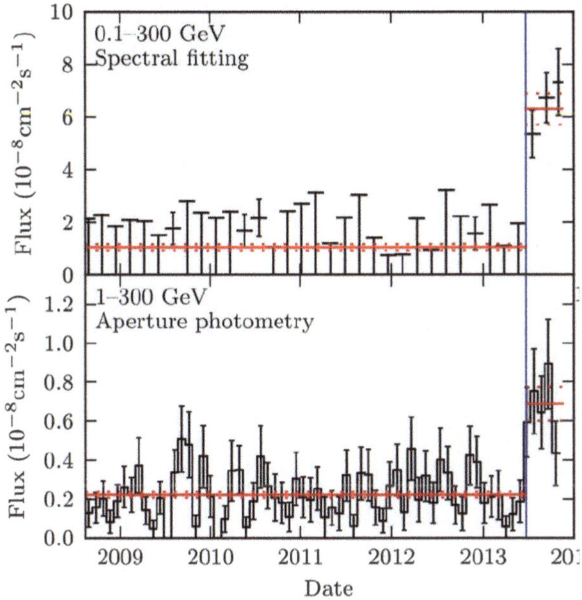

Fig. 4.6 Gamma-ray photon flux from PSR J1023+0038 (Stappers et al. 2014). *Top panel*: the 100 MeV to 300 GeV flux determined by spectral fitting. *Bottom panel*: the 1–300 GeV flux determined by aperture photometry, with Poisson uncertainties

Coti et al. 2014). In general, the new objects type detected by LAT show a spectral energy distribution characterised by a power law function plus and exponential cutoff at a few GeV. Radiation at TeV energies, if existing, should thus be produced by a new inverse Compton component. Observations at very-high energy (Aliu et al. 2016) of PSR J1023+0038 have resulted on upper limits, imposing limits on the maximum magnetic field strength in the synchrotron/inverse Compton region. The most promising results at very-high energy concerning this new type of sources are HESS J1808−204 (see Fig. 4.7) (Abdalla et al. 2016), TeV J2032+4130 (Albert et al. 2008) (in Fig. 4.8) and HESS J1747−248 (H.E.S.S. Collaboration et al. 2011) (in Fig. 4.9). The first is an extended very-high gamma-ray emission towards SGR1806−20 and the stellar cluster Cl* 1806−20, the second is an extended source that has been associated with an evolved pulsar which is gravitationally linked to a massive star companion, and the third TeV source is positionally in agreement with the globular cluster Terzan 5. The magnetar SGR 1806−20 is one of the most prominent and burst-active soft gamma repeaters. The emission detected at TeV energies is coincident with the synchrotron radio nebula G10.0−0.3 which is believed to be powered by LBV 1806−20. It is also in agreement with the position of the GeV source 3FGL J1809.2−2016c (Yeung et al. 2016). The GeV and TeV maps show a very good morphological agreement between the two sources, whereas the spectrum extends as a power-law function from a few hundreds of MeV to a few tens of TeV. The WR star LBV 1806−20 that belongs to the cluster Cl* 1806−20, which

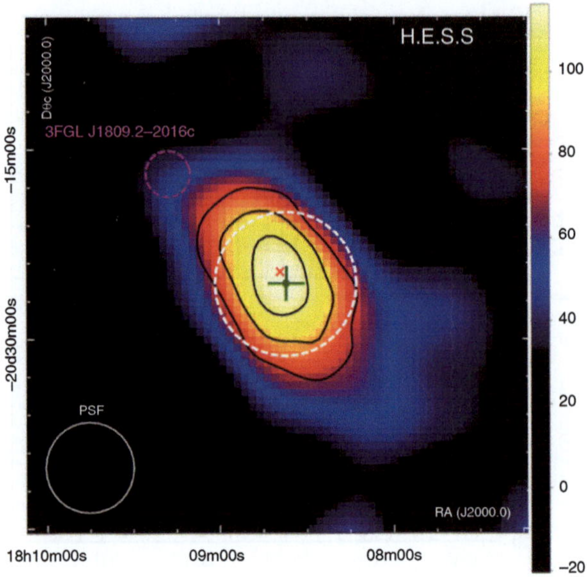

Fig. 4.7 H.E.S.S. exposure-corrected excess counts image of HESS J1808−204 towards the stellar cluster Cl* 1806−20 (*red cross*), containing SGR 1806−20 and LBV 1806−20 (Abdalla et al. 2016)

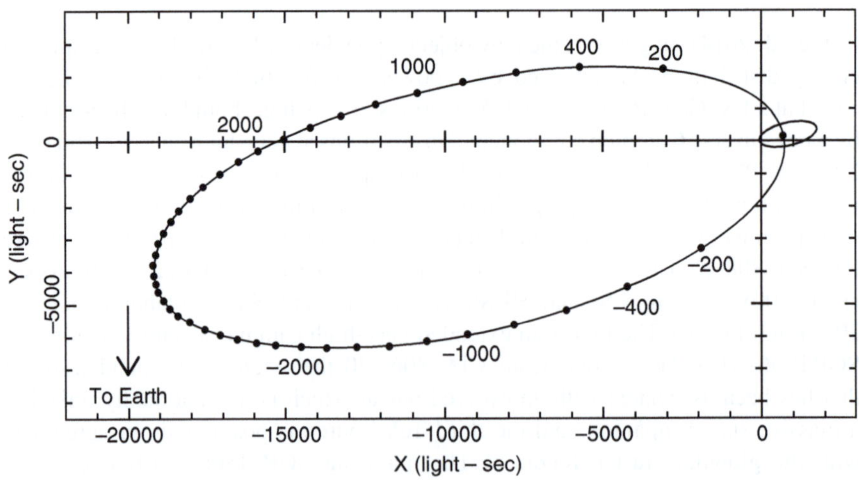

Fig. 4.8 Schematic diagram showing the approximate orbital motions of PSR J2032+4127 and its Be-star companion MT91 213 (Lyne et al. 2015)

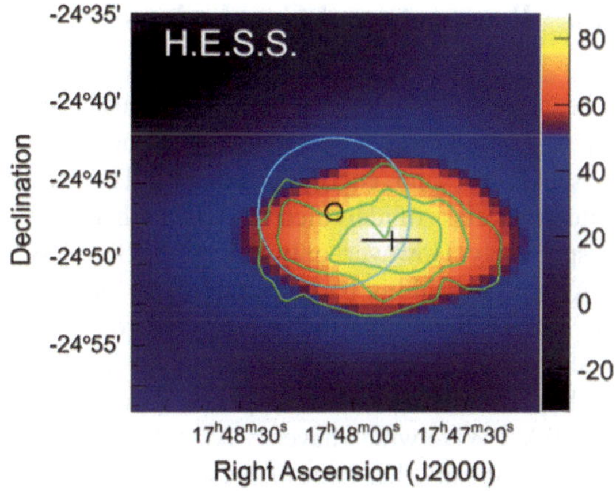

Fig. 4.9 Gamma-ray image of the Terzan 5 region (H.E.S.S. Collaboration et al. 2011). *Green lines* give 4–6 sigma significance contours. The *circles* show the half-mass radius of about 30 arcseconds (in *black*) and the larger tidal radius of 4.6 arcminutes (in *cyan*) of the GC

displays a kinetic stellar wind luminosity of L_W=10^{38} erg/s could easily account for the TeV luminosity via hadronic or leptonic mechanisms. At GeV energies, Yeung et al. (2016) have claimed variability from the direction of the magnetar. The association of the GeV and TeV emission with the magnetar, if confirmed, will entail the first TeV pulsar wind nebula powered by a magnetic-driven wind. H.E.S.S. has also detected very-high energy emission from the direction of the cluster Terzan 5 (H.E.S.S. Collaboration et al. 2011), which is especially rich in eclipsing binary systems among globular clusters. The origin of the emission was proposed to be related to the combined effect of the millisecond pulsars winds or to one particular one. Unfortunately, the TeV emission is displaced with respect the GeV one, and a clear association is still on hold. Searches for VHE emission from the globular clusters 47 Tucanae (Aharonian et al. 2009), M5, M15 (McCutcheon 2009), and M13 (Anderhub et al. 2009; McCutcheon 2009) have only resulted on upper limits. Finally, the unidentified source TeV J2032+4130 has been recently associated with the LAT pulsar PSR J2032+4127. PSR J2032+4127 is an energetic pulsar with a X-ray nebula associated (Murakami et al. 2011), and the TeV-GeV association could be easily established except, perhaps, for the pulsar age (0.11 Myr). However, radio observations have shown that the pulsar is in a highly-eccentric binary ($\sim$100 yrs) system with a 15-solar-mass Be star, MT91 213 (Lyne et al. 2015). That renders the association between the LAT pulsar and the TeV source an intriguing aspect, since it will imply either an extremely fast ($<$100 years) evolution of a high-energy nebula or a nebula created being disrupted only by the companion every pass on the periastron. The next passage which will be happening next year will shed some light on this interesting object.

4.6 Prospects and Conclusions

The association with energetic LAT pulsar powering TeV PWNe is supported by a combination of positional and morphological evidence, multi-wavelength observations, and energy arguments. The two populations constitute the most numerous class of identified Galactic high and very-high energy gamma-ray sources respectively. However, only half of the TeV PWNe candidates have an associated LAT pulsar. This is partially due to an observational bias (see Fig. 4.3) towards regions of the Galaxy which are not too contaminated by the diffuse Galactic background, or simply to the fact that the large size of the TeV sources often hinders a clear association with a unique GeV pulsar. However, it seems like luminosity of the TeV PWNe which have a LAT pulsar associated weakly depends on the pulsed GeV luminosity. The large dispersion (see Fig. 4.4) can be explained the dependency of the pulsed emission on the beaming factor and/or by the dependency of the PWN luminosity on many parameters, such photon field, age, etc.

At TeV energies, deep exposures unveiling cooling processes and complex morphology with sensitive instrument such the future CTA, will allow a clear determination of the key parameters that rule the high and very-high energy emission. Likewise, future observations will also allow a complete population study of the Galaxy (up to $\sim$15 pc, see de Oña-Wilhelmi et al. 2013) and an unbiased study of the correlation between LAT pulsar population and TeV PWN sources. These information will be crucial to develop time-dependent morphological models, which will finally reveal the intrinsic connections between the compact $\sim 10^6$ cm radius pulsars and the enormous tens-of-parsec extension nebulae.

References

Abdalla, H., Abramowski, A., Aharonian, F., et al.: arXiv:1606.05404 (2016)
Abdo, A.A., Ackermann, M., Ajello, M., et al.: Science **325**, 840 (2009)
Abdo, A.A., Ajello, M., Allafort, A., et al.: Astrophys. J. Suppl. Ser. **208**, 17 (2013)
Abeysekara, A.U., Linnemann, J.T.: Astrophys. J. **804**, 25 (2015)
Abramowski, A., Acero, F., Aharonian, F., et al.: Astron. Astrophys. **548**, A38 (2012)
Aharonian, F., Akhperjanian, A.G., Aye, K.-M., et al.: Astron. Astrophys. **432**, L25 (2005a)
Aharonian, F., Akhperjanian, A.G., Aye, K.-M., et al.: Astron. Astrophys. **435**, L17 (2005b)
Aharonian, F., Akhperjanian, A.G., Bazer-Bachi, A.R., et al.: Astron. Astrophys. **457**, 899 (2006a)
Aharonian, F., Akhperjanian, A.G., Bazer-Bachi, A.R., et al.: Astron. Astrophys. **460**, 365 (2006b)
Aharonian, F., Akhperjanian, A.G., Bazer-Bachi, A.R., et al.: Astron. Astrophys. **472**, 489 (2007)
Aharonian, F., Akhperjanian, A.G., Anton, G., et al.: Astron. Astrophys. **499**, 273 (2009)
Aharonian, F.A., Bogovalov, S.V., Khangulyan, D.: Nature **482**, 507 (2012)
Ahnen, M.L., et al.: Astron. Astrophys. **591**, A138 (2016)
Albert, J., Aliu, E., Anderhub, H., et al.: Astrophys. J. **675**, L25 (2008)
Aleksić, J., Ansoldi, S., Antonelli, L.A., et al.: J. High Energy Astrophys. **5**, 30 (2015)
Aliu, E., Archambault, S., Archer, A., et al.: arXiv:1609.01692 (2016)
Anderhub, H., Antonelli, L.A., Antoranz, P., et al.: Astrophys. J. **702**, 266 (2009)
Ansoldi, S., et al.: Astron. Astrophys. **585**, A133 (2016)

Archer, A., VERITAS Collaboration: APS April Meeting Abstracts (2015)
Atwood, W.B., et al.: Astrophys. J. **697** 1071 (2009)
Bednarek, W., Sitarek, J., Sobczak, T.: Mon. Not. R. Astron. Soc. **458**, 1083 (2016)
Breed, M., Venter, C., Harding, A.K.: arXiv, arXiv:1607.06480 (2016)
Bogovalov, S., Tsinganos, K.: Mon. Not. R. Astron. Soc. **357**, 918 (2005)
Caraveo, P.A.: Annu. Rev. Astron. Astrophys. **52**, 211 (2014)
Cerutti, B., Philippov, A.A., Spitkovsky, A.: Mon. Not. R. Astron. Soc. **457**, 2401 (2016)
Contopoulos, I.: Mon. Not. R. Astron. Soc. **463**, L94 (2016)
Coti Zelati, F., Baglio, M.C., Campana, S., et al.: Mon. Not. R. Astron. Soc. **444**, 1783 (2014)
de Oña Wilhelmi, E.: Am. Inst. Phys. Conf. **1357**, 213 (2011)
de Oña-Wilhelmi, E., Rudak, B., Barrio, J.A., et al.: Astropart. Phys. **43**, 287 (2013)
de Oña Wilhelmi, E., Papitto, A., Li, J., et al.: Mon. Not. R. Astron. Soc. **456**, 2647 (2016)
Del Zanna, L., Amato, E., Bucciantini, N.: Astron. Astrophys. **421**, 1063 (2004)
Del Zanna, L., Volpi, D., Amato, E., Bucciantini, N.: Astron. Astrophys. **453**, 621 (2006)
Dyks, J., Rudak, B.: Astrophys. J. **598**, 1201 (2003)
Fang, J., Zhang, L.: Astron. Astrophys. **515**, A20 (2010)
Gallant, Y.A., et al.: Am. Inst. Phys. Conf. **983**, 195 (2008)
Gelfand, J.D., Slane, P.O., Zhang, W.: Astrophys. J. **703**, 2051 (2009)
Gonthier, P.L., Story, S.A., Clow, B.D., Harding, A.K.: Astrophys. Space Sci. **309**, 245 (2007)
Grégoire, T., Knödlseder, J.: Astron. Astrophys. **554**, A62 (2013)
Grenier, I.A., Harding, A.K.: C. R. Phys. **16**, 641 (2015)
H.E.S.S. Collaboration, Abramowski, A., Acero, F., et al.: Astron. Astrophys. **531**, L18 (2011)
H.E.S.S. Collaboration, Abramowski, A., Acero, F., et al.: Astron. Astrophys. **548**, A46 (2012)
Johnson, T.J., Ray, P.S., Roy, J., et al.: Astrophys. J. **806**, 91 (2015)
Johnson, T.J., Smith, D.A., Kerr, M., den Hartog, P.R., Fermi Large Area Telescope Collaboration, Timing Consortium P., Search Consortium P.: Am. Astron. Soc. **223**, 153.02 (2014)
Kalapotharakos, C., Kust Harding, A., Kazanas, D., Brambilla, G.: Am. Astron. Soc. **227**, 423.01 (2016)
Kargaltsev, O., Rangelov, B., Pavlov, G.G.: arXiv, arXiv:1305.2552 (2013)
Kennel, C.F., Coroniti, F.V.: Astrophys. J. **283**, 710 (1984)
Klepser, S., Gallant, Y., Mayer, M., Valerius, K., For The H.E.S.S. Collaboration: Proc. 35th ICRC, Busan, South Korea, PoS(ICRC2017)731. arXiv:1707.04024 (2017)
Komissarov, S.S., Lyubarsky, Y.E.: Mon. Not. R. Astron. Soc. **344**, L93 (2003)
Komissarov, S., Lyubarsky, Y.: Astrophys. Space Sci. **293**, 107 (2004)
Li, J., Torres, D.F., de Ona Wilhelmi, E., Rea, N., Martin, J.: Astrophys. J. **831**, 19 (2016)
Lyne, A.G., Stappers, B.W., Keith, M.J., et al.: Mon. Not. R. Astron. Soc. **451**, 581 (2015)
Lyubarsky, Y.E.: Mon. Not. R. Astron. Soc. **329**, L34 (2002)
Martín, J., Torres, D.F., Rea, N.: Mon. Not. R. Astron. Soc. **427**, 415 (2012)
Mattana, F., Falanga, M., Götz, D., et al.: Astrophys. J. **694**, 12 (2009)
McCutcheon, M., for the VERITAS Collaboration: arXiv:0907.4974 (2009)
Mochol, I., Pétri, J.: Mon. Not. R. Astron. Soc. **449**, L51 (2015)
Murakami, H., Kitamoto, S., Kawachi, A., Nakamori, T.: Publ. Astron. Soc. Jpn. **63**, S873 (2011)
Muslimov, A.G., Harding, A.K.: Astrophys. J. **617**, 471 (2004)
Nolan, P.L., et al.: Astron. Astrophys. Suppl. Ser. **120**, 61 (1996)
Pétri, J.: Astron. Astrophys. **574**, A51 (2015)
Pedaletti, G., de Oña Wilhelmi, E., Torres, D.F., Natale, G.: J. High Energy Astrophys. **5**, 15 (2015)
Pierbattista, M., Harding, A.K., Gonthier, P.L., Grenier, I.A.: Astron. Astrophys. **588**, A137 (2016)
Porth, O., Komissarov, S.S., Keppens, R.: Mon. Not. R. Astron. Soc. **438**, 278 (2014)
Prosekin, A.Y., Kelner, S.R., Aharonian, F.A.: arXiv:1305.0783 (2013)
Ray, P.S., Parkinson, P.M.S.: Astrophys. Space Sci. Proc. **21**, 37 (2011)
Romani, R.W.: Astrophys. J. **754**, L25 (2012)
Romani, R.W., Yadigaroglu, I.-A.: Astrophys. J. **438**, 314 (1995)
Spitkovsky, A.: Astrophys. J. **648**, L51 (2006)
Stappers, B.W., Archibald, A.M., Hessels, J.W.T., et al.: Astrophys. J. **790**, 39 (2014)

Takata, J., Shibata, S., Hirotani, K., Chang, H.-K.: Mon. Not. R. Astron. Soc. **366**, 1310 (2006)
Takata, J., Li, K.L., Leung, G.C.K., et al.: Astrophys. J. **785**, 131 (2014)
Tam, P.H.T., Hui, C.Y., Huang, R.H.H., et al.: Astrophys. J. **724**, L207 (2010)
Tanaka, S.J., Takahara, F.: Mon. Not. R. Astron. Soc. **429**, 2945 (2013)
Tavani, M., AGILE Team: Nucl. Instrum. Methods Phys. Res. A **630**, 7 (2011)
Tchekhovskoy, A., Philippov, A., Spitkovsky, A.: Mon. Not. R. Astron. Soc. **457**, 3384 (2016)
Thompson, D.J., Bertsch, D.L., O'Neal Jr., R.H.: Astrophys. J. Suppl. Ser. **157**, 324 (2005)
Torres, D.F., Cillis, A., Martín, J., de Oña Wilhelmi, E.: J. High Energy Astrophys. **1**, 31 (2014)
Vallée, J.P.: Astrophys. J. **681**, 303–310 (2008)
VERITAS Collaboration, Aliu, E., Arlen, T., et al.: Science **334**, 69 (2011)
Viganò, D., Torres, D.F., Martín, J.: Mon. Not. R. Astron. Soc. **453**, 2599 (2015)
Watters, K.P., Romani, R.W.: Astrophys. J. **727**, 123 (2011)
Weekes, T.C., et al.: Astrophys. J. **342**, 379 (1989)
Yeung, P.K.H., Kong, A.K.H., Tam, P.H.T., et al.: Astrophys. J. **827**, 41 (2016)

Chapter 5
Prospects for Pulsar Wind Nebulae Observations with γ-Ray Astronomy Facilities: Cherenkov Telescope Array and Satellites

Giovanna Pedaletti

Abstract In the coming years, many facilities will be delivering high quality γ-ray data essential for the advancement of pulsar wind nebulae (PWN) studies. I will be reviewing the technical capabilities of various instruments: from satellites devoted to the MeV–GeV regime, to ground based observatories intended for energies >10 GeV. I will especially cover the prospects for CTA (Cherenkov Telescope Array), the next generation array of Imaging Air Cherenkov Telescopes, sensitive in the range from few tens of GeV to hundreds of TeV. Thanks to its substantial improvement in flux sensitivity, angular and energy resolution and instantaneous field of view with respect to the current Cherenkov arrays, CTA is expected to increase the number of detected PWN to the hundreds and allow a detailed morphological and spectral study of the most prominent ones. I also will cover some of the satellites planned to be launched in the next decade.

5.1 Introduction: An Overview of the Next *"gamma"* Decade

The 2020s will be an exciting decade for the most energetic photon window. The first revolution will come from the VHE (Very High Energy, 10 GeV $\lesssim E \lesssim$ 300 TeV) γ-ray regime. The Cherenkov Telescope Array (CTA) (Acharya et al. 2013) is currently in the preproduction phase as of 2016, but the first telescopes will be on site by the beginning of 2018 and full array operation is expected for the beginning of next decade. The CTA concept, its instrumental capabilities and some of the most exciting physics cases related to pulsar wind nebula (PWN) science will be summarized in Sect. 5.2. The extremely successful *Fermi* (Atwood et al. 2009) and AGILE (Tavani et al. 2009) satellite missions will probably come to an end during the CTA lifetime, but some follow up missions are being explored to continue covering the GeV regime and will be described in Sect. 5.4. The improved angular resolution of these missions will enhance the flux sensitivity and will be interesting

G. Pedaletti (✉)
DESY, Zeuthen, Germany
e-mail: giovanna.pedaletti@desy.de

D.F. Torres (ed.), *Modelling Pulsar Wind Nebulae*, Astrophysics and Space Science Library 446, DOI 10.1007/978-3-319-63031-1_5

to deepen the morphology studies of PWN in the GeV regime. Looking even a few years more into the future, with CTA operation in full swing, the new addition is expected in a lower gamma energy window closing what has been called the *"MeV gap"*. The satellite concept discussed here combine the Compton detection technique, sensitive up to tens of MeV, with the pair production at higher photon energies. The detector concepts, instrument capabilities and impact for PWN studies will be covered in Sect. 5.4.2. All the major instruments described in the following text are put in context of their relative instrument performances (only sensitivity and angular resolution are shown) in Fig. 5.1. Please note the exact definition of quantities in the caption of Fig. 5.1. Some of the currently operating facilities are also shown in Fig. 5.1 for comparison, but will not be discussed in the following chapter. Other chapters in this volume will be covering the state of the art in the γ-ray energy band, see Acero (2017); de Oña-Wilhelmi (2017); Zanin (2017). The chapter is not a complete review of all the facilities in the next decade. The recently launched CALET and DAMPE particle detectors with γ-ray detection capabilities are not covered, nor are the planned extensive air shower arrays for energies E$\gtrsim$10 TeV. For more information, an excellent review is presented in Knödlseder (2016).

5.2 The Future of VHE Astronomy: CTA

CTA is an international (world-wide) project for the development of the next generation ground-based γ-ray instrument (Acharya et al. 2013). CTA will be an open Observatory, with a foreseen lifetime of $\sim$30 years. The advances of CTA with respect to the present generation of Imaging Atmospheric Cherenkov Telescopes (IACTs) are the foreseen order of magnitude improvement in sensitivity, an improvement in angular resolution and wider energy coverage, the larger field of view (FoV) and the faster slewing capability. The detection of γ-rays with energies E $\gtrsim$ 10 GeV with ground based facilities is made possible thanks to the imaging atmospheric Cherenkov technique. VHE γ-rays interact with particles in the atmosphere on their way to the ground producing a cascade of particles, whose velocities are larger than the speed of light in the medium, leading to Cherenkov light emission. The showers produce flashes of Cherenkov light that can be imaged by IACTs. The shower images are then used to reconstruct the energy and direction of the original photon. Particle cascades can also be initiated by background cosmic rays and constitute the main source of background for VHE astronomy. Hadron-initiated showers are generally broader than those initiated by photons and gamma-hadron separation can be achieved thanks to the differences of the shower images. The Cherenkov flashes from showers illuminate an area at the ground level, the size of which depends on the energy of the primary γ-ray. An array of IACTs positioned inside this illuminated area makes for a stereoscopic system and allows for finer sampling and reconstruction of the Cherenkov light distribution with respect to a single imaging telescope. The maturity of the stereoscopic technique

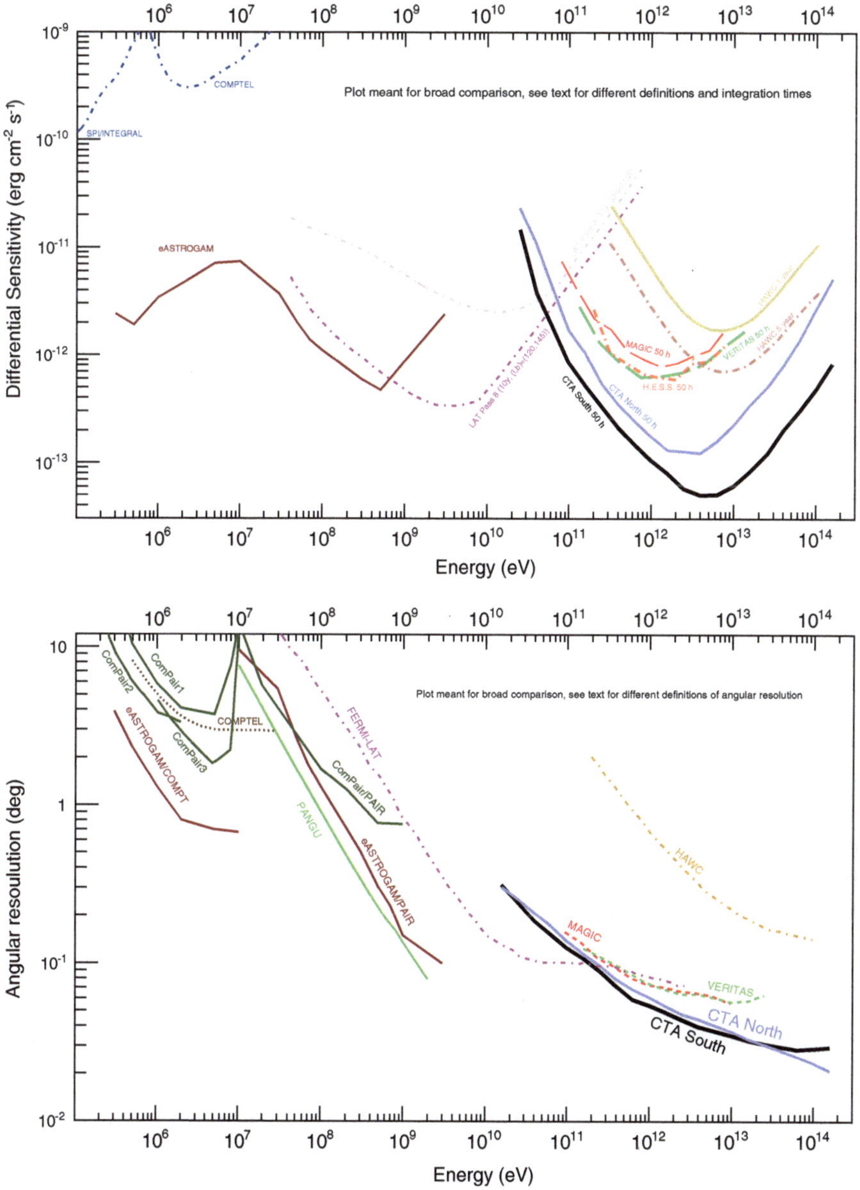

Fig. 5.1 Instrument performances for the major facilities described in the text, with a comparison to the state-of-the-art facilities. *Top*: Differential sensitivity. Shown here are the following sensitivities: 3σ sensitivity for Integral/SPI for a 10^6 s observation, from SPI Observer Manual (2017); COMPTEL for the full mission livetime ($\sim$9 years), from Knödlseder (2016); e-ASTROGAM 3σ sensitivity for continuum emission ($\Delta E = E$) of a point source located at high latitude, observed for 1 year effective exposure, from Tatischeff et al. (2016); Fermi-LAT sensitivity for a high and low galactic latitude 10 years exposure and 5σ sensitivity for MAGIC, H.E.S.S, VERITAS, and the Cherenkov Telescope Array (CTA) north and south (50 h observation), from CTA (2017) and references therein; HAWC sensitivity for a 1 year and 5 year full scan, from HAWC (2017).

and the huge discovery potential has been already demonstrated by the previously (HEGRA) and currently operational IACT arrays (H.E.S.S. MAGIC, VERITAS).

The CTA array is expected to have an unprecedented sensitivity down to ~50 GeV and above ~50 TeV, establishing a perfect link to water Cherenkov experiments at the highest energies and to satellites at low energies. In particular, the CTA gain in sensitivity is due to a combination of different sized telescopes in different parts of the light pool. Large telescopes with dish diameter ~23 m will be placed at the center of the array. Thanks in large part to their large mirror area, dim flashes from the low energy events are expected to be reconstructed. Tens of medium size telescopes, with dish diameter ~11 m, will be positioned so to cover a larger area on the ground, thus covering a large fraction of the light pool and enhancing the reconstruction of medium energy events. Finally, small size telescopes, with dish diameter ~4–7 m, will be placed in an even larger area on the ground, out to a radius of O(1 km). This will assure the increase of the effective area of the array for the bright but rare high energy events. A southern and a northern hemisphere site are foreseen as part of the CTA project, so that a full sky coverage is assured. An artistic impression of a possible CTA configuration is shown in Fig. 5.2 along with a comparison of the size of the three types of telescopes. A simplified structure is shown, with only single mirror telescopes. All the IACTs currently operating use single mirror telescopes, but CTA will comprise also dual mirror telescopes, with some prototypes already constructed. A detailed description of the different types of telescopes can be found in Acharya et al. (2013).

The performance of CTA has been the object of intense studies with detailed Monte Carlo simulations. The simulations start from the extensive air shower development in the atmosphere and include detailed reproduction of the optics and detector characteristics. Details can be found in Bernlöhr et al. (2013). The entire data analysis pipeline is then applied to the simulated signal (γ-ray) and background samples, to eventually obtain the instrument response functions (IRFs). Only protons are simulated for the background, due to computational limitations, as they anyway constitute the dominant background component.

Numerical values of the CTA IRFs can be downloaded from https://portal.cta-observatory.org/Pages/CTA-Performance.aspx. While simulations of even increased realism are still being run, the IRFs from the web site referenced above are expected to represent very closely the final performance.

Fig. 5.1 (continued) *Bottom*: Angular resolution. e-ASTROGAM: In the Compton domain (below 10 MeV), FWHM of the angular resolution measure for a source on axis. In the pair domain (above 10 MeV), it is the 68% containment radius for a 30°-off-axis point source, from Tatischeff et al. (2016). ComPair angular resolution from Moiseev et al. (2015), FWHM for Compton events and 68% containment radius for pair events (*Line 1*: Compton tracked events; *line 2*: Compton untracked events; *line 3*: Compton tracked events with a deeper non-baseline CZT calorimeter, see Sect. 5.4.2.2 for details). PANGU, 68% Point Spread Function at normal incidence, from Wu et al. (2016). Fermi-LAT 68% containment angle of the acceptance weighted PSF, from Fermi-LAT public performance (2017). 68% containment radius for MAGIC, VERITAS, CTA north and south and HAWC, from CTA (2017) and references therein

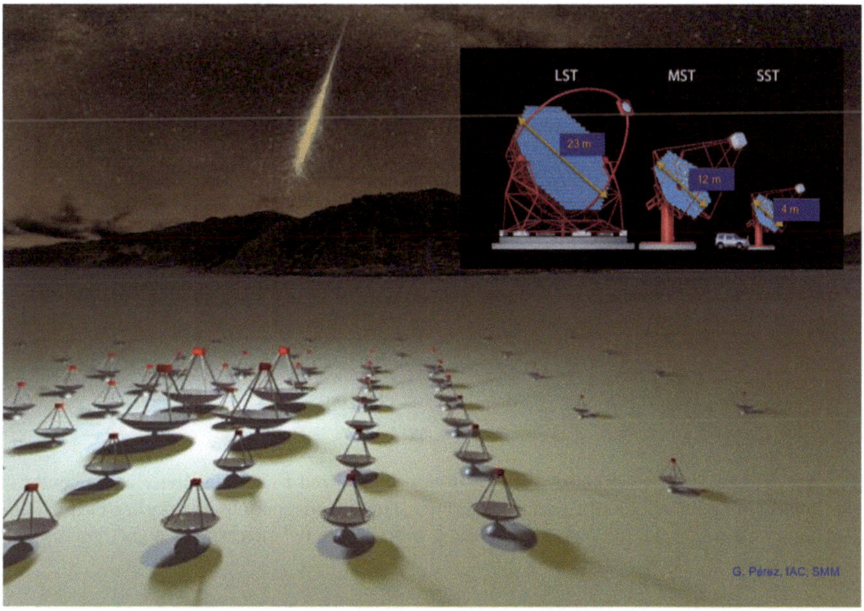

Fig. 5.2 Artistic impression of a possible CTA south configuration. Four LSTs are placed in the center of the array, with MSTs and SSTs placed in an evenly spaced grid to larger distances from the center of the array. Comparative sizes of the three types of telescopes are shown in the *inset*, see text for details

Figure 5.1 shows some of the performance characteristics in a broadband context, namely the differential flux sensitivity and the angular resolution. CTA is expected to deliver an impressive improvement in sensitivity in the southern array of one order of magnitude in the full energy range from tens of GeV to hundreds of TeV. The more modest improvement expected for the northern array is principally due to its smaller covered area and smaller number of telescopes. The northern array indeed is conceived to focus on the lower-energy range and no SSTs are planned in the norther site, thus the performance of the array at higher energies is especially affected. It is to be noted that IACTs operate mostly background free at the energies of tens of TeV, therefore the gain in sensitivity as a function of time is expected to be linear. The effective FoV (field-of-view) for each CTA observation will also be larger than currently achieved by IACT arrays and it will be exploited in the planned surveying of large portions of the sky. Figure 5.3 shows the degradation of point-source sensitivity as function of offset from the center of the FoV. This can be used as a proxy to understand the effective size of the FoV and its dependency on energy range. The FoV at the lowest reconstructed photon energy is limited, so survey capabilities are usually investigated for $E > 100\,\mathrm{GeV}$.

The angular resolution shown in Fig. 5.1 is taken from the public CTA IRFs. Those performance values are achieved for simulations targeted to obtain the best possible sensitivity. It is possible to instead tune the analysis towards the best

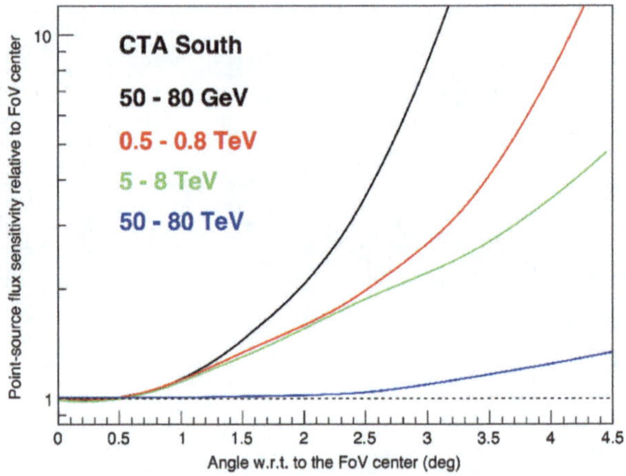

Fig. 5.3 Relative off-axis point-source sensitivity (normalized to the sensitivity at the center of the field of view) for four different energy bins [https://portal.cta-observatory.org/Pages/CTA-Performance.aspx]. Note the flatter dependency to the angle for increasingly higher energy bins

angular resolution, selecting, e.g., those reconstructed events imaged by several telescopes. The higher multiplicity ensures a larger precision in morphological reconstruction, accepting a loss of sensitivity. Therefore this approach is only appropriate for high fluxes and extended sources. This is shown in Fig. 9 of Acharya et al. (2013). This point is represented also in the comparison of the CTA north and south angular resolution. CTA north simulations lead to a better performance in angular resolution at the highest energies with respect to the southern array. This is due also to the fact that the events surviving at those energies in CTA north simulations are imaged only by MSTs, thus assuring a better reconstruction. But the simulated flux will be probably larger than considered in the CTA south simulation.

Another pillar of CTA capabilities will be the performance for fast transients at low energies: an increase of differential sensitivity of a factor of $\sim 10^4$ at energies $E < 100\,\text{GeV}$ for a 1 h time integration with respect to *Fermi*-LAT (Funk and Hinton 2013).

5.3 The Impact of CTA on the Physics of PWN

CTA will provide a goldmine of observations for PWN. I here will show three examples: the Galactic plane survey in Sect. 5.3.1; energy-dependent morphology capabilities in Sect. 5.3.2 and the observation of Crab and flares (Sect. 5.3.3). The study of specific PWN will not be covered here.

5.3.1 CTA Galactic Plane Survey: A Key Science Program

CTA will be an open observatory, with time available to scientists of all the participating countries. However, a certain percentage of observation time, decreasing every year, will be reserved to the CTA Consortium. This time will be mainly dedicated to the completion of the KSP (Key Science Programs). A complete overview of the KSPs and the expected impact to the VHE field is reported in CTA (2017).

One of the KSP cornerstones will be the Galactic Plane Survey (GPS), showcasing one of the unique features of CTA: *"the ability to produce the deepest surveys of the sky at VHE"; "the ability to open up the shortest timescale phenomenology at very-high energies"*. The current vision for the CTA GPS will guarantee the coverage of the full galactic plane. The currently operating experiments already surveyed a large fraction of the Galactic plane, discovering O(100) sources. A large fraction of the Galactic sources has been firmly identified as PWN, but the classification of the majority of the sources stays unidentified. For a comprehensive review of the state-of-the-art of PWN observations at TeV see Acero (2017) in this volume and Klepser (2016).

The observation time foreseen for the CTA GPS will allow us to reach a sensitivity level of 1.8 mCrab[1] units in the inner Galactic plane region ($-60° < l < 60°$) and 4 mCrab or better in the remaining scan CTA (2017). The numbers refer to point source sensitivity for a 5σ detection level. A double row strategy will be adopted, with a step in pointing of $\sim 2°$, ensuring a flatter acceptance on a large Galactic latitude range. The full sensitivity scan will be completed in a 10 years span. However, partial releases of the dataset during the years will already constitute a significant improvement over the currently available dataset. These early releases will also aid Guest Observers in proposal preparation.

From a simple extrapolation in the *"logN-logS"* plane (number of sources N above a certain emission level S), the target sensitivity of the CTA GPS will allow the detection of hundreds of sources (O(500), Renaud (2009)). A similar number could be expected for PWN alone, from population studies presented in Dubus et al. (2013). Expectations derived from the currently known population of Galactic sources in the Fermi-LAT catalogs is described also in CTA (2017). This level of sensitivity, for a source similar to, e.g., G21.5−0.9, would lead to detection even at a distance of ∼20 kpc, effectively ensuring complete coverage of a Galactic population of similar luminosity (de Oña-Wilhelmi 2013).

The CTA consortium has been simulating a possible realization of the CTA GPS. Figure 5.4 shows one such simulation effort, zoomed on the inner Galactic plane. The figure illustrates that source confusion will be a major technical challenge for the analysis of the CTA GPS dataset. In the inner Galactic plane ($-30° < l < 30°$) it is expected to have a source density of up to 3–4 sources per square degree. Many

[1] 1mCrab: $I(> 125\,\mathrm{GeV}) = 5.07 \times 10^{-13}\mathrm{cm}^{-2}\mathrm{s}^{-1}$.

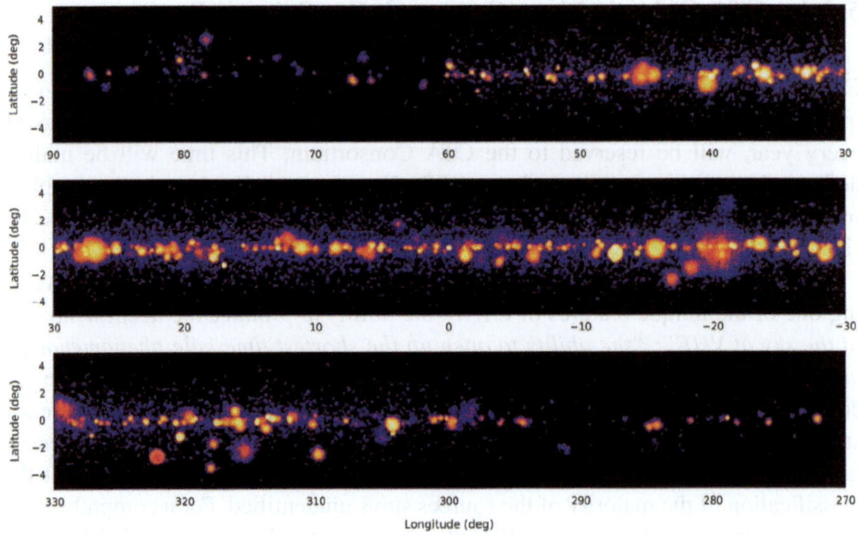

Fig. 5.4 Simulation of the CTA GPS, from CTA (2017), for the inner region of $(-80^\circ < l < 80^\circ; -5^\circ < b < 5^\circ)$

sources will have a considerable extension of degree scale which also contributes to the source confusion problem. The angular resolution of CTA ranges from $\sim$0.13° at 100 GeV to $\sim$0.05° at 1 TeV (cf. Fig. 5.1). Therefore source confusion will be reduced when exploring higher energies. This will be especially useful for the study of energy dependent morphology of interest for the identification of PWN (see Sect. 5.3.2). The excellent sensitivity, coupled with an energy resolution of better than $\sim$10% will also allow for a precise reconstruction of the inverse Compton peak for the detected PWN and their maximum energies, providing information of the total energy released by the parent pulsar in its life time, thanks to the longer cooling time of the particle population responsible for the emission at VHE energies, with respect to the population studied from the emission at X-ray energies. For more detailed discussion on the topic, see de Jager and Djannati-Ataï (2008).

A large asset to the study of the complex CTA GPS dataset will come also from the advancement in the mapping of the material content on the Galactic plane, especially with the NANTEN2 and MOPRA surveys, to study the environment in which the nebula is developing (Voisin et al. 2016).

The GPS will not be the only survey conducted with CTA that will be of potential interest to the study of PWN. Among the KSPs, the CTA Consortium is preparing also a deeper scan of the Galactic center and a scan of the Large Magellanic Cloud (CTA 2017).

5.3.2 CTA Energy Dependent Morphology and PWN Identification

Energy dependent morphology has been used already in the VHE regime as identification tool for the cooling mechanism in PWN (see e.g HESS J1825−137 and HESS J1303−631) (Aharonian et al. 2006; Abramowski et al. 2012). In a scenario where synchrotron is the dominant cooling mechanism, this will shape also the morphology of the source: the extension of the source at higher energies will be very compact. In a lower energy range, the older (more strongly cooled) electrons are still detectable and the size of the nebula is larger.

We have simulated a CTA observation of a source similar to HESS J1825−137, extrapolating the surface brightness distribution observed by H.E.S.S. to larger extensions. Also the spectral index distribution with offset from the pulsar position is extrapolated. The simulation is performed for an observation time of 10 h, similar to the expected coverage of the CTA GPS in the inner galactic plane. Figure 5.5

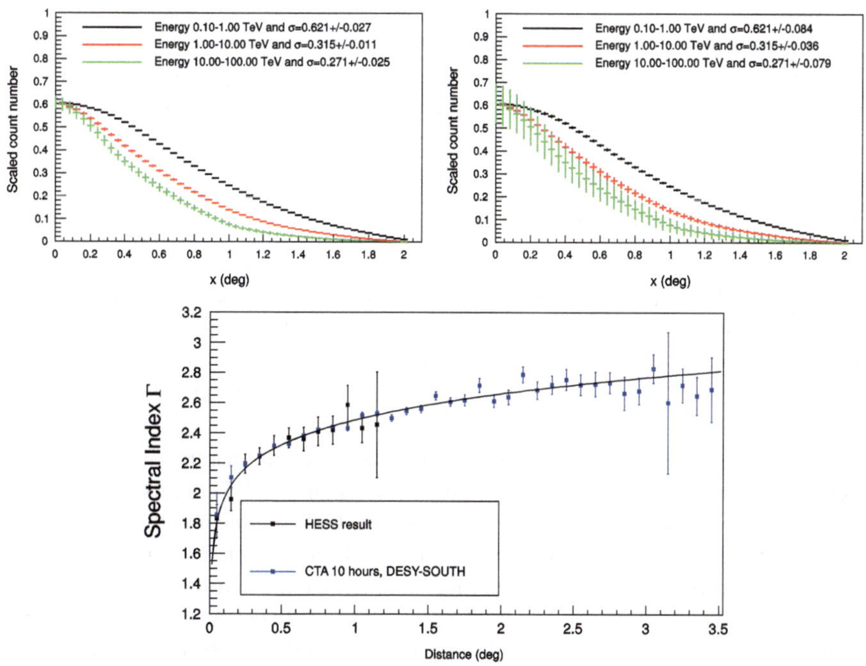

Fig. 5.5 Performance of CTA for energy dependent morphology. *Top Left*: Simulation of a 10 h observation on a source similar to HESS J1825−137, where the H.E.S.S. reconstructed surface brightness and spectral index evolution as a function of offset from the central position are extrapolated to higher angular distances. *Top Right*: Same as left, but the flux is reduced to 10%. *Bottom*: Reconstructed spectral index evolution as a function of offset, after feeding the extrapolated H.E.S.S. result on HESS J1825−137 (*solid line*) to a CTA simulation. Updated simulation with respect to what shown in de Oña-Wilhelmi (2013), using the public CTA instrument response functions

(top left) shows the expectation for such a simple simulation of energy dependent morphology. It can be seen how, dividing the energy spectrum of the source in three energy bins ([0.1–1],[1–10],[10–100] TeV), the different source extensions are clearly reconstructed. The original result on HESS J1825−137 was derived with 67 h of H.E.S.S. observation. The effect is still visible for the case shown in Fig. 5.5 (top right), where the integral flux of the source is reduced to 10% of the case shown in Fig. 5.5 (top left). In this latter simulation, the three different extensions are still easily reconstructed.

A 10 h CTA observation would provide enough precision also to reconstruct the spectral index of a source as bright as HESS J1825−137 to much larger extension than possible with current experiments (see Fig. 5.5, bottom). The spectral index is calculated in concentric wedges at increasing offset-angle from the center of the simulated FoV. The actual physical extension is determined from the balance of the age of the nebula, the propagation of the accelerated particles and their cooling. However, this is not an input to the simulation and the extension of the simulated nebula is only constrained by detectability. More simulations on the topic can be found in de Oña-Wilhelmi (2013), conducted with older sets of IRFs, but still relevant.

As mentioned above, the combination of energy dependent morphology and improved angular resolution as a function of energy, will be paramount to mitigate the source confusion in the CTA GPS and to aid in the identification of PWN.

5.3.3 CTA, Crab, and Its Flares

The Crab PWN has been adopted as the prototypical PWN, especially in VHE, being the first source detected at these energies (Weekes et al. 1989). Although the Crab is the most studied source at VHE, the highest energy part of its spectrum is still not firmly established, with small inconsistencies among current and older experiments (MAGIC coll 2015). CTA will allow a precise reconstruction of the high energy spectrum cut-off and thus of the maximum energy of the parent electrons. For a detailed review of studies of the Crab Nebula see Zanin (2017) in this volume.

One of the most spectacular surprises from the GeV regime was the discovery of flares from the Crab PWN. The Crab Nebula, thanks to its high flux and stability, has been used for many years as non-thermal standard candle and for cross-calibration among instruments. However, data of both Fermi-LAT and AGILE showed some major *variations* in the Crab Nebula flux in the 100 MeV–100 GeV energy range. These could be attributed to daily-timescales flares of the Crab Nebula itself, see Abdo et al. (2011); Tavani et al. (2011) and especially Bühler and Blandford (2014) for a detailed review. The spectral energy distribution (SED) of the Crab Nebula is characterized by a classical two humped structure. The low energy non-thermal emission is reconstructed as synchrotron emission, while the high energy emission is modeled through inverse Compton (IC) where the high energy electrons scatter off the surrounding soft photon field. The transition between the emission from the two

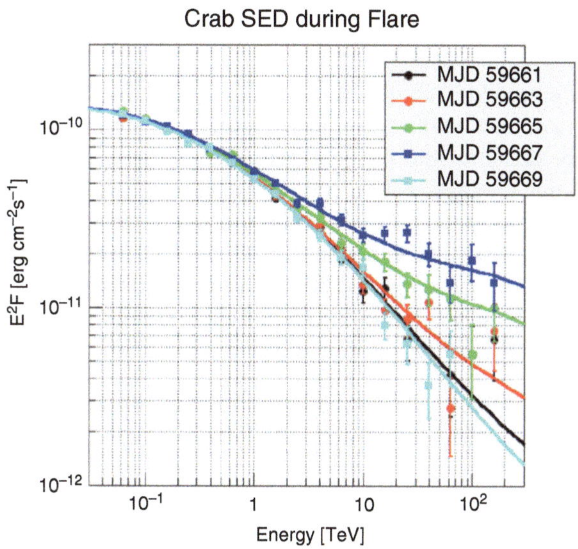

Fig. 5.6 Simulations of CTA observation of flares in the Crab Nebula direction. Each simulated day corresponds to a 4 h observation at good zenith angle ($\sim 20^{\circ}$). The spectrum in input for the simulation is taken from Kohri et al. (2012), with a Lorentz factor $\Gamma = 70$. Simulation from CTA (2017)

emission mechanisms happens at hundreds of MeV. Thc cmission of Crab Nebula during the flare can be modeled as an additional synchrotron emission component, with a higher cut-off with respect to the *"quiescent"* state, therefore requiring a modification of the average conditions. If a related IC emission appears also during flaring states, it should modify the *"quiescent"* Crab Nebula spectrum at multi-TeV energies. Up to date no corresponding variation has been detected at multi-TeV energies with IACTs.

A simulation of a possible CTA observation during such flares is shown in Fig. 5.6. The simulation starts from the conditions seen in the Crab Nebula flare of April 2011 (Bühler et al. 2012), modeled in the paper of Kohri et al. (2012) as related to Doppler boosted emission of a small blob. In the simulation, a Lorentz factor of $\Gamma = 70$ is assumed. The simulation is described in more detail in CTA (2017) as part of the study of the CTA technical capabilities for transients in the KSP framework. The flare is followed in the simulation for 10 days, with a 4 h observation time per night. The variability at tens of TeV would be clearly detectable.

For more simulations of Crab observation with CTA see de Oña-Wilhelmi (2013).

5.4 Space-Borne Detectors

The minimum detectable energies for IACTs is determined by dominating background systematic uncertainties. Below this threshold, detection techniques focus on the direct detection of the incident photon by space-borne detectors, like the currently operating Large Area Telescope on board the *Fermi* satellite and the GRID instrument on board of the AGILE satellite. A large improvement of the effective area of space-borne detectors with respect to *Fermi*-LAT would translate in a much larger load that could not be carried by the current vectors for space deployment. Therefore an improvement in flux sensitivity could be better implemented thanks to an improvement in angular resolution. More considerations on how to improve performances with respect to the current generations are nicely touched upon in Charles et al. (2014). The currently operating pair-conversion telescopes do not offer polarimetric capabilities, but see Giomi et al. (2016) for an estimation of Fermi-LAT sensitivity for γ-ray polarization. Polarimetry is included in some concepts for future detectors that will be described in Sect. 5.4.1. For a review of polarimetry in pair-conversion regime, see Bernard (2013). Section 5.4.2 will be devoted to describe satellite concepts that combine Compton and pair-production detection techniques.

5.4.1 Improvement in Angular Resolution and Polarimetric Capabilities

PANGU (*PAir-productioN Gamma-ray Unit*) is a γ-ray telescope concept optimized for pair-production detection technique. The energy range of interest is the "sub-GeV" regime (10 MeV $\lesssim E \lesssim$ 1 GeV). The concept is currently evolving from a small payload of ~60 kg (Wu et al. 2014) meant for the recent ESA-CAS Call for Joint Small Science Mission. The same mission concept, with an enlarged acceptance, is now being prepared as a proposal for the upcoming ESA M5 mission with a payload of ~200 kg. The latter proposal will be described here, from Wu et al. (2016). PANGU will have a wide field of view (~ 4.5 sr) and will provide imaging, spectral, timing and polarization capabilities.

A small angular resolution will be achieved thanks to the high precision tracker. The tracker will be composed of layers of silicon strip detectors without any interleaved passive material. The passive material layers are usually employed to increase the pair-conversion efficiency at the expense of angular resolution. A further improvement is given by the thinness of the strips (150 μm). To balance the loss in pair-conversion efficiency, PANGU will consist of many detectors layers (100 layers, arranged in 50 couples spaced by 15 mm). The design does not necessarily include a calorimeter. The energy estimation is done through the reconstruction of the Multiple Coulomb Scattering angles in the tracker, developed for PANGU (see Wu et al. (2016) for more details). Figure 5.1 (bottom) shows the expected improvement in angular resolution, with a PSF of $< 1°$ at 100 MeV (in comparison to the several degrees achieved from *Fermi*-LAT).

The polarization fraction is estimated from the distribution of the azimuth angle of the converted electron-positron pair. This distribution could be smeared by multiple scattering in the tracks. However, thanks to the good track angular resolution in PANGU, polarization measurements in the "sub-GeV" energy range will actually be possible. From current simulations, the convolution of the photon polarization fraction with the polarization asymmetry of the detection process, could be detected down to $\sim$10% up to 1 GeV (provided enough statistics), see Wu et al. (2016).

GAMMA400 is a ROSKOSMOS mission. It consist of an alternative concept for a satellite mission employing pair-conversion detection technique. It owes it name to the initial concept (*Gamma Astronomical Multifunctional Modular Apparatus with the maximum gamma-ray energy of 400 GeV*). The concept has however evolved starting from energy of $\sim$20 MeV to several hundreds GeV. An excellent angular resolution of few tenths of degree at 1 GeV and $< 1'$ at $\sim$100 GeV would be achieved by the precise reconstruction of the conversion point in the tracker and the reconstruction of the shower axis in the first calorimeter separated by 50 cm. It will perform pointed observations. Given that the project is currently in a re-assessment phase, no detailed information on performance are included here. For a summary of design concept and performances already investigated, see Topchiev et al. (2015).

Time Projection Chamber (**TPC**) gaseous detectors will also provide an excellent angular resolution and polarimetric capabilities. In this kind of detector, traversing charged particles ionize the detector material. The direction reconstruction is done before the Multiple Coulomb Scattering dominates the track development, avoiding the smearing of the azimuthal distribution of direction reconstruction essential for polarimetry. Two concept for Argon based TPC are explored.

The *Hermetic ARgon POlarimeter* (**HARPO**) (Bernard et al. 2014) is a high pressure gaseous TPC mission, sensitive to energies of $E \sim 3$ MeV up to hundreds of GeV. An excellent angular resolution of 0.4° at 100 MeV is expected. The *Advanced Energetic Pair Telescope* (**AdEPT**) is a TPC-based satellite concept expected to be sensitive to the medium energy range of $E \sim 5$–200 MeV. It requires a low density of the detector material. Thanks to its Three-Dimensional Track Imager will provide a 5-plus fold improvement in angular resolution of AdEPT below $\sim$200 MeV compared with *Fermi*-LAT and minimum detectable polarization less than 10% for a 10 mCrab source in 10^6 s (Hunter et al. 2014).

5.4.2 Combining the Compton and the Pair-Production Detection Technique

The MeV range is complicated because of several reasons. Several processes compete at these energies and Compton scattering dominates below $\sim$10 MeV over

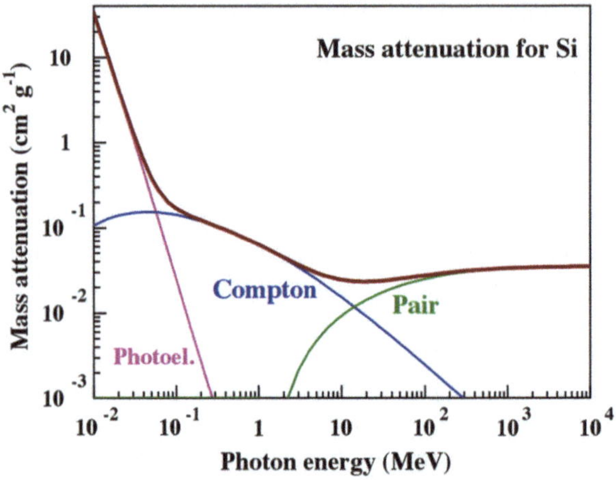

Fig. 5.7 Mass attenuation factors for Si, highlighting the regime where Compton scattering and pair production dominate, from Tatischeff (2017)

pair production, see Fig. 5.7, which shows the example of mass attenuation for Si as a function of energy. In addition, there is also a very rich background, exactly at MeV energies, from the induced radioactivity of the detector material itself.

Two mission concepts are planning to merge the Compton detection technique with the pair conversion technique in the same satellite mission and in a single instrument: e-ASTROGAM and ComPair.

5.4.2.1 e-ASTROGAM

e-ASTROGAM (Tatischeff et al. 2016) is a satellite mission that will be submitted to the M5 Medium-size call for ESA (European Space Agency) missions, thus expected to flight in ∼2029 and with 3 nominal years of operations. It merges two previous concepts: AstroMeV for the MeV range and Gamma-Light for the GeV range. It will be sensitive to energies 0.2 MeV $< E <$ 3 GeV.

The e-ASTROGAM telescope will consist of three components: a Silicon Tracker where the incident photon Compton scatters or pair produces; a Calorimeter needed to reconstruct the original energy of the incident photon from the deposited energy of the secondary particles; an anticoincidence veto for the background induced by charged particles. The schematic design is shown in Fig. 5.8. The tracker is very "light". 56 layers of detectors are expected, with each layer composed of 100 double sided strip detectors (similar technology as already used in *Fermi*-LAT and AGILE), only 500 μm thin. This ensures a precise 3-D position resolution of the converted pairs and of the recoil electrons from Compton scattering and an excellent spectral resolution. The calorimeter will be similar in concept to what already in use in *Fermi*-LAT, but with a very low noise readout at both ends. γ-ray events

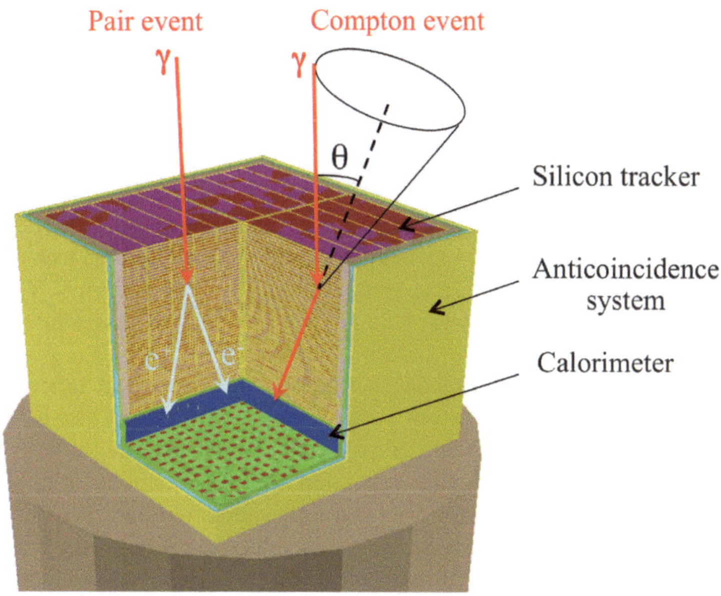

Fig. 5.8 Schematic design of e-ASTROGAM (the upper surface of anticoincidence detector is not shown for simplicity). Diagrams of the pair conversion and Compton detection principles are also overlaid. From Tatischeff et al. (2016)

can be triggered in the calorimeter independently from the tracker, useful for short transients.

Figure 5.1 shows the expected sensitivity for the continuum emission (ΔE=E) of a point source located at high latitude, observed for 1 year effective exposure and the expected angular resolution. It is possible to appreciate the impressive improvement of roughly two orders of magnitude in sensitivity at 1 MeV with respect to the last existing facility, COMPTEL, retired in the year 2000 at the end of the Compton Gamma Ray Observatory mission ($\sim$9 year duration). The improvement in angular resolution with respect to COMPTEL is of a factor 4 at 5 MeV, and of a factor 3.5 with respect to *Fermi*-LAT at 1 GeV. This is thanks to the lightness of the tracker, where the passive material is minimized, reducing the effect of multiple Coulomb scattering. The energy resolution in the Compton domain is expected to go from a $\sim$5% at 0.3 MeV (FWHM after event reconstruction and selection on the angular resolution measure) to $\sim$1% at 10 MeV; the expected spectral resolution is within 20–30% in the pair production domain. The field of view is expected to be 2.9 sr for the Compton domain and $>$2.5 sr for the pair production domain. Polarimetry will be possible both in the Compton and in the pair conversion domain thanks to the precise direction reconstruction of secondary particles track, so that the azimuthal distribution of these tracks, paramount for polarimetry, is not blurred. The minimum detectable polarization at the 99% confidence level is expected as low as 0.7% for a Crab-like source in 1 Ms (statistical uncertainties only), at the low energies

(0.2 MeV $< E <$ 2 MeV, hence in the Compton regime) (Tatischeff et al. 2016). Detailed information on performance and simulations can be found in Tatischeff et al. (2016) and Tavani (2016).

5.4.2.2 ComPair

ComPair (Moiseev et al. 2015) is a satellite concept that, like e-ASTROGAM, will integrate the Compton detection technique with the pair conversion technique in a single instrument, devoted to the energies of 0.2 MeV $< E <$ 500 MeV. The design is however focused towards a larger field of view of $\sim$3.5 sr with moderate angular and energy resolution.

The design is classical in components (tracker, calorimeter and anticoincidence detector) but with innovative twists in components and integration. Also in ComPair, conversion efficiency is sacrificed removing the passive material in the tracker to benefit the improvement of angular resolution and the possibility of polarization measurement. The angular resolution performances are shown in a broadband context in Fig. 5.1. The performances are calculated separately in the Compton regime for tracked and untracked events. Tracked events are those for which both the scattered photons and the recoiling electron direction can be identified, in contrast to untracked events for which only the scattered photons is measured, thus leading to a circle of possible direction for the incident photon (see Compton technique representation in Fig. 5.8).

The calorimeter is meant to combine solid-state detectors CdZnTe-strip (CZT) and inorganic scintillators CsI(Tl). The CZT calorimeter will provide good spatial and energy resolution needed for good Compton performance. The CsI calorimeter will add depth to contain higher energy showers from pair events. For tracked events and a deeper CZT calorimeter, the angular resolution is expected to improve further (see line 3 of ComPair in Fig. 5.1). The baseline design corresponds to line 1 of ComPair in Fig. 5.1.

Therefore the event selection will play on a payoff between reducing the effective area for the tracked events (thus with a better PSF) and the larger effective area for the untracked events with a worse direction reconstruction, depending on the physics case investigated. The expected sensitivity improvement is of: factor 20–30 in comparison to COMPTEL ($E <$ 30 MeV); factor 5–10 in comparison to EGRET (30 MeV $< E <$ 100 MeV); similar sensitivity to Fermi at $E \sim$ 100 MeV, but a factor of few worse at $E \sim$ 500 MeV, see Moiseev et al. (2015). Polarimetry is possible, even if the thickness of the strip of the tracker is not optimal: 1 month of ComPair observations would measure an assumed 20% polarization of Crab, with a 30% accuracy (Moiseev et al. 2015).

This is a quickly evolving field, where concept for satellites are evolving quickly. The ComPair design is now evolving into the AMEGO satellite project.[2]

[2]See: https://asd.gsfc.nasa.gov/amego/index.html.

5.5 The Importance of the MeV–GeV Range for the Physics of PWN

The space-borne detector concepts described in Sect. 5.4 are in an earlier project phase with respect to CTA. Therefore, while the science objectives are already very clear and detailed, not many detailed simulations of specific physics case are published. Only a general overview of the possibilities is given in the following.

The SED of PWN is usually modeled by a two humped structure, with the lower energy emission component interpreted as synchrotron emission and the higher energy one as IC emission. The transition between the emission from the two emission mechanisms happens at tens or hundreds of MeV. Due to the limited sensitivity of the existing datasets in this energy range, this is not measured with great accuracy. Measuring the cut-off energy of the synchrotron emission would help constraining the maximum energy of the electrons in the PWN, the magnetic field in the inner region of the PWN and the particle transport in the nebula.

This might be challenging as the expected fluxes from other PWN in this energy regime are much lower than the Crab nebula one, for which the cut-off energy is measured. However, Fig. 5.9 shows the example of an assumed SED for two young PWN: MSH 15-52 with a characteristic age of ~1500 years and G21.5−09 with a

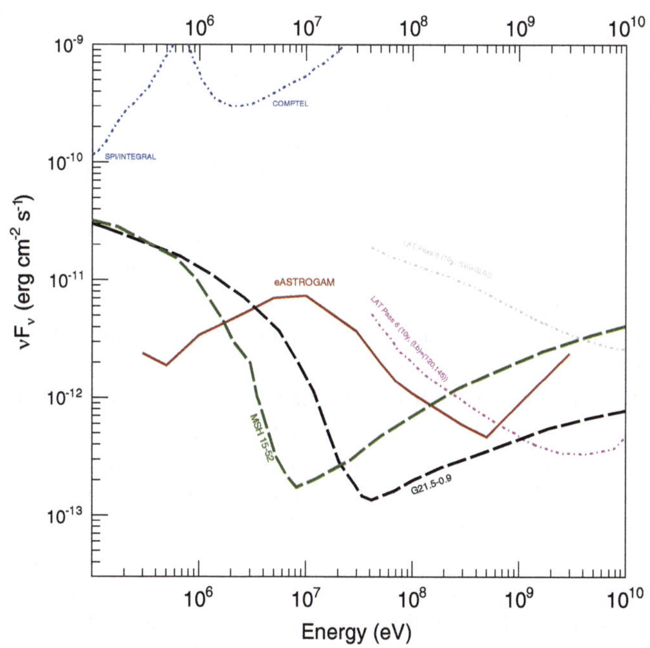

Fig. 5.9 Example of SED levels for young PWN in the MeV–GeV range. The model SED for MSH 15-52 and G21.5−09 are taken from Torres et al. (2014). The sensitivities are shown as in Fig. 5.1

characteristic age of 4860 years (but the age of 870 years is used in the modeling, derived from the PWN expansion rate). The SED are taken from the best fit SED in the modelling of Torres et al. (2014), obtained from the available MWL dataset. For both cases, the energy range 0.1 MeV $< E <$ 10 GeV is not covered by existing dataset, but will be within the reach of observations with the upcoming facilities described in Sect. 5.4.2 above. In the example of e-ASTROGAM, the sensitivity is shown for 1 year of effective observation, effectively achievable after three years of observations, thanks to the large field of view. A measurement of the higher end of the synchrotron spectrum will be possible also for older PWN, such as Vela PWN, see Kargaltsev et al. (2015). Study of the cooling mechanism could be aided by the measurement of a possible evolution of the spectrum in different regions of the PWN emission. The greater angular resolution expected in the forthcoming facilities will make this possible also at the GeV range and below. The improved angular resolution will also make easier PWN detection above the interstellar diffuse background.

The physics case of flares in the Crab nebula at GeV energies has been already discussed in Sect. 5.3.3 above, in the context of CTA possible detections in the associated IC component of a flaring state. The high energy cut-off in the synchrotron spectrum is determined by the balance of acceleration process and energy losses. If the maximum energy is limited only by synchrotron energy losses, it is to be noted that the synchrotron emission has an intrinsic cut-off at $E_{\rm sync,co} \sim 160$ MeV (Aharonian 2004). This value can decrease depending on the ratio of the effective electric field with respect to the magnetic field, that is usually much smaller than unity. This is compatible with the cut-off for the *"quiescent"* state of the Crab Nebula ($E_{\rm co} \sim 100$ GeV, Abdo et al. (2010)), but not with the flaring states observed in Abdo et al. (2011); Tavani et al. (2011). A simultaneous coverage and follow-up of possible flares (now detected in *Fermi*-LAT data with a frequency of more than once a year) also at MeV energies would help in constraining the flare evolution and interpretation. Similar flares might happen in other sources similar to the Crab Nebula, but might not be detected due to sensitivity limitation, hence an improvement in sensitivity, especially below 100 MeV, is extremely important. However, it is not possible to quantify the brightness and cadency of flares expected in other PWN.

Polarization information will be advantageous in the study of PWN. It could constrain the geometry of the magnetic field in the acceleration region and its homogeneity. A high degree of polarization has been measured in the Crab Nebula with INTEGRAL/SPI (0.1 MeV $< E <$ 1 MeV, Dean et al. 2008) and INTEGRAL IBIS (0.2 MeV $< E <$ 0.8 MeV, Forot et al. 2008), expected in the case of synchrotron emission. The electric vector polarization angle is found to be closely aligned with the pulsar spin axis, suggesting a common acceleration region responsible for the optical emission close to the pulsar that shows a similar polarization angle. The forthcoming facilities described above would enlarge the energy range for which the measurement is possible both for the *"quiescent"* and *"flaring"* Crab Nebula. Polarimetric information at GeV energies, so far never obtained, would be extremely beneficial for the disentanglement of emission models proposed.

5.6 Summary

I have summarized the main technical characteristics of the facilities that will be online in the 2020 decade, ranging nine orders of magnitude, from energies of $E \sim 0.1\,\text{MeV}$ to $E \sim 100\,\text{TeV}$. For a complementary review, including also other instruments, see Knödlseder (2016).

The highest part of the energy range, above tens of GeV, will be covered with CTA, that will boost the PWN population studies, thanks to a large commitment of observation time in the scan of the Galactic plane. CTA will also provide unprecedented detailed spectral and morphological information of the detected objects. At the end of the decade, the major step forward will happen in the MeV–GeV range, with space-borne detectors. The greatest improvement in this regime will be driven by the angular resolution of the instruments. This will lead to an improvement in point-source sensitivity, in morphology capabilities and especially will make polarimetry possible also at GeV energies.

Acknowledgements I wish to thank the Sant Cugat forum organizers for the very stimulating atmosphere and discussions during the workshop. Special thanks to Gernot Maier and Vincent Tatisheff for providing detailed numbers on performances for the current instrumentation in the HE-VHE regime and for the e-ASTROGAM mission, respectively. While most of the CTA material has been previously published and only summarized here, I greatly benefit from being a CTA member for the use of the material and the helpful internal reviewing process of the CTA sections. CTA gratefully acknowledge support from the agencies and organizations listed under Funding Agencies at this website: http://www.cta-observatory.org/.

References

Abdo, A.A., et al.: Astrophys. J. **708**(2) 1254 (2010)
Abdo, A., et al.: Science **331**(6018), 739–742 (2011)
Abramowski, A., et al.: Astron. Astrophys. **548A**, 46H (2012)
Acero, F.: Gamma-ray observations of PWNe (GeV to TeV). In: Torres, D.F. (ed.) Modelling Pulsar Wind Nebulae. Astrophysics and Space Science Library, vol. 446. Springer, Cham (2017). doi:10.1007/978-3-319-63031-1_5
Acharya, B.S., Actis, M., Aghajani, T., et al.: Astropart. Phys. **43**, 3–18 (2013)
Aharonian, F.: Very High Energy Cosmic Gamma Radiation, 1st edn., pp. 126–130. World Scientific, Singapore (2004)
Aharonian, F., et al.: Astron. Astrophys. **460**, 365 (2006)
Atwood, W.B., et al.: Astrophys. J. **697**, 1071 (2009)
Bernard, D.: Nucl. Instrum. Methods Phys. Res. A **729**, 765B (2013)
Bernard, D., et al.: In: Proc. SPIE 9144 (2014)
Bernlöhr, K., Barnacka, A., Becherini A., et al.: Astropart. Phys. **43**, 171–188 (2013)
Bühler, R., Blandford, R.: Rep. Prog. Phys. **77**, 6 (2014)
Bühler, R., et al.: Astrophys. J. **749**, 1 (2012)
Charles, E.: In: Proc. SPIE 91440E-130 (2014)
CTA Consortium: Science with CTA, CTA Consortium (2017)
de Jager, O.C., Djannati-Ataï, A.: In: Becker, W. (ed.) Springer Lecture Notes on Neutron Stars and Pulsars: 40 Years After Their Discovery (2008)

de Oña-Wilhelmi, E.: Astropart. Phys. **43**, 287–300 (2013)
de Oña-Wilhelmi, E.: Observations of note relating Fermi pulsars and nebulae. In: Torres, D.F. (ed.) Modelling Pulsar Wind Nebulae. Astrophysics and Space Science Library, vol. 446. Springer, Cham (2017). doi:10.1007/978-3-319-63031-1_5
Dean, A.J., et al.: Science **321**(5893) 1183–1185 (2008)
Dubus, G., et al.: Astropart. Phys. **43**, 317 (2013)
Fermi-LAT public performance page: http://www.slac.stanford.edu/exp/glast/groups/canda/lat_Performance.htm (2017)
Forot, M., et al.: Astrophys. J. **688**, 29 (2008)
Funk, S., Hinton, J.: Astropart. Phys. **43**, 348 (2013)
Giomi, M., et al.: In: Gamma2016 Symposium Proceedings. arXiv: 1610.06729 (2016)
HAWC public performance page: http://www.hawc-observatory.org/observatory/sensi.php (2017)
Hunter, S.D., et al.: Astropart. Phys. **59**, 18–28 (2014)
Kargaltsev, O., et al.: Space Sci. Rev. **191**, 391–439 (2015)
Klepser, S.: In: Gamma2016 Symposium Proceedings (2016)
Knödlseder, J.: C. R. Phys. **17**(6) 663–678 (2016)
Kohri, K., et al.: Mon. Not. R. Astron. Soc. **424**, 2249 (2012)
MAGIC coll: J. High Energy Astrophys. **5**, 30 (2015)
Moiseev, A., et al.: ICRC2015, arXiv:1508.07349 (2015)
Renaud, M.: In: Proceedings of 44th Recontres de Moriond (2009)
SPI Observer Manual, http://integral.esac.esa.int/AO9/AO9_SPI_om.pdf (2017)
Tatischeff, V., et al.: In: Proc. SPIE 9905. Numbers on sensitivities and angular resolution updated from a private communication from V. Tatischeff (2016)
Tatischeff, V., for the e-ASTROGAM Collaboration: In: 28th Texas Symposium on Relativistic Astrophysics (2017)
Tavani, M., et al.: Astron. Astrophys. **502**, 3, 995 (2009)
Tavani, M., et al.: Science **331**, 736T (2011)
Tavani, M., for the e-ASTROGAM Collaboration: In: Gamma2016 Symposium Proceedings (2016)
Topchiev, N.P., et al.: In: ICRC2015 Proceedings, arXiv:1507.06246 (2015)
Torres, D.F., et al.: JHEAp **1–2**, 31–62 (2014)
Voisin, F., et al.: Mon. Not. R. Astron. Soc. **458**, 2813 (2016)
Weekes, T.C., et al.: Astrophys. J. **342**, 379W (1989)
Wu, X., et al.: In: Proc. SPIE 9144-130 (2014)
Wu, X., et al.: In: Proc. SPIE 9905, 99056E (2016)
Zanin, R.: Crab observational status: nebulae, pulsations and flares. In: Torres, D.F. (ed.) Modelling Pulsar Wind Nebulae. Astrophysics and Space Science Library, vol. 446. Springer, Cham (2017). doi:10.1007/978-3-319-63031-1_5

Chapter 6
Crab Observational Status: Nebulae, Pulsations, and Flares

Roberta Zanin

Abstract The Crab, both pulsar and nebula, is one of the most studied astrophysical objects. This chapter reviews our knowledge of the system, as observed at different frequencies. It pays special attention to the flare phenomenology, and the information carried by polarization measurements.

6.1 Introduction

The Crab nebula is the remnant of the supernova explosion that occurred on July 4, AD 1054. Chinese astronomers extensively reported a bright "guest star" visible in daytime for 23 days and at night for almost 2 years (Stephenson and Green 2003). References about this exceptional star were also found in Japan and in an Arabic work from Constantinople. Rediscovered by the English amateur John Bevis in 1731, it was then listed as first entry in the catalogue of spurious comets by Charles Messier. It was named after the drawing by Lord Rosse describing the observations using the 36-in. reflector at Birr Castle in 1844. The Crab was the first astronomical object to be identified as supernova explosion by Hubble in 1928 thanks to the detection of its expansion (Hubble 1928). Six years later Baade and Zwicky suggested the supernova explosion as the link between ordinary stars and neutron stars (NS) (Baade and Zwicky 1934). In 1942 Minkowski associated one of the two central stars in the nebula with the explosion in 1942 (Minkowski 1942) and in 1967 Pacini proposed this star to be a highly magnetized spinning NS, a pulsar (Pacini 1967). This scenario was finally confirmed 1 year later, in 1968, with the detection of radio pulsations from the central object (Staelin and Reifentein 1968).

The Crab lies in the Perseus Arm of the Milky Way galaxy, at a distance of 2 kpc (Trimble 1973). The presently visible nebula is detected at all wavelengths, from radio frequencies up to very-high-energies (VHE). It is the archetypal pulsar wind

R. Zanin (✉)
MPIK, Heidelberg, Germany
e-mail: Roberta.Zanin@mpi-hd.mpg.de

D.F. Torres (ed.), *Modelling Pulsar Wind Nebulae*, Astrophysics and Space Science Library 446, DOI 10.1007/978-3-319-63031-1_6

nebula (PWN) that is powered by the powerful pulsar PSR B0531+21, located at its center ($\dot{E} = 1.3 \times 10^{38}$ erg s^{-1}). The rotational energy lost by the pulsar goes into a relativistic wind of particles, primarily electrons and positrons (pulsar wind). When this wind first impacts on the surrounding medium, a reverse shock is launched towards the pulsar, that is the termination shock (TS). At the TS particles are slowed down, randomized and heated and the power-law nebular particle spectrum is thought to be produced. This bubble of (shocked) hot relativistic particles emitting synchrotron radiation is the PWN. The best available models describing the PWN emission and morphology are based on magneto-hydro-dynamic (MHD) description of the pulsar outflows, introduced (Kennel and Coroniti 1984)

The Crab, both pulsar and nebula, is one of the most studied astrophysical objects, having the most precise and detailed spectral and morphological measurements, thanks to its high luminosity and proximity. As such, it is one of the best laboratories to study the pulsar physics, as well as the processes of both particle acceleration in relativistic shocks, and transport of particles downstream. These basic astrophysical processes occur in many other classes of astrophysical sources, like active galactic nuclei or gamma-ray bursts, but whose observational conditions are much more complex (for instance, because of the further distance, the flux variability, or the lower brightness).

The Crab nebula is also used as flux/spectral calibration target by various instruments because of its brightness across the electromagnetic spectrum and stability over time. The latter property has been only recently challenged by the observation of flaring activity in γ rays (Tavani et al. 2011; Fermi Collaboration 2011) and long-term flux variations in the X-ray energy band (Wilson-Hodge et al. 2011). This discovery together with the recent detection of VHE radiation from the pulsar (VERITAS Collaboration 2011; MAGIC Collaboration 2012, 2016) turned on the interest of the astrophysical community on the Crab. On one hand hundreds of theoretical paper were published providing a possible explanation for these unexpected results. On the other hand similar phenomena are now being searched in other sources attempting to establish them as general properties of this class of sources. So far, there is only a claim of flux variability from the PWN N157B, the Crab-twin in the Large Magellanic Cloud (Saito and Uchyama 2016).

6.2 Crab Pulsar

As leftover of the supernova explosion SN 1054, the Crab pulsar is one of the few pulsars whose age is perfectly known. Discovered at radio frequencies by Staelin and Reifenstein in 1968 (Staelin and Reifentein 1968) thanks due its giant pulses (see Sect. 6.2.4), it was the first pulsar to be detected at optical wavelengths (Cocke et al. 1969).

6.2.1 The Rotational Frequency

The Crab pulsar has a period $P = 33.6\,\text{ms}$, which is slowing down with $\dot{P} = 4.21 \times 10^{-13}$. It looses rotational energy with a rate of $5 \times 10^{38}\,\text{erg}\,\text{s}^{-1}$ (i.e. the spin-down luminosity L_{spin}), under the assumption that the moment of inertia of the neutron star is $I \sim 10^{45}\,\text{g}\,\text{cm}^{-2}$. Just 1% of this energy goes into collimated electromagnetic radiation, which sweeping the field of the Earth, produces the observed pulsations, whereas the majority of the pulsar's spin-down luminosity is carried away by a combination of the magnetic radiation and the relativistic pulsar wind and later released by the synchrotron nebula.

The slowdown of the rotational frequency ν follows a power-law function $\dot{\nu} = -k\nu^n$, where k depends on the mass, the inertia momentum, and the angle between the spin and the magnetic-axis, while n is the braking index ($n = 3$ for ideal dipolar magnetic torque). The estimation of the braking index mainly depends on the measurement of the rotational frequency and its first, and second derivatives. Despite this monotonic slowdown, there are unpredictable changes of the rotational frequency due to timing noise or glitches. Timing noise is a random change in the rotational frequency most likely due to magnetospheric instabilities (Lyne et al. 2010), while glitches are almost instantaneous increases in the rotation rate whose origin is related to the NS interior (Espinoza et al. 2011). The glitches last for seconds and usually occur in bunches. They produce an increase in the slowdown rate due to the fact that the rotation rate never recovers completely after the glitch (Lyne et al. 2015). From 1968 to 2013, 24 glitches were detected in the Crab pulsar with a 11-year period of increased glitch activity. The 24 observed glitches produced an overall increase of 6% in the rotation rate (Lyne et al. 2015).

The braking index of the Crab pulsar between glitches is 2.519 ± 0.002 (Lyne et al. 1993). This lower value of n is thought to be related to the change in the magnetic field, in particular in the angle between the spin- and the magnetic-axis (Lyne et al. 2015).

6.2.2 Pulse Profile

The Crab pulsar shows a two-peak mean pulse profile almost at all wavelengths: the main peak at low radio frequencies (0.1–5 GHz), by definition set at phase 0, is usually referred to as P1, whereas P2 or interpulse is located $\sim 0.4°$ away. This double-peaked structure remains almost phase-aligned over all energy bands (Moffett and Hankins 1996; Kuiper et al. 2001), even though the amplitude and the width of the two peaks varies with the energy.

The radio pulse profile is much more complex though, with two additional low-frequency components and two distinct high-frequency (>5 GHz) ones (refer to Eilek and Hankins (2016) for a review). The two low-frequency components precede P1 by 0.10 and 0.05°, respectively. The latter is usually refers to as precursor and appear in the 0.3–0.7 GHz range, while the other one, LFC (low-frequency component) is more evident at 1.4 GHz (Moffett and Hankins 1996). The two high-frequency components (HFC1, HFC2) follows P2 by ~0.2° and ~0.35°, respectively (Moffett and Hankins 1999). The interplay between the two main peaks itself is frequency-dependent. The interpulse disappears at 2.7 GHz to appear back at 4.5 GHz and in the 4.5–8.4 GHz it becomes the dominant component. At 8.4 GHz the profile shows only three components: P2, HFC1 and HFC2 with P1 missing. Then between 8.4 GHz and 2.2 μm, the profile evolves to two components with much broader peaks and P1 back dominating over P2 (Moffett and Hankins 1996). In addition, the interpulse behaves very differently at low and high-frequencies. In particular, at high frequencies its position is shifted to earlier phases by 0.018° and its single pulses show significant spectral and timing variability (Moffett and Hankins 1996). These characteristics may indicate that it is a different feature from the low-frequency P2 (Moffett and Hankins 1999).

Figure 6.1 shows the pulse profile from radio frequencies up to gamma rays. Above μm the profile shows only the two main peaks, sometimes connected by the *bridge* emission. The flux ratios P2/P1 follows a repetitive pattern, increasing with energy up from μm to few MeV, decreasing up to 100 MeV, and increase again up to TeV energies (Kuiper et al. 2001; MAGIC Collaboration 2014) (see Fig. 6.2). The *bridge*/P1 ratio behaves similarly, as pointed out in MAGIC Collaboration (2014). No theoretical explanation has been put forward so far to explain such behaviour. At high energies P2 becomes dominant at ~50 GeV (MAGIC Collaboration 2014), whereas the *bridge* appeared between 10 and ~150 GeV. At these energies the main peaks are much narrower than at 1 GeV. However, given the small statistics available, any asymmetry in the pulse shape cannot be estimated (MAGIC Collaboration 2012).

6.2.3 *Polarization*

The pulsed emission is found to be polarized, even though at smaller level with respect to the nebular one. At optical frequencies, the Crab pulsar is one of few pulsars from which polarization has been detected (Smith et al. 1988). Once subtracted the continuous nebular emission, it turns out to be on average just slightly polarized with a polarization degree, P.D. = $(5.2 \pm 0.3)\%$ and a polarization angle, P.A. = $(105.1 \pm 1.6)°$ (Slowikowska et al. 2009; Moran et al. 2013) which hints to be aligned to the pulsar proper motion $(110 \pm 2 \pm 9)°$. The P.A. changes with the phase, as shown in the two bottom panels in Fig. 6.4: it spans from P.A. ~40° to P.A. ~170° during P1, it decreases during the bridge and swings from P.A. 90° to P.A. 180° during P2 (Slowikowska et al. 2009). These variations in polarization are consistent in shape with a beam of synchrotron radiation beamed from both poles of an orthogonal rotator. At radio wavelengths, all the components are moderately

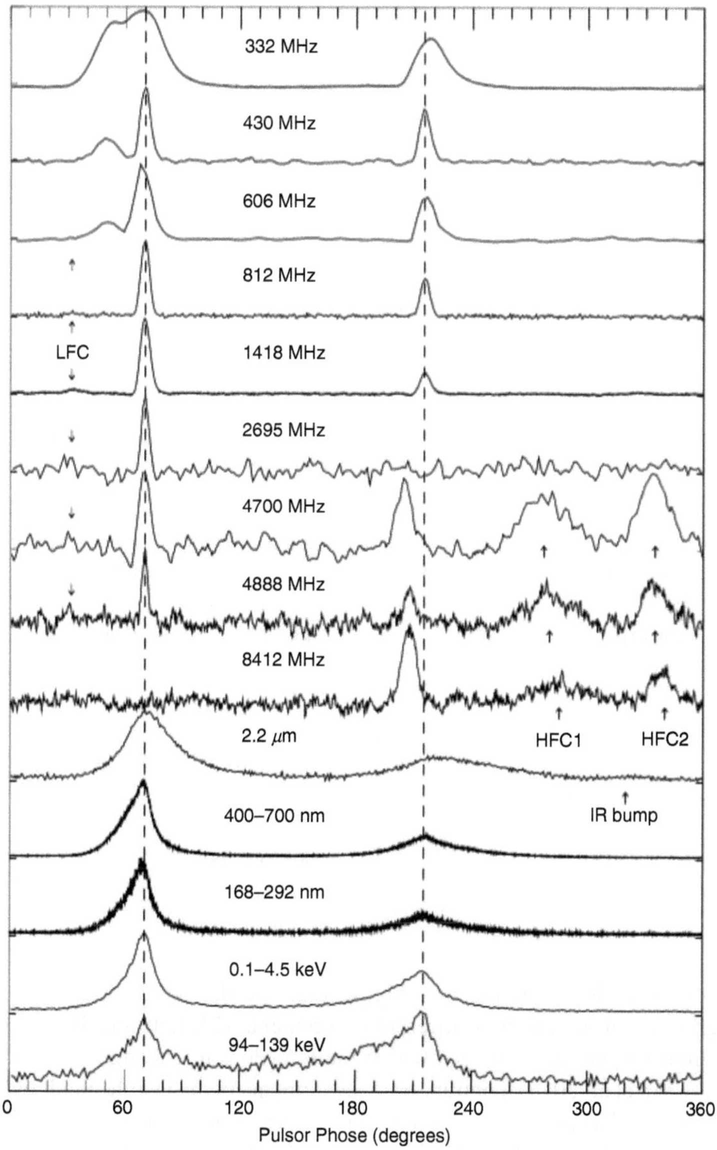

Fig. 6.1 Summary of Crab pulsar profiles taken from Moffett and Hankins (1996)

linear polarized with properties which are frequency-dependent, and summarized in Table 6.1. P1 shows a P.D. around 20% between few hundreds of MHz and 9 GHz with almost a constant P.A. of ~120°, while P2 has similar polarization properties as P1 at 1.5 GHz, but at 5 GHz it has a P.D. higher than 50% with more than 90° rotation with respect to its P.A. at 1.5 GHz. The latter, together with the

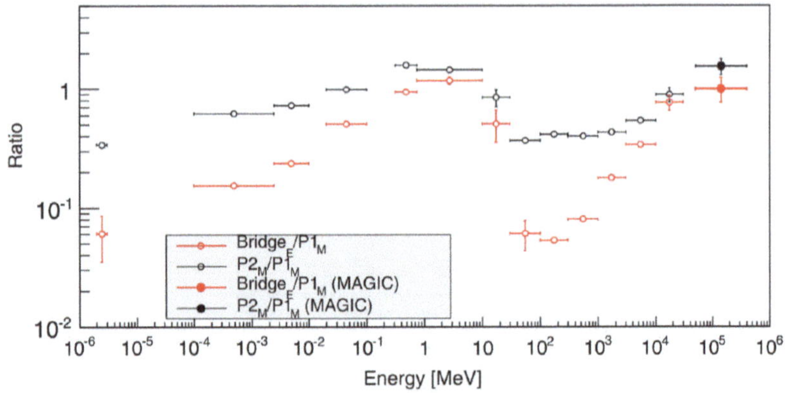

Fig. 6.2 P2/P1 (*black markers*) and Bridge/P1 (*red markers*) flux ratio. The subscripts $_E$ and $_M$ define the interval of phases from where the flux was extracted, but are irrelevant for the purpose of this review. Taken from MAGIC Collaboration (2014)

Table 6.1 Polarization parameters for the different components of the pulse profile at different radio frequencies, and in the optical energy band (Manchester 1971; Moffett and Hankins 1999; Slowikowska et al. 2015)

	400 MHz		1.5 GHz		4.9 GHz		Optical	
	P.D. [%]	P.A. [°]	P.D. [%]	P.A. [°]	P.D. [%]	P.A. [°]	P.D. [%]	P.A. [°]
Pulsar							5.2 ± 0.3	105.1 ± 1.6
P1	20	140	22.98 ± 0.30	120	25	120	5–20	40–170
P2	–	–	21.3 ± 1.0	120	50–100	120 ± 90	3–15	90–180
Precursor	80	140	–	–	–	–	–	–
LFC	–	–	98.2 ± 6.7	160	–	–	–	–
HFC1	–	–	–	–	50–100	180	–	–
HFC2	–	–	–	–	50–100	180	–	–

shift position of P2 at frequencies higher than 5 GHz, indicates that P2 at low- and high-frequencies are two different components (Moffett and Hankins 1999). Recent high-resolution results show that the P.A. does not change across the pulse (Slowikowska et al. 2015), contrary to measurements in the visible band. In addition to the linear polarization, a weak circular polarization is measured at a level of -1.3% and -3.2% for P1 and P2, respectively (Slowikowska et al. 2015). The LFC shows, at 1.5 GHz, an almost complete linear polarization and high circular polarization with 20% P.D. suggesting that it originates at a different location and via a different mechanism than the stronger P1 and P2. In particular, whereas the LFC is thought to come from a low-altitude region, the two main peaks seem to be high-altitude caustic components.

So far there are no available measurements of X-ray polarization for the Crab pulsar, neither for the averaged nor for phase-resolved emission. The OSO-8 experiments, that had the capabilities to perform this measurement, were not granted with the needed observation time (Weisskopf 2009). At γ rays there is no significant indication of polarization from the peaks (Forot et al. 2008).

6.2.4 Giant Pulses

One of the peculiarity of the Crab pulsar is that most of its radio emission arises from giant pulses (GPs): narrow and bright pulses with a flux $10^2 - 10^7$ higher than the average one. GPs are much narrower than a typical pulse: they last from few ns to few μs (Hankins et al. 2003) with the narrower at higher energy. GPs come in bunches of pulses and the bunch duration is of at most 100 μs. Their spectrum is well-represented by a power-law distribution, distinct from the Gaussian shape of the ordinary pulse population (Knight et al. 2006). GPs have been observed to occur during the main pulse and the interpulse. In particular, GPs from P1 and the low-frequency P2 share the same temporal and spectral characteristics, while those in the high-frequency P2 are different (Hankins and Eilek 2007). Despite their pulse profile differences the emission processes at low- and high-frequencies are related, as shown by the positive correlation between low- and high-frequency P2 GPs within one spin period (Mickaliger et al. 2012). P1 GPs have a characteristic time of few μs, and a strength which is highly variable with the weaker the more common. Hankins and collaborators suggested that these microbursts are a superposition of shorter ns single pulses (Eilek and Hankins 2016; Hankins et al. 2016).

The nature of the giant pulses is still far from being understood. Certainly possible correlations with flux increase at different wavelengths is important for testing theories on the giant pulse formation. It is generally believed that they are related either to changes in the beaming directions, or to variations in the pair creation rate, or to changes in the coherence in the magnetospheric plasma. In the first two scenarios, one would expect a correlation between the GPs and the higher-energy pulsed emission, while in the third scenario there should be no correlation since the higher-energy emission is of incoherent origin. The only positive results so far were obtained in the optical band where the optical main pulses concurrent with GPs at 1.4 GHz were on average 3% brighter than the others (Shearer et al. 2003). While, at higher energies, no flux increase was observed simultaneously to radio GPs, neither in soft X-rays (*Chandra* Bilous et al. 2012) nor in gamma rays (*EGRET* Ramanamurthy and Thompson 1998, *Fermi*-LAT Bilous et al. 2011, VERITAS VERITAS Collaboration 2012). These results indicate that variations in the pair production rates or in the beaming directions are not the dominant mechanisms in the production of GPs. In the 0.1–5 GeV the limit on the flux of the pulsar concurrent with the GPs is of less than four times the averaged gamma-ray flux (Bilous et al. 2011).

6.2.5 Spectral Measurements

Figure 6.3 shows the spectral energy distribution of the phase-averaged pulsed emission which comprises two main components. The first one extends from optical to ∼10 MeV with a plateaux of maximum luminosity from ∼10 KeV to ∼1 GeV (Sollerman et al. 2000). Beyond 1 MeV the spectrum steepens up to ∼70 MeV where the second component becomes dominant. The latter has a hard power-law spectrum with a photon index ∼2 and a rather sharp spectral break at a few GeV (Kuiper et al. 2001; Fermi Collaboration 2010). The optical to soft γ-ray emission is believed to be synchrotron radiation of the secondary pairs. While at high energies the pulsed emission is expected to come from synchro-curvature radiation by electron-positron pairs moving along curved paths inside the light cylinder. The maximum photon energy is limited by either magnetic and gamma-gamma pair absorption or radiation losses, that result in spectral cutoffs at around a few GeV. The Crab pulsar is one of the few pulsar where the synchrotron component has higher luminosities than the high-energy one (Kuiper and Hermsen 2015).

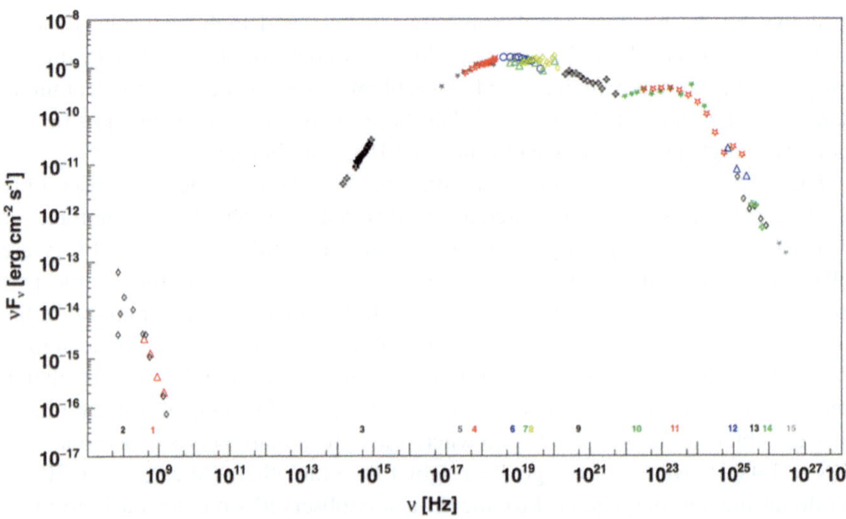

Fig. 6.3 Broadband SED of the Crab pulsar. Data taken from Bülher and Blandford (2014). Numbers below the sed indicate the reference from which the corresponding data come from: (1) Thompson et al. (1999), (2) Thompson et al. (1999), (3) Tziamtzis et al. (2009), (4) *BeppoSax*-MECS (Kuiper et al. 2001), (5) *BeppoSax*-LECS (Kuiper et al. 2001), (6) *BeppoSax*-PDS (Kuiper et al. 2001), (7) Gris (Kuiper et al. 2001), (8) *Osse* (Kuiper et al. 2001), (9) *COMPTEL* (Kuiper et al. 2001), (10) *EGRET* (Kuiper et al. 2001), (11) *Fermi*-LAT (Fermi Collaboration et al. 2013), (12) MAGIC-mono (MAGIC Collaboration 2011), (13) MAGIC-stereo (MAGIC Collaboration 2012), (14) VERITAS (VERITAS Collaboration 2011), (15) MAGIC-stereo (MAGIC Collaboration 2016)

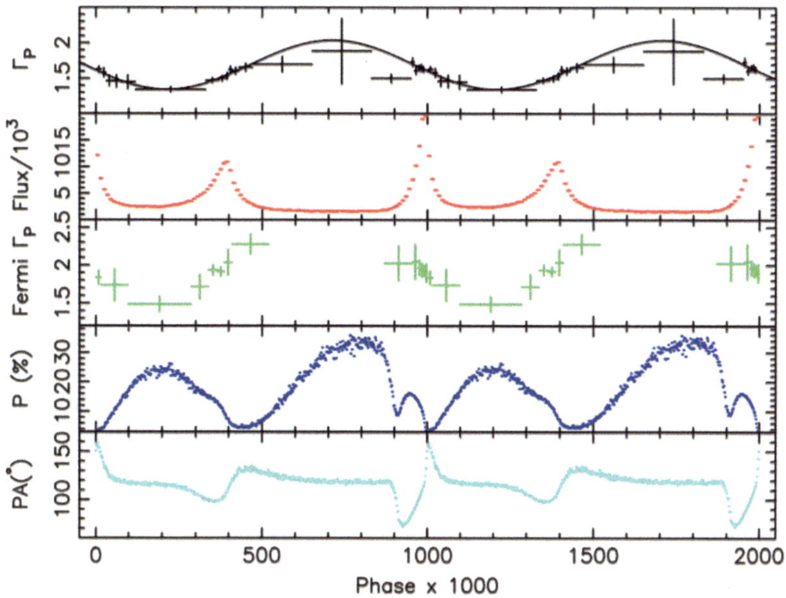

Fig. 6.4 From top to bottom. The P.A. and P.D. for the Crab pulsar at optical frequencies as a function of rotational phase. Figure taken from Weisskopf et al. (2011)

The phase-averaged SED in optical to hard X-rays is smooth, with a varying from ~0.1 at optical frequencies, through −0.5 (soft X-rays) to −0.7 (hard X-rays) (Sollerman et al. 2000; Carraminana et al. 2000). In X-ray highly significant variations of the photon index as a function of the pulse phase have been detected by several instruments (Pravdo et al. 1997; Massaro et al. 2000; Weisskopf et al. 2011; Madsen et al. 2015). The hardest index occur in the *bridge* emission to reach the maxima in the *off-pulse* region, as shown in the top panel of Fig. 6.4 (Weisskopf et al. 2011). This change behaviour of the photon index just before P1 coincides with a rapid change in the optical polarization (see Sect. 6.2.3) suggesting them as correlated phenomena. These variations of the photon index trace the changes of the pair spectrum across the open field lines, thus providing a valuable tool to test theoretical models.

At high energies all the 160 known gamma-ray pulsars, detected by *Fermi*-LAT, agree with the synchro-curvature scenario (Fermi Collaboration et al. 2013), but one, the Crab pulsar. The spectrum of the Crab pulsar as seen by *Fermi*-LAT (0.1–100 GeV) is well parametrized by a power-law function with a sub-exponential cutoff (b<1). The latter, preferred by more than 3σ on the pure exponential cutoff, is not so unusual for young gamma-ray pulsars (McCann 2015). This is interpreted as the overlapping of beams coming from different regions and/or with different cutoff energies. The best-fit value for the photon index is $\gamma = 1.59 \pm 0.01$, for the energy break is (5.09 ± 0.63) GeV and for the curvature index b = 0.43 ± 0.01 (Buehler et al. 2012). The variations of the photon index with the pulse phase at HE highly resembles those at X-rays and are illustrated in the middle panel in Fig. 6.4 (Fermi Collaboration 2010).

Recent measurements at VHE showed a totally unexpected power-law component emerging above the synchro-curvature cutoff and extending above 100 GeV (VERITAS Collaboration 2011; MAGIC Collaboration 2011, 2012). The steep power-law function with a photon index $\sim$3.5 is excluding the spectral break at more than 6σ level (VERITAS Collaboration 2011). The phase-resolved spectral analysis shows the new component is a feature of both peaks and *bridge*. P2 is significantly harder than P1 with a difference in the photon indices, obtained by a joint MAGIC-*Fermi*-LAT fits, of 0.5 ± 0.1 (MAGIC Collaboration 2016). Whereas the steeper P1 ($\Gamma = 3.5 \pm 0.1$) spectrum cannot be detected beyond 600 GeV, P2 extends at higher energies, up to 1.5 TeV without any sign of cutoff (MAGIC Collaboration 2016). A lower limit on the spectral cutoff is estimated at 700 GeV. Compared to P1, the bridge emission at HE is fainter but with a similar photon index. Indeed it fades away already at 150 GeV (MAGIC Collaboration 2014).

6.2.6 Lessons from the Crab Pulsar Observations

The study of the pulse profile can give deep insight on the regions where the pulsed emission is produced. At high energies, for instance, population studies, made possible by the discovery of the 160 gamma-ray pulsars by *Fermi*-LAT, established that the great majority of high-energy pulsars have a double-peaked light curve. This shows that high-energy emission must originate at high altitudes (Romani and Watters 2010; Pierbattista et al. 2012), either in gaps inside the magnetosphere (Cheng et al. 1986; Romani and Yadiglarogu 1995; Muslinov and Harding 2004)—where unscreened electric fields can form—or in current sheets existing both within and beyond the light cylinder—where current flows along the separatrix between the closed and open magnetic field lines (Contopoulos et al. 1999; Kalapotharakos et al. 2012; Contopoulos et al. 2014; Cerutti et al. 2015). (For a more detailed description on the recent advances on the understanding of the pulsar magnetosphere and its interplay with the pulsar wind region, please refer to Chap. 7). Explaining the double-peaked profile with low-altitude emission models, in fact, would require a geometry with both viewing and inclination angles close to 90° that is too special to meet the observed numbers.

The VHE emission is not likely to be produced via synchro-curvature by particles streaming along the magnetic field lines inside the light cylinder. In the dipole magnetic field approximation, the production of TeV photons via synchro-curvature would require an unrealistic curvature radius, more than one order of magnitude larger than the estimated values $\sim$2 R_{LC} (where R_{LC} is the radius of the light cylinder) (Vigan et al. 2015). This component is likely to be dominated by inverse Compton scattering off either photons in particle cascading inside the magnetospheric gaps (Lyutikov et al. 2012; MAGIC Collaboration 2011; Hirotani 2015) or pulsed photons that escaped the light cylinder (Bogovalov and Aharonian 2000; Aharonian et al. 2012; Petri 2012; Mochol and Petri 2015). In this second scenario, the VHE pulsed emission would provide the first direct detection of the cold pulsar wind region—where particles are stationary in their comoving frame. The new advances in the particle-in-cell simulations of the pulsar magnetosphere show that particles can be accelerated to relativistic energies in the current sheets outside the light cylinder due to magnetic reconnection and emit synchrotron radiation (Petri 2012; Cerutti et al. 2015). In a specific work on the Crab pulsar, Mochol and collaborator estimated that this pulsed synchrotron emission could reach up to few hundreds of GeV and that at higher energies a synchrotron-self-Compton (SSC) component could emerge up to few TeV (Mochol and Petri 2015). What it is still missing though is a simultaneous computation of the magnetospheric and beyond-the-light-cylinder emission. So far none of the existing model can explain simultaneously the spectrum and the pulse profile with its narrow peaks at VHE. In addition there is no comprehensive theoretical model reproducing all the observable of the pulsed emission i.e. the broadband spectrum with its spectral changes across the pulse-phase in X- and γ rays, and the pulse profile with its energy-dependent peak amplitude and width.

The radio pulsed observations of the Crab taught us two important lessons. First, radio emission in the Crab pulsar does not come from the low-altitude gaps, as generally believed for the majority of radio pulsars. The almost-alignment of the radio main peaks with the high-energy ones suggest that they come from the same region, i.e. at high altitudes. This is confirmed also by the fact that we see seven radio components throughout the whole rotational period. Second, despite the related location, radio emission is believed to be produced in a much more compact region than gamma-ray emission given its much narrower main peaks, and the absence of the *bridge* emission (Eilek and Hankins 2016). Also the radio flux variability on timescales down to ns (see Sect. 6.2.4) points towards a needed compact production site, and an overall non-steady emission region. The random appearance of the single giant pulses suggests the sporadic existence of a mechanism which drives the coherent radio emission within the emission region (Eilek and Hankins 2016).

6.3 Crab Nebula

Most of what we know about PWNe comes from the Crab. In fact most of the theoretical models have been drawn to match the Crab observations. As we will discuss in Sect. 6.3.6, the accurate spectral and morphological measurements of the Crab at all wavelengths provide a powerful tool to test all the theoretical predictions.

6.3.1 The "Halo" and the Cage

The Crab nebula is expanding into the supernova ejecta which are themselves freely expanding (Chevalier 1977). Low expansion velocity of the ejecta suggests that the remnant is the result of a Type-II explosion with a progenitor star of 8–12 $M_{\odot}$ (MacAlpine and Satterfield 2008).

The remnant as yet remains still undetected, and it is often referred to as "halo". Its existence was inferred by the need to justify the discrepancy between the estimated energy of the visible nebula ($\sim 7\times10^{49}$erg) and the typical energy of a core-collapsed supernova ($\sim 1\times10^{51}$erg). Chevalier indeed suggested that this missing energy was contained in the halo (Chevalier 1977). However, an alternative explanation has been proposed, that SN 1054 was an under-luminous supernova (Smith 2013; Yang and Chevalier 2015).

The Crab nebula is not only expanding, but it has been accelerated since the original explosion. This acceleration is expected: as the PWN is expanding into the remnant that is itself expanding with a velocity that is proportional to the radius, it has to move even faster to catch them (Chevalier 1984). The expansion rate is of 1500 km/s (Bietenholz et al. 2001). By extrapolating backwards in time the obtained expansion rate, the convergence year is 1255 ± 27, implying an accelerating expansion (Bietenholz and Nugent 2015).

The synchrotron-emitting nebula is confined into an cage of thermal filaments of ionized gas that are the ejecta of the supernova explosion, possibly mixed with some material from the pre-supernova (Fesen et al. 1992). The filaments are interpreted as the result of the Rayleigh-Taylor (R-T) instabilities occurring in the interaction between high-pressure, but low-density nebular plasma and the still freely expanding supernova ejecta (Hester et al. 1996). MHD simulations show that this instability causes finger-like structure pointing inwards into the center of the nebula (Porth et al. 2014), as it is seen by the high-resolution optical images by the *Hubble Space Telescope* (see Fig. 6.5). Fingers have typically arc-second dimensions and are usually connected each other by arc-like bridges of emission. The filaments are also expanding but are significantly less accelerated than the synchrotron nebula (Bietenholz and Nugent 2015) (convergence year is 1091 ± 34). This result supports the idea the the filaments are due to R-T instabilities at the interface between non-thermal and thermal emission.

Visible in the optical and IR band, the filaments are of special interest because their emission line spectra contain chemical information about the progenitor star (Davidson and Fesen 1985). The obtained results suggest that the ejecta are nitrogen-depleted, but carbon-rich (Temim et al. 2006). Filaments also contain freshly formed SN dust as shown by their optical extinction (Sankrit et al. 1998). In particular, the spatial correlation between the dust emission and the thermal filaments indicates that the SN-ejecta dust is fully concentrated in the filaments (Temim et al. 2012). The dust is well described by carbon and silicate compositions with a total estimated mass of 0.01–0.001 $M_{\odot}$ (Temim et al. 2012). However, the

Fig. 6.5 Composite image of the Crab nebula showing the X-ray image in *light-blue* as detected by *Chandra*, the IR image in *red* by the *Spitzer Space Telescope* and the optical image in *yellow* by the *Hubble Space Telescope*. Credit: X-ray: NASA/CXC/ASU/J.Hester et al.; Optical: NASA/ESA/ASU/J.Hester and A.Loll; Infrared: NASA/JPL-Caltech/Univ. Minn./R.Gehrz

latter value was questioned by Gomez et al. (2012) who are suggesting one order of magnitude larger dust mass.

6.3.2 Morphological Measurements

The detailed images of the Crab Nebula taken at different wavelengths reveal a complex and energy dependent morphology. The size of the nebula shrinks for increasing energies because of the "synchrotron burn-off" effect, i.e. the radiative lifetime of the propagating electrons being shorter for high energy than for low energy particles. The composite X-ray/IR/optical image of the Crab nebula, in Fig. 6.5, shows clearly this energy dependent behaviour. In the optical band, the nebula has an elliptical shape of 2.9×4.0 pc (5' $\times$ 6.8'), whereas in the X-ray band the nebular radius is about 0.4 pc (42"). In particular, recent *NuSTAR* observations of the Crab revealed a significant shrinking of size for increasing energies within the 3–78 keV band following a power-law function with an index of 0.094 ± 0.018 (Madsen et al. 2015).

The 3D nebular structure is almost symmetric about the spin axis of the pulsar. The observed axis-symmetry goes from South-East (SE) to NorthWest (NW) and is titled by 27° with respect to the plane of the sky. Such a symmetry is evident at X-rays where *Chandra* images, with a 0.8" PSF, clearly shows a torus with bipolar jets (Weisskopf et al. 2000). The Doppler brightening of the flows defines the direction of the axis titling, thus having the jet, brighter since beamed toward us, to the SE and the dimmer counter-jet to the NW. The jets have been seen to decrease in size with energy and evolve with time. In particular, a still-ongoing study of *Chandra* images taken at different times has shown that the tail of the jet wags on timescales of few years (Weisskopf et al. 2011) which could be explained by either precession of the jet or by the motion of kinks along the jet. Most likely the counter-jet has similar behaviour but it is masked by the foreground torus. On the other hand, the wide range covered by the *NuSTAR* detector allowed to measure the size decrease with energy finding that the counter-jet size falls off more rapidly with the energy than the jet (Madsen et al. 2015). This suggests that the particle transport mechanism in the counter-jet is different from the jet and the torus.

The most striking feature of the X-ray image is the inner ring usually identified with the TS. It has an elliptical shape, likely a circular ring in projection, with a semimajor axis of 14" (0.14 pc) and a semiminor axis of 7" (0.07 pc) (Weisskopf et al. 2000). The center of this ellipse is located 0.09" (10 light days) away from the pulsar implying that the ellipse does not lie on the equatorial plane of the pulsar, but at +4.5° latitude in a pulsar-centered coordinate system (Weisskopf et al. 2012). The inner ring, contrasty to the torus, has a surface brightness that is rather uniform in azimuth angle, indicating that the relativistic beaming is less significant than for the torus.

Along the axis, at 0.65" from the pulsar to the SE, a small compact and bright feature of elongated shape (with 0.5" = 0.005 pc length), called *knot*, pops up (Hester et al. 1995). It is highly variable both in position and brightness, but still it is a persistent feature. Its shape is not purely elliptical, but it has a curvature (like a "smile"), that, together with the brightness, is related to the *knot*-pulsar separation (Rudy et al. 2015). The fact that it has no counterpart on the symmetric axis suggests that the wind characteristics are not symmetrical. The *knot* has been explained both as a shock in the jet flow (Lou 1998), a Doppler beaming in the high-latitude wind (Del Zanna et al. 2006) or just downstream the TS (Komissarov and Lyutikov 2011). Under the assumption that particles can be accelerated only at TS, the *knot* has been suggested to be the production site of most of the synchrotron gamma-ray emission at $\sim$100 MeV, given their short cooling times and the strong Doppler beaming in the vicinity of the TS (Komissarov and Lyutikov 2011).

The X-ray torus is populated by series of elliptical ripples with width of 0.2", usually referred to as wisps. The name was given by J. Scargle who was the first one to study the dynamic substructures of the inner Crab Nebula at optical wavelengths (Scargle 1969). The *wisps* propagate outwards with velocities projected on the sky ranging from 0.1 (v/c) to 0.4 (v/c), and the larger the further they are from the pulsar. They form and peel off from the region commonly associate with the TS

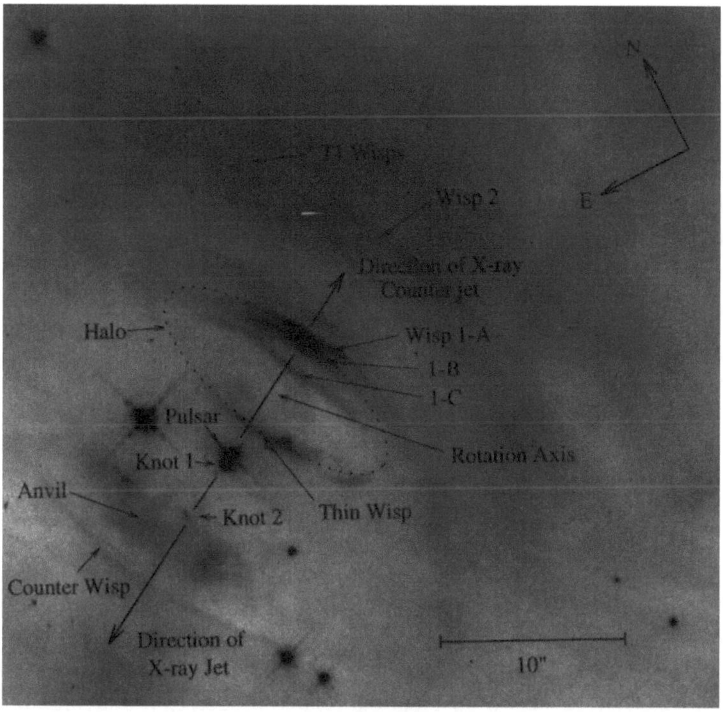

Fig. 6.6 Image of the inner Crab nebula, close to the pulsar, by the *Hubble Space Telescope* with all the features labelled. Credits to Hester, J. et al. (1995)

once per year (Schweitzer et al. 2013) with a life time of month time scales. In terms of MHD modeling, the *wisps* are interpreted as asymmetric enhancement emissivity immediately downstream the TS. Due to the anisotropy of the pulsar wind luminosity, and the related oblate shape of the TS (Del Zanna et al. 2004), asymmetric regions form inside the nebula and their emissivity towards the observer is higher, because of either a local compression of the magnetic field or Doppler boosting.

All these X-ray features appear also in the optical, and IR bands filling roughly the same volume, although with variability on different timescales. The better resolution of the *Hubble Space Telescope*, on scales down to 0.2", allowed a more precise classification of the existing substructures (Hester et al. 1995). The are listed and illustrated in Fig. 6.6. We list in the following the features that are persistent although with varying position and brightness (Hester et al. 1995):

- *thin wisp*, i.e. the closest to the pulsar, at 1.8" (0.02 pc) distance from the pulsar. It looks like closing a ringlike structure around the axis of symmetry, called the *halo*. The image of the *thin wisp* shows that it breaks into different *knots* and clumps.

- *wisp 1* at 7.3" from the pulsar to the NW, it lies on the above-described *halo*, thus above the equatorial plane. The *wisp 1* is visible also in X-rays as a superimposition of two/three structures. It has a *counter-wisp* in the located at 8.4" to the SE, which, however, lies on another plane.
- *wisp 2* is more an amorphous structure at 14" from the pulsar, remarkably changing shape along the years. It lies on the equatorial plane.
- *anvil* sets along the jet axis with an amorphous shape. It is interpreted as the counterpart to the SE of the above-mentioned *halo*. It hosts another *knot*, fainter but similar in dimension to the main *knot*.

The NW *wisps* are brighter than those to the SE due to Doppler beaming of the flow. A simultaneous study in the optical and X-ray bands of the brightest *wisps* to the NW showed that the position of the optical and X-ray *wisps* is similar but not coincident, with the X-ray ones located closer to the pulsar (Schweitzer et al. 2013). This would suggest that they are not produced by the same particle distributions. In addition the authors found the optical *wisps* being more Doppler boosted than the X-ray ones.

The *wisps* are also visible at radio frequencies (Bietenholz et al. 2001) moving outwards with projected velocities smaller than *c*/3. Their correspondence with the optical *wisps* is not exact mainly in terms of brightness: the *wisps* on the NW are not significantly brighter than the SE ones and the brightest optical *wisps* close to the pulsar do not have counterparts at radio frequencies. This lack of exact correspondence suggests two different but related processes producing optical and radio *wisps* (Bietenholz et al. 2004). Besides the motion of the *wisps*, the radio emission is variable on timescales larger than 1 year. In addition, striation is also detected in images spanning $\sim$3 years with an arclike geometry, usually not centered on the pulsar (Bietenholz et al. 2004). Such effect suggests the existence of wavelike motions in the body of the nebula due to wavelike disturbances excited by the *wisp* instabilities and propagating away from the injection region in the synchrotron emitting plasma.

The symmetric about the SE-NW axis is visible also at larger scales in the nebula. In fact, Projected on the plane of the sky, the elliptical optical nebula has the major axis aligned to the spin axis of the pulsar. However, the nebular radio extent agrees well with the optical one everywhere, but in the NW part where the radio emission extends beyond the edge of the thermal filaments. Such a symmetry break is most likely due to the variations in the density of the ejecta into which the synchrotron nebula is expanding. For shock velocities below 195 km/s the cooling time is smaller than the age of the nebula allowing the filament formation through R-T instabilities, whereas for higher velocities the already cooled gas prevents the growth of the filaments (Sankrit et al. 1998).

At γ rays, MHD simulations shows that the size of the inverse Compton nebula is expected to decrease with energy (Volpi et al. 2008). In addition, they distinguish the jet-torus structure, and also the *knot*, even though with a lower brightness contrast than in X-rays. The strong similarity between the X-ray and the gamma-ray maps is due to the fact that these photons are produced in the same region of the nebula, relatively close to the pulsar. Unfortunately, the point spread function of the current generation of gamma-ray telescopes is still too large to be able to resolve single features. Therefore, the comparison between the data and the simulations can be based only on the size of the nebula, and on possible asymmetric shrinking. At VHE, HEGRA was excluding a size larger than 2' (1.2 pc) for energies below 10 TeV (HEGRA Collaboration 2004) in agreement with the MAGIC 95% confidence level upper limits on the extension at 2.4' (1.4 pc) between 150 and 400 GeV and at 2.2' (1.3 pc) above 500 GeV (MAGIC Collaboration 2008). These estimations are totally dominated by the systematic uncertainties in the photon direction reconstructions.

6.3.3 Spectral Measurements

The emission from the Crab nebula is mostly non-thermal of synchrotron and inverse Compton origin, except for the thermal peak due to the dust that pops up in the IR band at ~50 μm. The non-thermal broad-band spectrum consists of two components. The primary emission mechanism at most energies is synchrotron radiation of particles gyrating in the nebular magnetic field. The synchrotron SED peaks in the optical through X-rays and cuts off at about a hundred of MeV, close to critical energy. At higher energies, the dominant emission mechanism is the inverse Compton scattering by the same particles off different photon fields, mainly the cosmic microwave background (CMB), the synchrotron photons (SSC emission), the FIR photons produced by the reprocessing of the nebular radiation by the dust (Atoyan and Aharonian 1996) and the line emission from the filaments (Meyer et al. 2010). The inverse Compton component peaks between 10 and 100 GeV, but its cut off is still matter of debate. The spectral steepening at the high end is indeed expected because of a combination of the maximum energies of the parent population and the transition to the Klein-Nishina regime.

The Crab nebula is an extreme accelerator, one of the few known Pevatrons in our Galaxy: synchrotron photons at ~100 MeV are expected to be produced by particles with ~1 PeV energy in a 0.1 mG. It is very efficient at converting the energy of the shocked wind into synchrotron radiation with a conversion efficiency of about 10%. It is not that efficient as gamma-ray emitter though mainly due to the large magnetic field B $\left(\frac{L_{synch}}{L_{IC}} = \frac{U_B}{U_{ph}}\right.$, where L_{synch} is the luminosity of the synchrotron emission, L_{ph} is the luminosity of the inverse Compton radiation, and U_B and U_{ph} are the energy densities of the magnetic field and of the seed photon fields, respectively). Its L_{IC} is just 0.05% of the spin-down flux, that is, however, high enough to allow the detection of VHE γ rays. The inverse Compton component, if considered together with the synchrotron one, enables constraint of properties (under the assumption that B is uniform) of the magnetic-field strength and particle number density, which cannot be achieved using only synchrotron measurements (de Jager et al. 1996).

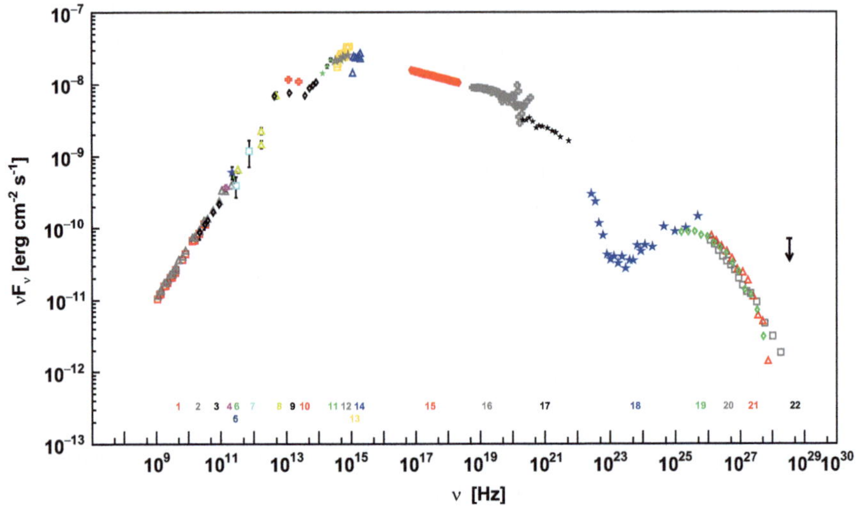

Fig. 6.7 SED of the Crab nebula. Numbers on the bottom indicate the reference where data come from: (1) Baars et al. (1977), (2) Maca-Prez et al. (2010), (3) Wieland et al. (2011), (4) Arendt et al. (2011), (5) Bandiera et al. (2002), (6) Mezger et al. (1986), (7) Wright et al. (1979), (8) Green et al. (2004), (9) Temim et al. (2006), (10) Marsden et al. (1984), (11) Garsdalen (1979), (12) Veron-Cetty and Woltjer (1993), (13) Hennessy et al. (1992), (15) Kirsch et al. (2005), (16) Jourdain and Roques (2009), (17) Kuiper et al. (2001), (18) Buehler et al. (2012), (19) MAGIC Collaboration (2015), (20) HEGRA Collaboration (2004), (21) H.E.S.S. Collaboration (2006), (22) Borione et al. (1997)

From the observational point of view, the broad-band SED of the Crab nebula, showed in Fig. 6.7, has now reached a great level of accuracy. At radio frequencies the nebular integrated emission is a power-law spectrum ($S_\nu = x^\alpha$) with index α being around the canonical -0.30 (Baars et al. 1977; Maca-Prez et al. 2010) from 10 MHz to the submillimeter domain. Bietenholz et al. (1997) showed that the spectral index is quite constant across the nebula. In particular, between 300 MHz and 5 GHz there is no evidence of spectral steeping near the edge of the nebula or in the jets (Bietenholz et al. 1997). The IR spectral measurements fit the extrapolation of the radio spectrum (Green et al. 2004; Arendt et al. 2011), thus excluding the existence of two different synchrotron components suggested by Bandiera et al. (2002) to justify a spectral hardening at $\sim$1 mm (Bandiera et al. 2002). In the optical domain the photon index is of -0.6 in the inner nebula (Veron-Cetty and Woltjer 1993). The fit to the integrated radio-to-optical spectrum suggest a spectral break in the far-IR at $\sim$420 μm (Arendt et al. 2011), but missing data in this energy range prevent a more precise estimation.

In the X-ray band, recent measurements by *NuSTAR* described the nebular emission between 3 and 78 keV as a mildly broken power law with the two photon indices being $\Gamma_1 = (2.123 \pm 0.003)$ and $\Gamma_2 = (2.134^{+0.001}_{-0.004})$ (Madsen et al. 2015). The photon index remains constant, $\sim$2.14, up to 300 keV whereas it softens to

2.27 in the 0.75–30 MeV range (Kuiper et al. 2001), continuing to soften until 700 MeV beyond which the inverse Compton component sets in. Above 100 MeV the synchrotron emission is well described by a (3.59 ± 0.07) power law by using 33 months of *Fermi*-LAT data (Buehler et al. 2012). The combination of *Fermi*-LAT (0.1–300 GeV) and *COMPTEL* (0.75–30 MeV) data (Kuiper et al. 2001) allowed the estimation of an energy cutoff E_{cutoff} = (92 ± 12) MeV (Fermi Collaboration 2010).

The synchrotron component shows a trend of softer spectra with increasing distance from the pulsar. This is interpreted as synchrotron cooling of the highest energy electrons as convected or diffuse away from the inner nebula. The IR and optical spectral morphological maps show a significant increase of the spectral index with larger values further away from the pulsar: in IR it spans from 0.3 in the inner nebula to 0.6 at the edge of the nebula towards SE, or 1.0 towards NW (Temim et al. 2012), whereas in optical it lies in the 0.6–0.8/1.0 range following the same asymmetry seen in IR (Veron-Cetty and Woltjer 1993). However, the spectrum in the inner region of the nebula is rather uniform from IR to X-rays. Indeed the small substructures show no spectral variations and the spectral index of the torus appeared to be constant up to its edge where it steepens abruptly (Mori et al. 2004). This indicates that advection dominates over diffusion time scales for X-ray producing particles. At radio frequencies the spectrum is harder in the inner nebula (Green et al. 2004; Arendt et al. 2011), but it is more difficult to disentangle between different regions, as radio emission comes from larger volumes.

The new generation of gamma-ray detectors, *Fermi*-LAT between 0.1 and few hundreds of GeV on one side, and the imaging atmospheric Cherenkov telescopes (IACT, 0.05<E<∼100 TeV) at higher energies led to the measurement of the nebular inverse Compton spectrum with high statistical precision and a complete coverage of its peak. The rising edge of the inverse Compton component is parametrized by a smoothly broken power law with a curvature index of $\beta = 0.2$:

$$\phi(E) = \phi_0(\frac{E}{100\,\text{MeV}})^{-\gamma_1} \left(1 + \frac{E}{E_b}^{\frac{\gamma_2-\gamma_1}{\beta}}\right)^{\beta} \tag{6.1}$$

At low energies the photon index, γ_1 is 1.48 ± 0.07 and softens to $\gamma_2 = 2.19 \pm 0.17$ above the energy break $E_b = (13.9 \pm 5.8)$ GeV (Buehler et al. 2012).

The falling edge of the inverse Compton component falls in the domain of the Cherenkov telescopes—IACTs, water Cherenkov telescopes, and air shower arrays. The first detection of the Crab at VHE occurred in 1989 by the pioneering Whipple telescope (Weeks et al. 1989), and was later confirmed by a number of other experiments (ASGAT (Goret et al. 1993) THEMISTOCLE (Baillon et al. 1993); STACEE (Oser et al. 2001); CELESTE (de Naurois et al. 2002); HEGRA (HEGRA Collaboration 2000, 2004); Hillas et al. (1998); H.E.S.S. (H.E.S.S. Collaboration 2006); MAGIC (MAGIC Collaboration 2008, 2012, 2015), VERITAS (Meagher 2015)). Recently the MAGIC collaboration published the Crab nebula spectrum covering almost three decades in energy between 50 GeV and 30 TeV (MAGIC Collaboration

2015). The energy spectrum is significantly curved both at low- and high-energies and well represented, in the considered energy range, by a log-parabola:

$$\phi(E) = \phi_0(\frac{E}{E_0})^{-\alpha+\beta \log(E/E_0)} \tag{6.2}$$

where the photon index $\alpha = 2.47 \pm 0.02_{stat} \pm 0.05_{sys}$, the curvature parameter $\beta = -0.24 \pm 0.01_{stat} \pm 0.07_{sys}$ and $\phi_0 = 3.23 \pm 0.03_{stat} \pm 0.35_{sys}\,\mathrm{TeV^{-1}\,cm^{-2}\,s^{-1}}$. At low energy it reveals the long-anticipated inverse Compton peak below 100 GeV (MAGIC Collaboration 2008) and its combination with the *Fermi*-LAT data allows for its precise estimation. A joint *Fermi*-LAT/MAGIC fit results in a peak position of $(53 \pm 3^{+31}_{-13})$ GeV and shows that the peak is rather flat between 10 and 200 GeV.

Over specific energy sub-ranges other functional forms can provide good fits of the measured VHE emission from the Crab Nebula. In particular, above few hundreds of GeV it was described by either a single power law until 80 TeV (HEGRA Collaboration 2004) or a power law with exponential cutoff at (14 ± 2) TeV (H.E.S.S. Collaboration 2006). However, the discrepancy between the available results, as well as the large systematic uncertainties of the IACTs above some tens of TeV, prevents a common accord on the position of this spectral softening. Recently the MAGIC collaboration presented some new results obtained with a total independent data sample taken under observational conditions that are more favorable to the detection of photons at the highest energies. The obtained energy spectrum, covering the 0.8–80 TeV energy range, shows a sub-exponential steepening with an estimated cutoff at 30–50 TeV (Zanin 2015), in agreement with the measurements obtained by using the entire 8 year of data by the already dismissed water Cherenkov detector Milagro (Milagro Collaboration 2012). Above 100 TeV there are only flux upper limits produced by air shower arrays (CASA-MIA (Borione et al. 1997), the Tibet air shower array Amenomori et al. 2015), but they are still few orders of magnitude higher than the expected emission.

6.3.4 The Gamma-Ray Flares

For almost 100 year it is known that the Crab is a dynamic source. Its substructures show brightness variations on timescales of months (Scargle 1969). However, the flux integrated over larger volumes of the nebula is changing at a few percent per year (Wilson-Hodge et al. 2011), below the systematic uncertainties of many detection techniques.

Flux variability is indeed expected from the high end of the synchrotron component (de Jager et al. 1996), whose emission is produced by PeV electrons having a lifetime of 1 year. Nevertheless, the discovery of flaring activity on short timescales in the energy band between 100 MeV and few GeV came as a surprise (Tavani et al. 2011; Fermi Collaboration 2011).

As October 2016 *Fermi*-LAT has detected 11 flaring events since its launch in August 2008, that are listed in Table 6.2. That corresponds to a rate of 1.4 event per year. The enhanced activity, with respect to the average flux, typically lasts for ∼1 week, between few days and a couple of weeks. The increase of the flux was up a factor 30 larger than the average flux of $F(E > 100\,\text{MeV}) = (6.1 \pm 0.2)\times10^{-7}\,\text{cm}^{-2}\,\text{s}^{-1}$, with significant variations on few hours scales. The complex structure of the brightest flare, occurred in April 2011, is shown in Fig. 6.8, as example (Buehler et al. 2012; Striani et al. 2013). In this particular event, the flux doubled in less 8 hrs at the raising edges of the two main bursts. Such rapid flux variations constrain the emitting region to be as compact as R

Table 6.2 List of flares events as October 2016 with duration and spectral information when available as well as with their corresponding references

Starting date	Duration (days)	Flux increase	Photon index	Reference
10.2007				Tavani et al. (2011)
26.01.2009	16	3.8 ± 0.5	4.3 ± 0.3	Fermi Collaboration (2011)
18.09.2010	4	5.5 ± 0.8	2.7 ± 0.2	Fermi Collaboration (2011) and Tavani et al. (2011)
09.04.2011	9	∼30	1.27 ± 0.12	Buehler et al. (2012) and Striani et al. (2013)
02.07.2012		∼2		ATel4239
01.03.2013	14	∼20	1.70 ± 0.3	Mayer et al. (2013)
15.10.2013	5	∼4		ATel 5485
06.03.2014	4–5	∼2		ATel 5971
12.08.2014	9	∼3		ATel 6401
28.12.2015	5–6	∼2		ATel 8519
30.09.2016	16–17	∼3		ATel 9586, 9588

The flux increase corresponds to the peak emission

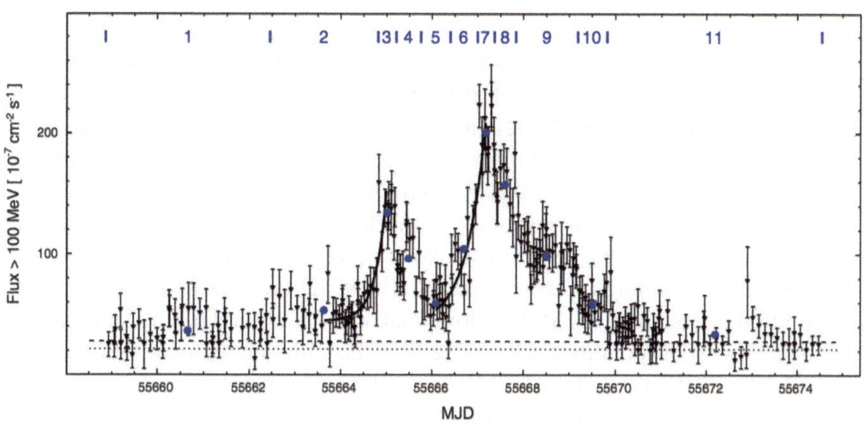

Fig. 6.8 Integral flux of the Crab nebula above 100 MeV as a function of the time during the 2011 April flare. The binning of the plot is 9 min. Taken from Buehler et al. (2012)

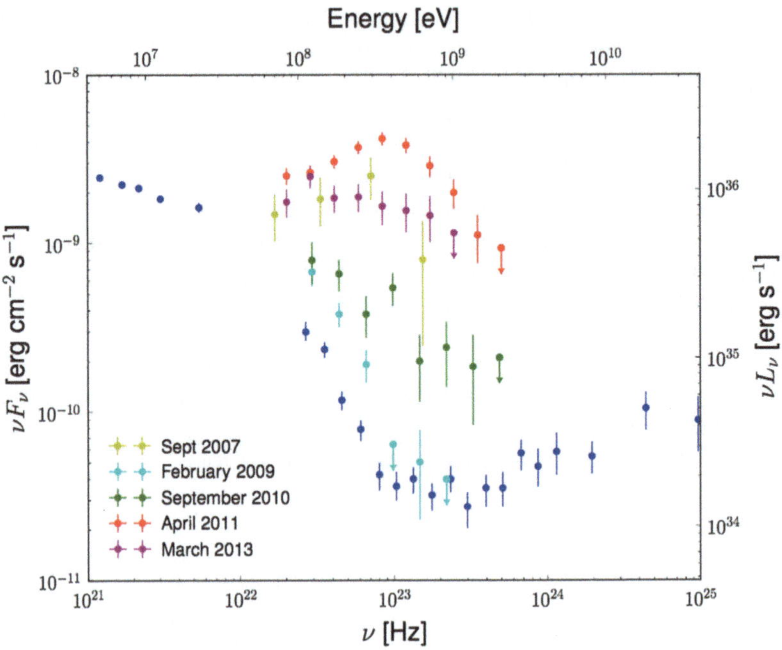

Fig. 6.9 SED at the maximum flux level for five Crab flares. Taken from Bülher and Blandford (2014)

$\leq c\dot{t}_d \sim 10^{-4}$ pc. Unfortunately this value is orders of magnitude smaller than the PSF of the current gamma-ray detectors. It is still unclear if the flares are really distinct spectral components, but their SEDs differ between flares (see Fig. 6.9) (Bülher and Blandford 2014). The February 2009 flare shows a flux increase but no spectral changing with respect to the baseline Crab spectrum. While in April 2011 flare the energy spectrum of this emerging component is evolving during the flare. Well-represented by a power law with an exponential cutoff, it has a constant photon index $\gamma = 1.27 \pm 0.12$ and a cutoff E_C that increases with the luminosity L_{synch}, following the relation: $L_{synch} \propto E_C^{(3.42 \pm 0.86)}$ (Buehler et al. 2012) (see Fig. 6.10). At the maximum flux, the spectral cutoff is $E_C = (375 \pm 26)$ MeV and $L_{synch} = 4 \times 10^{36}$ erg s^{-1}, that is approximately 1% of the spin-down power of the pulsar. The spectral properties of the flares challenged the understanding of the PWN physics: first the diffuse shock acceleration fails in producing hard spectra with a ~1.3 photon index; secondly in MHD the maximum energy of the synchrotron emission is 160 MeV, below what actually observed. The latter implies that either the MHD conditions are not valid in the emission region, or the emission is relativistically boosted towards the line of sight (a Doppler factor of 3–4 would be needed to match the data) (Yuan et al. 2011; Bednarek and Idec 2011; Komissarov and Lyutikov 2011; Lyutikov et al. 2012; Clausen-Brawn and Lyutikov 2012).

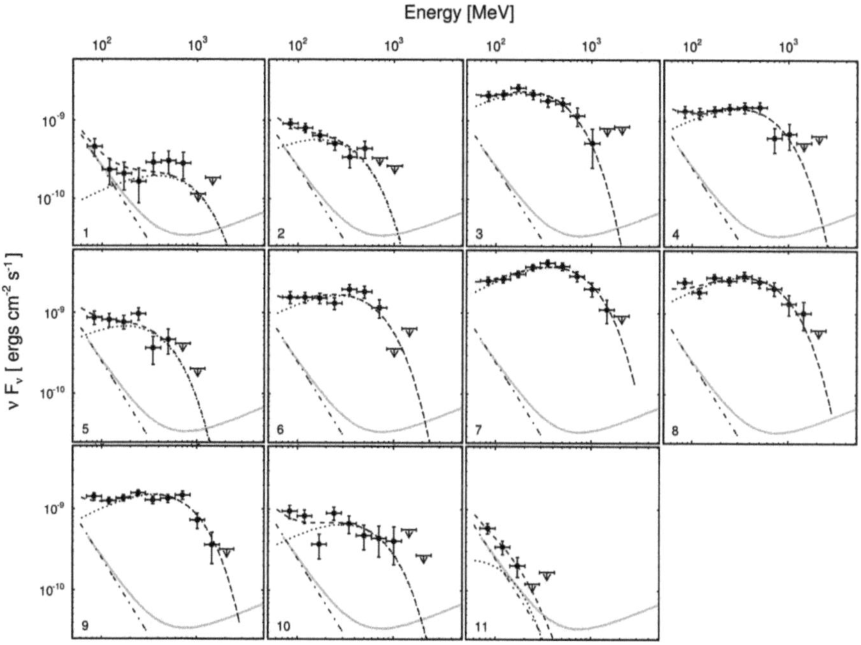

Fig. 6.10 Evolution of the SED during the 2011 April flare. Credits to Buehler et al. (2012)

The origin of the Crab flares is still unclear. Many efforts were conducted in the past years to look for counterparts at other wavelengths, especially at radio, optical and X-rays frequencies where their high resolution instruments could allow to pinpoint the region of the flux enhancement. To date, no correlated flux variations have been detected in the radio (Lobanov et al. 2011; Weisskopf et al. 2013), infrared (Rudy et al. 2015), optical (Rudy et al. 2015), X-ray (Weisskopf et al. 2013; Rudy et al. 2015) bands. Also the inverse Compton component remained constant (H.E.S.S. Collaboration 2014; VERITAS Collaboration 2014). The *knot* was immediately suggested as possible emission site (Tavani et al. 2011; Komissarov and Lyutikov 2011) for three reasons: (1) the tight constraints on the size of the emission region due to the rapid flux increases; (2) the high flaring luminosity which implies that the emission must originate close to the pulsar; (3) the relativistic flow lies at high altitudes. However, the available data on the *knot* do not show strong correlation with the gamma-ray flares (Rudy et al. 2015). This result favours a non MHD process at the origin of the gamma-ray flares, and magnetic reconnection seems the most likely one. Recent particle-in-cell simulations showed that magnetic reconnection can accelerate particles beyond the synchrotron burnoff (Cerutti et al. 2013) as well as produce spectra as hard as $\sim$1.3 (Sironi and Spitkovsky 2011), and explain flux variations on hour timescales thanks to the beaming of high energy particles (Cerutti et al. 2012). The gamma-ray flare are therefore likely connected to explosive reconnection events triggered by current instabilities. More theoretical discussion on this topic can be found in Chap. 10.

6.3.5 *Polarization Measurements*

Polarization provides the best available tool to study the magnetic field structure, thus the geometry of the emission regions. The Crab nebula was the first source from which polarization was measured (Shklovsky 1953). It is indeed highly polarized as expected from a synchrotron emitting source. At large scales the linear polarization is rather uniform with a polarization angle P.A. $\sim$155° indicating a magnetic field that is aligned to NE-SW, almost perpendicular to the pulsar spin axis (at a position angle of 124°$\pm$0.1° Ng and Romani 2004). This large-scale polarization is detected from radio (e.g. at 9 mm, P.A. = 154.8 $\pm$ 2.0 (Flett and Henderson 1979)) through millimeter (e.g. at 3.3 mm P.A. = 153.7 $\pm$ 0.4 Aumont et al. 2010), and optical (Oort and Walraven 1957), soft X-rays (P.A. = 155.8 $\pm$ 1.4 Weisskopf et al. 1978) to hard X-rays (P.A. = 149.2 $\pm$ 16 Chauvin et al. 2015) frequencies. These results confirm the expectation of a constant P.A. as the synchrotron emission from radio to submillimeter band is believed to be produced by the same particle population accelerated by the same magnetic field (see Sect. 6.3.3). Since the P.D. varies rapidly inside the source, its comparison requires spatially coincident measurements and that is not our case. The optical measurements, for instance, were extracted from a region of 0.5', while those at the 2.6–5.2 keV within a radius of 3° from the pulsar.

At optical wavelengths the high resolution *Hubble Space Telescope* images allowed for a morphological study of the polarized emission, hence of the magnetic field structure in the inner nebula region close to the pulsar. The polarized map shows the torus and the *wisps* as prominent features (P.D.$\sim$40%) with the *knot* being the brightest one (P.D. = (59 $\pm$ 1.9)% Moran et al. 2013). The P.A. of both *knot* and *wisps* is $\sim$125° (Slowikowska et al. 2009; Moran et al. 2013) perfectly aligned with the pulsar spin axis. This indicates that these structures have polarization properties different from those from the rest of the inner nebula. Table 6.3 resumes the polarization properties of the different detected features. In the last years the Galway Astronomical Stokes Polarimeter (GASP) opened the new window of the optical polarimetry allowing to measure circular polarization, that is directly dependent on the magnetic field (Legg and Westfold 1968; de Brca et al. 2015). As such, the circular polarization can provide estimates of the magnetic field strength in the emission region. Recent observations of the Crab show a -1.2% circular polarization at 3σ level (Moran et al. 2016). Still more statistics is needed before a definitive estimation of the magnetic field strength can be made, and phase-resolved measurements will clear up which is the source of the circular polarization, if the pulsar or the *knot*.

In the soft gamma-ray band both the IBIS (0.2–5 MeV) and SPI (0.1–1 MeV) detectors on board of the *INTEGRAL* satellite measured strong polarization from the nebula with a P.A. $\sim$122° that is consistent with the spin-axis of the pulsar and

Table 6.3 Results for P.D. and P.A. of the different nebular substructures obtained by the *Hubble Space Telescope* (Moran et al. 2013)

	P.D. [%]	P.A. [°]
knot	59.0 ± 1.9	124.7 ± 1.0
thin wisp	36.7 ± 1.4	127.1 ± 1.0
wisp 1	39.8 ± 1.6	124.7 ± 1.2
counter-wisp	40.6 ± 1.5	130.3 ± 1.1

a more uncertain P.D. spanning the 46% and 72% range (Dean et al. 2008; Forot et al. 2008). The remarkable similarity of the gamma-ray polarization with that of the structures in the inner nebula suggests that the gamma-ray emission originate at the same site close to the pulsar. The high-level of polarization (more than 60% of the maximum physical P.D.) indicates the magnetic field is rather uniform on the associated scales. In case of gamma-ray synchrotron emission, the maximum achievable degree of polarization is ∼77%, obtained from the expression $\Pi = (p+1)/(p+7/3)$ taken from Lei et al. (1997), where p indicates the index of the power-law distribution of the parent electrons that will be $p = 2\Gamma - 1 \sim 3.5$, and Γ the photon index.

Changes in the polarization parameters were recently searched both in optical and γ rays as tracers of potential variations in the geometry of the local magnetic field. This could shed some light on the mechanism producing the gamma-ray flares (Sect. 6.3.4). Moran and collaborators compared 2005 with 2012 *Hubble Space Telescope* Crab images and 2003–2007 with 2012–2014 *INTEGRAL* data, finding that both wavebands showed constant P.D., but similar shift in the P.A.: optical $24.2 \pm .1.6°$, γ rays $35 \pm 16°$ (Moran et al. 2016). These results suggest a change in the magnetic field orientation close to the pulsar giving more support to the magnetic reconnection scenario. However, optical polarimetric observations of the Crab nebula carried out simultaneously with the gamma-ray flare are necessary to claim a correlation between the changes in the P.A. and the increase of the high-energy flux (Moran et al. 2016).

A high-resolution X-ray polarization maps would probe the magnetic field structure close to the TS where higher energy particles are accelerated. The importance of this still unexplored window is demonstrated by the plethora of missions that have been recently proposed (Soffitta et al. 2013; Weisskopf et al. 2014). In particular, recent simulations performed for the *XIPE* satellite show that this kind of instruments will be able to determine the magnetic field orientation and the level of turbulence in the torus, in the jet and at various distance from the pulsar (de Oña-Wilhelmi et al. 2016). At the light of the recent advances of the MHD modeling where the turbulence of the magnetic field plays a central role (see Sect. 6.3.6) the launch of one of these missions could lead to a real breakthrough in the understanding of the PWNe.

6.3.6 Lessons on the Crab Nebula

An averaged magnetic field of the Crab nebula of 300 μG is inferred from the IR break frequency (Marsden et al. 1984) and it is in agreement with the estimations in the original MHD models (200–300 μG) (Kennel and Coroniti 1984). On the other hand a fit to the available gamma-ray data suggests an averaged magnetic field of $\sim$150 μG under its uniformity assumption (Meyer et al. 2010; MAGIC Collaboration 2015) or even lower ($<$100 μG) in the context of an evolving magnetic field with time (Martin et al. 2012; MAGIC Collaboration 2015). However, the wide breadth of the inverse Compton peak, revealed by the recent joint *Fermi*-LAT and MAGIC data, cannot be reproduced by models assuming a constant magnetic field, hence implying that the magnetic field structure plays an important role in the production of the observed emission (MAGIC Collaboration 2015).

Asymmetric relativistic MHD models have been so successful because they can reproduce both the overall jet-torus structure, and the observed optical substructures, like *wisps* and *knot* (Del Zanna et al. 2004; Komissarov and Lyubarky 2004). The interpretation of the jet-torus morphology is based on the idea that the pulsar wind is anisotropic (Bogovalov and Aharonian 2000). A non-null angle between the magnetic and the spin axis, θ, produces the corrugation of the current sheets forming beyond the light cylinder and this results in a latitude dependence ($\propto \sin^2\theta$) of its energy flux (that is usually refer to as *striped wind*). In this scenario the downstream flow first converges towards the equator and are then diverted along the symmetry axis by magnetic-hoop-stresses. Ensuring a post-shock collimation requires high levels of magnetization of the wind, that is evaluated by the σ parameter defined the ratio between the Poynting and the kinetic energy. For the Crab nebula, observational morphological results provide $\sigma = 0.03$, hence an average magnetic field of $\sim$50 μG, that is a factor 3–4 lower than estimated. In fact, accounting for the synchrotron emission would require a larger number of emitting particles which, in turn, would result in a significant overestimation of the inverse Compton flux (Volpi et al. 2008). On the other hand, in order to reproduce the broadband spectrum, the MHD models imply a magnetization parameter larger than 1, and this would result in too bright jets. This paradox seems to be an artefact of the axial symmetry imposed by the 2D simulations. Recently few groups have started running 3D relativistic MHD simulations which have demonstrated that kink instabilities can develop and that would twist the magnetic field, hence reducing the associated hoop stress. These results suggest that with high magnetization ($\sigma > 1$) both spectral and morphological properties may be recovered. Such a high magnetization has indeed important consequences at the level of understanding the acceleration mechanisms at work. The Fermi I type acceleration seems not be an option anymore and possible alternatives are magnetic reconnection (Sironi and Spitkovsky 2011) or resonant absorption of ion-cyclotron waves in a ion-doped plasma (Amato and Arons 2006). Still, this is one of the most important open-questions about the PWN physics.

The broadband synchrotron spectrum with its spectral break in the IR band suggests that there are two distinct particle populations (Atoyan and Aharonian 1996; Meyer et al. 2010). In fact, assuming an average magnetic field of 300 μG, only electrons with an energy larger than a few hundreds of GeV could be cooled down during the 1000 year Crab lifetime. The resulting electron spectrum is broken power law with photon indices $\gamma_e \sim 1.5$ and $\gamma_e \sim 2.2$.

While it is generally accepted that the electrons, that are responsible for the synchrotron emission above the IR break (called *wind electrons*), are freshly accelerated electrons presently injected downstream the TS (Atoyan and Aharonian 1996), there are three possible scenarios describing the origin of the radio emitting electrons (called *radio electrons*):

- relic particles, injected for a short time after the SN explosion. In this case the expected pair multiplicity $k = 10^4$, and the Lorentz factor $\gamma = 10^6$ would match the values predicted by the MHD models (Kennel and Coroniti 1984).
- relic particles but continuously accelerated by turbulence inside the nebula.
- ongoing injected particles as the X-rays ones. In this case we expect a much higher pair multiplicity, $k = 10^{5-6}$ and a lower wind Lorentz factor ($\gamma = 10^5$) than predicted by the available magnetospheric models.

A recent work addresses this problem by means of 2D MHD simulations leading to the conclusion that while we can exclude the first scenario, it is not possible to discriminate between the other two based on the emission morphology or the time variability. Also the existence of the radio wisp which could naively be taken as proof of ongoing acceleration was appearing in both scenarios (Olmi et al. 2014).

References

Aharonian, F., Bogovalov, S., Khangulyan, D.: Abrupt acceleration of a 'cold' ultrarelativistic wind from the Crab pulsar. Nature **482**, 507 (2012)

Amato, E., Arons, J.: Heating and nonthermal particle acceleration in relativistic, transverse magnetosonic shock waves in Proton-Electron-Positron plasmas. Astrophys. J. **653**, 325 (2006)

Amenomori, M., et al.: Search for gamma rays above 100 TeV from the Crab nebula with the Tibet air shower array and the 100 m^2 muon detector. Astrophys. J. **813**, 98 (2015)

Arendt, R.G., et al.: The radio-2 mm spectral index of the Crab nebula measured with GISMO. Astrophys. J. **734**, 54 (2011)

Atoyan, A.M., Aharonian, F.: On the fluxes of inverse Compton gamma-rays expected from the Crab nebula. Astron. Astrophys. Suppl. **453**, 456 (1996)

Aumont, J., et al.: Measurement of the Crab nebula polarization at 90 GHz as a calibrator for CMB experiments. Astron. Astrophys. **514**, A70 (2010)

Baade, W., Zwicky, F.: Remarks on super-Novae and cosmic rays. Phys. Rev. **46**, 76 (1934)

Baars, J.W.M., Genzel, R., Pauliny-Toth, I.I.K., Witzel, A.: Astron. Astrophys. **61**, 99 (1977)

Baillon, P., et al.: Gamma ray spectrum of the Crab nebula in the multi TeV region k20. Astropart. J. **1**, 341 (1993)

Bandiera, R., Neri, R., Cesaroni, R.: The Crab nebula at 1.3 mm. Evidence for a new synchrotron component. Astron. Astrophys. **386**, 1044 (2002)

Bednarek, W., Idec, W.: On the variability of the GeV and multi-TeV gamma-ray emission from the Crab nebula. Mon. Not. R. Astron. Soc. **414**, 2229 (2011)

Bietenholz, M.F., Nugent, R.L.: New expansion rate measurements of the Crab nebula in radio and optical. Mon. Not. R. Astron. Soc. **454**, 2416 (2015)

Bietenholz, M.F., et al.: The radio spectral index of the Crab nebula. Astrophys. J. **490**, 291 (1997)

Bietenholz, M.F., Frail, D.A., Hester, J.J.: The Crab nebula's moving wisps in radio. Astrophys. J. **560**, 254 (2001)

Bietenholz, M.F., Frail, D.A., Hester, J.J., Bartel, N.: The Crab nebula's wisps in radio and optical. Astrophys. J. **615**, 794 (2004)

Bilous, A.V., et al.: Correlation of *Fermi* photons with high-frequency radio giant pulses from the Crab pulsar. Astrophys. J. **728**, 110B (2011)

Bilous, A.V., McLaughlin, M.A., Kondratiev, V.I., Ransom, S.M.: Correlation of *Chandra* photons with the radio giant pulses from the Crab pulsar. Astrophys. J. **749** 24 (2012)

Bogovalov, S., Aharonian, F.: Very-high-energy gamma radiation associated with the unshocked wind of the Crab pulsar. Mon. Not. R. Astron. Soc. **313**, 504 (2000)

Borione, A., et al.: A search for ultra-high-energy gamma-ray emission from the Crab nebula and pulsar. Astrophys. J. **481**, 313 (1997)

Buehler, R., et al.: Gamma-ray activity in the Crab nebula: the exceptional flare of 2011 April. Astrophys. J. **749**, 26 (2012)

Bülher, R., Blandford, R.: The surprising Crab pulsar and its nebula: a review. Rep. Prog. Phys. **77**, 6 (2014)

Carraminana, A., Cadez, A., Zwitter, T.: Optical spectrum of main-, inter- and Off-pulse emission from Crab pulsar. Astrophys. J. **542**, 974 (2000)

Cerutti, B., Werner, G.R., Uzdensky, D.A., Begelman, M.C.: Beaming and rapid variability of high-energy radiation from relativistic pair plasma reconnection. Astrophys. J. **754**, 33 (2012)

Cerutti, B., Werner, G.R., Uzdensky, D.A., Begelman, M.C.: Simulations of particle acceleration beyond the classical synchrotron burnoff limit in magnetic reconnection: an explanation of the Crab flares. Astrophys. J. **770**, 147 (2013)

Cerutti, B., Philippov, A., Parfrey, K., Spitkovsky, A.: Particle acceleration in axisymmetric pulsar current sheets. Mon. Not. R. Astron. Soc. **448**, 606 (2015)

Chauvin, H.G., et al.: Observation of polarised hard X-ray emission from the Crab by the PoGOLite pathfinder. Mon. Not. R. Astron. Soc. **456**, L84–L88 (2016)

Cheng, K.S., Ho, C., Rudermann, M.:Energetic radiation from rapidly spinning pulsars. I – Outermagnetosphere gaps. II – VELA and Crab. Astrophys. J. **300**, 500 (1986)

Chevalier, R.A.: Was SN 1054 a type II supernova? In: Schramm, D.N. (ed.) Supernovae. Astrophysics and Space Science Library, vol. 66, p. 53. Springer (1977)

Chevalier, R.A.: The interaction of Crab-like supernova remnants with their surroundings. Astrophys. J. **280**, 797 (1984)

Clausen-Brawn, E., Lyutikov, M.: Crab nebula gamma-ray flares as relativistic reconnection minijets. Mon. Not. R. Astron. Soc. **426**, 1374 (2012)

Cocke, W.J.,Disney, M.J., Taylor, D.J.: Discovery of optical signals from pulsar NP 0532. Nature **221**, 525–527 (1969)

Contopoulos, J., Kazanas, D., Fendt, C.: The axisymmetric pulsar magnetosphere. Astrophys. J. **511**, 351 (1999)

Contopoulos, J., Kalapotharakos, C., Kazanas, D.:A new standard pulsar magnetosphere. Astrophys. J. **781**, 46 (2014)

Davidson, K., Fesen R.A.: Recent developments concerning the Crab nebula. ARAA **23** 119 (1985)

de Búrca, D., Shearer, A.: Circular polarization of synchrotron radiation in high magnetic fields. Mon. Not. R. Astron. Soc. **450**, 533 (2015)

de Jager, O., et al.: Gamma-ray observations of the Crab nebula: a study of the synchro-compton spectrum. Astrophys. J. **457**, 253 (1996)

de Naurois, M., et al.: Measurement of the Crab flux above 60 GeV with the CELESTE Cerenkov telescope. Astrophys. J. **566**, 343 (2002) de Oña-Wilhelmi, E., et al: Unveiling the magnetic structure of VHE SNRs/PWNe with *XIPE*, the X-ray imaging-polarimetry explorer. In: Proceedings of the Gamma 2016 Symposium, Heidelberg (2016, July)

Dean, A.J., et al.: Polarized gamma-ray emission from the Crab. Science **321**, 1183 (2008)

Del Zanna, L., Amato, E., Bucciantini, N.: Axially symmetric relativistic MHD simulations of pulsar wind nebulae in supernova remnants. On the origin of torus and jet-like features. Astron. Astrophys. **421**, 1063 (2004)

Del Zanna, L., Volpi, D., Amato, E., Bucciantini, N.: Simulated synchrotron emission from pulsar wind nebulae. Astron. Astrophys. **453**, 621 (2006)

Eilek, J.A., Hankins, T.: Radio emission physics in the Crab pulsar. J. Plasma Phys. **82**, 3 (2016)

Espinoza, C.M., Lyne, A.G., Stappers, B., Kramer, M.: A study of 315 glitches in the rotation of 102 pulsars. Mon. Not. R. Astron. Soc. **414**, 1679 (2011)

Fermi Collaboration: Fermi Large Area Telescope Observations of the Crab pulsar and nebula. Astrophys. J. **708**, 1254 (2010)

Fermi Collaboration: Gamma-ray flares from the Crab nebula. Science **331**, 739 (2011)

Fermi Collaboration, et al.: The second Fermi large area telescope catalog of gamma-ray pulsars. Astrophys. J. Suppl. **208**, 59 (2013)

Fesen, R.A., Martin, C.L., Shull, J.M.: The synchrotron bays of the Crab nebula – a magnetic structure associated with a presupernova circumstellar disk. Astrophys. J. **399**, 599 (1992)

Flett, A.M., Henderson, C.: Observations of the polarized emission of Taurus A, Cas A and Cygnus A at 9-mm wavelength. Mon. Not. R. Astron. Soc. **189**, 86 (1979)

Forot, M., Laurent, P., Grenier, I.A., Gouiffes, C., Lebrun, F.: Astrophys. J. **688**, L29 (2008)

Garsdalen, G.L.: Near infrared observations of the Crab nebula. Astron. Soc. Pac. **91**, 436 (1979)

Gomez, H. L., et al.: A cool dust factory in the Crab nebula: a Herschel study of the filaments. Astrophys. J., **760**, 1 (2012)

Goret, P., Palfrey, T., Tabary, A., Vacanti, G., Bazer-Bachi, R.: Observations of TeV gamma rays from the Crab nebula. Astron. Astrophys. **270**, 401 (1993)

Green, D.A., Tuffs, R.J., Popescu, C.C.: Far-infrared and submillimetre observations of the Crab nebula. MNRAS **355**, 1315 (2004)

H.E.S.S. Collaboration: Observations of the Crab nebula with HESS. Astron. Astrophys. **457**, 899 (2006)

H.E.S.S. Collaboration: H.E.S.S. observations of the Crab during its March 2013 GeV gamma-ray flare. Astron. Astrophys. **562**, L4 (2014)

Hankins, T., Eilek, J.A.: Radio emission signatures in the Crab pulsar. Astrophys. J. **670**, 693 (2007)

Hankins, T., Kern, J.S., Weatherall, J.C., Eilek, J.A.: Nanosecond radio bursts from strong plasma turbulence in the Crab pulsar. Nature **422**, 6928 (2003)

Hankins, T., Eilek, J.A., Jones, G.: The Crab Pulsar at Centimeter Wavelengths II: Single Pulses. arxiv:1608.08881

HEGRA Collaboration: The Energy Spectrum of TEV gamma rays from the Crab nebula as measured by the HEGRA system of imaging air Cerenkov telescopes. Astrophys. J. **539**, 317 (2000)

HEGRA Collaboration: The Crab nebula and Pulsar between 500 GeV and 80 TeV: observations with the HEGRA stereoscopic air Cerenkov telescopes. Astrophys. J. **614**, 897 (2004)

Hennessy, S.G., et al: Ultraviolet imaging telescope observations of the Crab nebula. Astrophys. J. **395**, 13 (1992)

Hester, J.J., et al.: WFPC2 studies of the Crab nebula. I. HST and ROSAT imaging of the synchrotron nebula. Astrophys. J. **448**, 240 (1995)

Hester, J., et al.: WFPC2 studies of the Crab nebula. III. Magnetic Rayleigh-Taylor instabilities and the origin of the filaments. Astrophys. J. **456**, 225 (1996)

Hillas, A.M., et al.: The spectrum of teravolt gamma rays from the Crab nebula. Astrophys. J. **503**, 744 (1998)

Hirotani, K.: Three-dimensional non-vacuum pulsar outer-gap model: localized acceleration electric field in the higher altitudes. Astrophys. J. **798**, 40 (2015)

Hubble, E.P.: Novae or temporary stars. Astron. Soc. Pac. Leafl. **14**, 55 (1928)
Jourdain, E., Roques, J.P.: The high-energy emission of the Crab nebula from 20 keV to 6 MeV with integral SPI. Astrophys. J. **704**, 17 (2009)
Kalapotharakos, C., Kazanas, D., Harding, A., Contopoulos, J.: Toward a realistic pulsar magnetosphere. Astrophys. J. **749**, 1 (2012)
Kennel, C.F., Coroniti, F.V.: Magnetohydrodynamic model of Crab nebula radiation. Astrophys. J. **283**, 710 (1984)
Kirsch, M.G., et al.: Crab: the standard x-ray candle with all (modern) x-ray satellites. Proc. SPIE **5898**, 22 (2005)
Knight, H.S., Bailes, M., Manchester, R.N., Ord, S.M., Jacoby, B.A.: Green bank telescope studies of giant pulses from millisecond pulsars. Astrophys. J. **640**, 941 (2006)
Komissarov, S.S., Lyubarky, Y.E.: Synchrotron nebulae created by anisotropic magnetized pulsar winds. Mon. Not. R. Astron. Soc. **349**, 779 (2004)
Komissarov, S.S., Lyutikov, M.: On the origin of variable gamma-ray emission from the Crab nebula. Mon. Not. R. Astron. Soc. **414**, 2017 (2011)
Kuiper, L., Hermsen, W.: The soft γ-ray pulsar population: a high-energy overview. Mon. Not. R. Astron. Soc. **449**, 3827 (2015)
Kuiper, L., et al.: The Crab pulsar in the 0.75–30 MeV range as seen by CGRO COMPTEL. A coherent high-energy picture from soft X-rays up to high-energy gamma-rays. Astron. Astrophys. **378**, 918 (2001)
Legg, M.P., Westfold, K.C.: Elliptic polarization of synchrotron radiation. Astrophys. J. **154**, 499 (1968)
Lei, F., Dean, A.J., Hills, G.L.: Compton polarimetry in gamma-ray astronomy. Space Sci. Rev. **82**, 309 (1997)
Lobanov, A.P., Horns, D., Muxlow, T.W.B.: VLBI imaging of a flare in the Crab nebula: more than just a spot. Astron. Astrophys. **533**, 10L (2011)
Lou, Y.: Wisps and knots in the central Crab nebula. Mon. Not. R. Astron. Soc. **294**, 443 (1998)
Lyne, A.G., Pritchard, R.S., Graham-Smith, F.: Twenty-three years of Crab pulsar rotational history. Mon. Not. R. Astron. Soc. **265**, 1003 (1993)
Lyne, A.G., Hobbs, G., Kramer, M., Stairs, I., Stappers, B.: Switched magnetospheric regulation of pulsar spin-down. Science **329**, 408 (2010)
Lyne, A.G., et al.: 45 years of rotation of the Crab pulsar. Mon. Not. R. Astron. Soc. **446**, 857 (2015)
Lyutikov, M., Balsara, D., Matthew, C.: Crab GeV flares from the corrugated termination shock. Mon. Not. R. Astron. Soc. **422**, 3118 (2012)
Lyutikov, M., Otte, A.N., McCann, A.: The very high energy emission from pulsars: a case for inverse Compton scattering. Astrophys. J. **754**, 33 (2012)
MacAlpine, G.M., Satterfield, T.J.: The Crab nebula's composition and precursor star mass. Astron. J. **136**, 2152 (2008)
Macías-Pérez, J.F., Mayet, F., Aumont, J. Desert, F.X.: Global spectral energy distribution of the Crab nebula in the prospect of the Planck satellite polarization calibration. Astrophys. J. **711**, 417 (2010)
Madsen, K.K., et al.: Broadband X-ray imaging and spectroscopy of the Crab nebula and pulsar with NuSTAR. Astrophys. J. **801** 66 (2015)
MAGIC Collaboration: VHE gamma-Ray Observation of the Crab nebula and its Pulsar with the MAGIC Telescope. Astrophys. J. **674**, 1037 (2008)
MAGIC Collaboration, et al.: Observations of the Crab pulsar between 25 and 100 GeV with the MAGIC I telescope. Astrophys. J. **742**, 43 (2011)
MAGIC Collaboration: Phase-resolved energy spectra of the Crab pulsar in the range of 50–400 GeV measured with the MAGIC telescopes. Astron. Astrophys. **540**, 69 (2012)
MAGIC Collaboration: Detection of bridge emission above 50 GeV from the Crab pulsar with the MAGIC telescopes. Astron. Astrophys. **565**, L12 (2014)
MAGIC Collaboration: Measurement of the Crab nebula spectrum over three decades in energy with the MAGIC telescopes. J. High Energy Astrophys. **5**, 30 (2015)

MAGIC Collaboration: Teraelectronvolt pulsed emission from the Crab pulsar detected by MAGIC. Astron. Astrophys. **585**, 133 (2016)

Manchester, R.N.: Observations of the Crab pulsar at 410 and 1664 MHz. Astrophys. J. **163**, L61 (1971)

Marsden, P.L., et al.: Far-infrared observations of the Crab nebula. Astrophys. J. **278**, L29 (1984)

Martín, J., Torres, D.F., Rea, N.: Time-dependent modelling of pulsar wind nebulae: study on the impact of the diffusion-loss approximations. Mon. Not. R. Astron. Soc. **427**, 415 (2012)

Massaro, E., Cusumano, G., Litterio, M., Mineo, T.: Fine phase resolved spectroscopy of the X-ray emission of the Crab pulsar (PSR B0531+21) observed with BeppoSAX. Astron. Astrophys. **361**, 695 (2000)

Mayer, M., et al.: Rapid gamma-ray flux variability during the 2013 March Crab nebula flare. Astrophys. J. **775**, 37 (2013)

McCann, A.: A stacked analysis of 115 pulsars observed by *Fermi*-LAT. Astrophys. J. **804**, 86 (2015)

Meagher, K., for the VERITAS Collaboration: Six years of VERITAS observations of the Crab nebula. In: Proceedings of the 34th International Cosmic Ray Conference, id. 792 (2015)

Meyer, M., Horns, D., Zechlin, H.S.: The Crab nebula as a standard candle in very high-energy astrophysics. Astron. Astrophys. **523**, A2 (2010)

Mezger, P.G., Tuffs, R.J., Chini, R., Kreysa, E., Gemuend, H.-P.: Maps of Cassiopeia a and the Crab nebula at lambda 1.2 MM. Astron. Astrophys. **167**, 145 (1986)

Mickaliger, M.B., et al: A giant sample of giant pulses from the Crab pulsar. Astrophys. J. **760**, 64 (2012)

Milagro Collaboration: Observation and spectral measurements of the Crab nebula with Milagro. Astrophys. J. **750**, 1 (2012)

Minkowski, R.: The Crab nebula. Astrophys. J. **96**, 199 (1942)

Mochol, I., Petri, J.: Very high energy emission as a probe of relativistic magnetic reconnection in pulsar winds. Mon. Not. R. Astron. Soc. **449**, L51 (2015)

Moffett, D.A., Hankins, T.: Multifrequency radio observations of the Crab pulsar. Astron. J. **468**, 779 (1996)

Moffett, D.A., Hankins, T.H.: Polarimetric properties of the Crab pulsar between 1.4 and 8.4 GHZ. Astrophys. J. **522**, 1046 (1999)

Moran, P., et al.: Optical polarimetry of the inner Crab nebula and pulsar. Mon. Not. R. Astron. Soc. **433**, 2564 (2013)

Moran, P., et al.: A recent change in the optical and gamma-ray polarization of the Crab nebula and pulsar. Mon. Not. R. Astron. Soc. **456**, 2974 (2016)

Mori, K., et al.: Spatial variation of the X-ray spectrum of the Crab nebula. Astrophys. J. **609**, 186 (2004)

Muslinov, A.G., Harding, A.K.: High-altitude particle acceleration and radiation in pulsar slot gaps. Astrophys. J. **606**, 1143 (2004)

Ng, C.Y., Romani, R.:Fitting pulsar wind tori. Astrophys. J. **610**, 479 (2004)

Olmi, B., Del Zanna, L., Amato, E., Bandiera, R., Bucciantini, N.: On the magnetohydrodynamic modelling of the Crab nebula radio emission. Mon. Not. R. Astron. Soc. **438**, 1518 (2014)

Oort, J.H., Walraven, T.: Polarization and the radiating mechanism of the Crab nebula. In: Proceedings from 4th IAU Symposium, p. 197 (1957)

Oser, S., et al.: High-energy gamma-ray observations of the Crab nebula and pulsar with the solar tower atmospheric Cerenkov effect experiment. Astrophys. J. **547**, 949 (2001)

Pacini, F.: Energy emission from a neutron star. Nature **216**, 567 (1967)

Petri, J.: High-energy emission from the pulsar striped wind: a synchrotron model for gamma-ray pulsars. Mon. Not. R. Astron. Soc. **424**, 2023 (2012)

Pierbattista, M., Grenier, I.A., Harding, A.K., Gonthier, P.L.: Constraining gamma-ray pulsar gap models with a simulated pulsar population. Astron. Astrophys. **545**, A42 (2012)

Porth, O., Komissarov, S., Keppens, R.: Three-dimensional magnetohydrodynamic simulations of the Crab nebula. Mon. Not. R. Astron. Soc. **438**, 278 (2014)

Pravdo, S.H., Angelini, L., Harding, A.K.: X-Ray spectral evolution of the Crab pulse. Astrophys. J. **491**, 808 (1997)
Ramanamurthy, P.V., Thompson, D.J.: Search for short-term variations in the E > 50 MeV gamma-ray emission of the Crab Pulsar. Astrophys. J. **496**, 863 (1998)
Romani, R., Watters, K.P.: Constraining pulsar magnetosphere geometry with gamma-ray light curves. Astrophys. J. **714**, 810 (2010)
Romani, R., Yadiglarogu, I.A.: Gamma-ray pulsars: emission zones and viewing geometries. Astrophys. J. **438**, 314 (1995)
Rudy, A., et al.: Characterization of the Inner Knot of the Crab: the site of the gamma-ray flares? Astrophys. J. **811**, 24 (2015)
Saito, S., Uchyama, Y.: AIP Conference Proceedings: 6th International Symposium on High-Energy Gamma-Ray Astronomy, July 2016, Heidelberg (2016)
Sankrit, R., et al.: WFPC2 studies of the Crab nebula. II. Ionization structure of the Crab filaments. Astrophys. J. **504**, 344 (1998)
Scargle, J.D.: Activity in the Crab nebula. Astrophys. J. **156**, 401 (1969)
Schweitzer, T., et al. Characterization of the optical and X-ray properties of the north-western wisps in the Crab nebula. Mon. Not. R. Astron. Soc. **433**, 3325 (2013)
Shearer, A., et al.: Enhanced optical emission during Crab giant radio pulses. Science **301**, 493 (2003)
Shklovsky, I.S.: Radioastronomiia. Dokl. Akad. Nauk SSSR **91**(3), 475 (1953). If non-Russian, see translation in Selected Papers on Cosmic Ray Origin Theories by Stephen Rosen, Dover, p. 150 (1969)
Sironi, L., Spitkovsky, A.: Acceleration of particles at the termination shock of a relativistic striped wind. Astrophys. J. **741**, 39 (2011)
Slowikowska, A., Kanbach, G., Kramer, M., Stefanescu, A.: Optical polarization of the Crab pulsar: precision measurements and comparison to the radio emission. Mon. Not. R. Astron. Soc. **397**, 103 (2009)
Slowikowska, A., et al.: High-time-resolution measurements of the polarization of the Crab pulsar at 1.38 GHz. Astrophys. J. **799**, 90 (2015)
Smith, N.: The Crab nebula and the class of type IIn-P supernovae caused by sub-energetic electron capture explosions. Mon. Not. R. Astron. Soc. **434**, 102S (2013)
Smith, F.G., Jones, D.H.P., Dick, J.S.B., Pike, C.D: The optical polarization of the Crab Pulsar. Mon. Not. R. Astron. Soc. **233**, 305 (1988)
Soffitta, P., et al.: *XIPE*: the X-ray imaging polarimetry explorer. Exp. Astron. **36**, 523 (2013)
Sollerman, J., et al.: Observations of the Crab nebula and its pulsar in the far-ultraviolet and in the optical. Astrophys. J. **537**, 861 (2000)
Staelin, D.H., Reifentein, E.C.: Pulsating radio sources near the Crab nebula. Science **162**, 1481 (1968)
Stephenson, F.R., Green, D.A.: Was the supernova of AD1054 reported in European history? J. Astron. Hist. Herit. **6**, 46 (2003)
Striani, E., et al.: Variable gamma-ray emission from the Crab nebula: short flares and long "waves". Astrophys. J. **765**, 52 (2013)
Tavani, M., et al.: Discovery of powerful gamma-ray flares from the Crab nebula. Science **331**, 736 (2011)
Temim, T., et al.: Spitzer space telescope infrared imaging and spectroscopy of the Crab nebula. Astron. J. **132**, 1610 (2006)
Temim, T., et al.: Properties and spatial distribution of dust emission in the Crab nebula. Astrophys. J. **753**, 72 (2012)
Thompson, D.J., et al.: Gamma Radiation from PSR B1055-52. Astrophys. J. **516**, 297 (1999)
Trimble, V.: The distance to the Crab nebula and NP 0532. Publ. Astron. Soc. Pac. **85**, 507 (1973)
Tziamtzis, A., Lundqvist, P., Djupvik, A.A.:The Crab pulsar and its pulsar-wind nebula in the optical and infrared. Astron. Astrophys. **497**, 167 (2009)
VERITAS Collaboration: Detection of pulsed gamma rays above 100 GeV from the Crab pulsar. Science **334**, 69 (2011)

VERITAS Collaboration: Search for a correlation between very-high-energy gamma rays and giant radio pulses in the Crab pulsar. Astrophys. J. **760**, 136 (2012)

VERITAS Collaboration: A search for enhanced very high energy gamma-ray emission from the 2013 March Crab nebula flare. Astrophys. J. **781**, L1 (2014)

Veron-Cetty, M.P., Woltjer, L.: Spettrophotometry of the continuum in the Crab nebula. Astron. Astrophys. **270**, 370 (1993)

Viganò, D., Torres, D.F., Hirotani, K., Pessah, M.E.: Compact formulae, dynamics and radiation of charged particles under synchro-curvature losses. Mon. Not. R. Astron. Soc. **447**, 1164 (2015)

Volpi, D., Del Zanna, L., Amato, E., Bucciantini, N.: Non-thermal emission from relativistic MHD simulations of pulsar wind nebulae: from synchrotron to inverse Compton. Astron. Astrophys. **485** 337 (2008)

Weeks, T., et al.: Observation of TeV gamma rays from the Crab nebula using the atmospheric Cerenkov imaging technique. Astrophys. J. **342**, 379 (1989)

Weisskopf, M.: High energy polarization and the Crab: historical remarks. In: Proceedings of the Workshop: Polarimetry days in Rome: Crab Status, Theory and Prospects, Rome July 2008 (2009)

Weisskopf, M.: New results from an old friend: the Crab nebula and its pulsar. In: Proceedings of Workshop on "The Extreme and Variable High Energy Sky", Cagliari (2011)

Weisskopf, M.C., Silver, E.H., Kestenbaum, H.L., Long, K.S., Novick, R.: A precision measurement of the X-ray polarization of the Crab nebula without pulsar contamination. Astrophys. J. **220**, L117 (1978)

Weisskopf, M., et al.: Discovery of spatial and spectral structure in the X-ray emission from the Crab nebula. Astrophys. J. **536**, L81 (2000)

Weisskopf, M., et al.: Chandra Phase-resolved X-ray spectroscopy of the Crab pulsar. Astrophys. J. **743**, 139 (2011)

Weisskopf, M., Elsner, R.F., Kolosziejczak, J.J., O'Dell, S.L., Tennant, A.F.: Unraveling the geometry of the Crab nebula's "inner ring". Astrophys. J. **746**, 41 (2012)

Weisskopf, M., et al.: Chandra, Keck, and VLA observations of the Crab nebula during the 2011-April gamma-ray flare. Astrophys. J. **765**, 56 (2013)

Weisskopf, M., et al.: Transforming our understanding of the X-ray universe: the imaging X-ray polarimeter explorer (IXPE). In: Proceedings of the American Astronomical Society, HEAD Meeting, id.116.15 (2014)

Wieland, J.L., et al.: Seven-year Wilkinson Microwave Anisotropy Probe (WMAP) observations: planets and celestial calibration sources. Astrophys. J. Suppl. **192**, 2 (2011)

Wilson-Hodge, C.A., et al.: When a standard candle flickers. Astrophys. J. **727**, 40 (2011)

Wright, E.L., Harper, D.A., Hildebrand, R.H., Keene, J., Whitcomb, S.E.: Millimetre and submillimetre measurements of the Crab nebula. Nature **279**, 703 (1979)

Yang, H., Chevalier, R.A.: Evolution of the Crab nebula in a low energy supernova. Astrophys. J. **806**, 153 (2015)

Yuan, Q., et al: A statistical model for the gamma-ray variability of the Crab nebula. Astrophys. J. **730**, L15 (2011)

Zanin, R., for the MAGIC Collaboration: The most precise measurements of the Crab nebula inverse Compton spectral component. In: Proceedings of the 34th International Cosmic Ray Conference, id. 831 (2015)

Chapter 7
Pulsar Striped Winds

Iwona Mochol

Abstract According to magnetohydrodynamic (MHD) models, the rotational energy of a rapidly spinning neutron star is carried away by a relativistic wind and deposited at a large distance, in the nebula, downstream of the wind termination shock. The energy transport in the outflow is mediated by Poynting flux, but it is not clear how the energy stored in the fields is transferred into the energized population of emitting particles. The most plausible dissipation mechanisms are thought to be related to the "striped" structure of the wind, in particular, to the existence of a current sheet, prone to reconnection events. In this model the current sheet is a natural place for internal dissipation and acceleration of particles responsible for pulsed, high-energy emission. Moreover, reconnection is a promising scenario for explaining annihilation of fields at the shock and conversion of their energy into the kinetic energy of particles. The shock structure, however, is likely to differ in the low-density plasmas, in which non-MHD effects intervene. In this regime, the striped wind can dissipate its energy via an electromagnetic precursor of the shock.

7.1 Introduction

7.1.1 General Picture

According to current models the entire post-supernova system is tightly coupled. The central engine—a rapidly spinning neutron star—is responsible for creation of particles, their acceleration and finally outflow in the form of a relativistic wind. The energy in the wind is transported by Poynting flux and released to the plasma at large distances from the star, in the nebula, where the energized particles radiate away the acquired energy and perform $p\mathrm{d}V$ work on the surroundings. The global evolution of the system is usually modelled using relativistic MHD. Although it

I. Mochol (✉)
Observatoire Astronomique de Strasbourg, 11 Rue de l'Université, F-67000 Strasbourg, France

Institute of Nuclear Physics, Polish Academy of Sciences, ul. E. Radzikowskiego 152, 31-342 Kraków, Poland
e-mail: iwona.mochol@astro.unistra.fr

D.F. Torres (ed.), *Modelling Pulsar Wind Nebulae*, Astrophysics and Space Science Library 446, DOI 10.1007/978-3-319-63031-1_7

allows us to get a glimpse on how the system is globally interconnected, we still do not understand the details of the coupling mechanisms and the processes that lead to the release of the magnetic energy.

The abundant pair plasma, transported by the wind, is thought to be created in electromagnetic cascades in the closest neighbourhood of the star, in the magnetosphere. The conventional boundary of this region is defined by the light cylinder of a cylindrical radius $r_{\rm L} = cP/(2\pi)$ (where P is the rotational period of a pulsar), at which the corotation velocity of the particles with the star would be equal to the speed of light. The corotating plasma is coupled to the magnetic field lines, which in the dipole-like geometry close within the light cylinder. The field lines which cross the light cylinder carry the currents that brake the central rotator via Lorentz forces.

This rotational energy of the central neutron star is transported outwards by the wind, a mixture of plasma and electromagnetic fields. The outflow is terminated at the roughly standing shock, where its ram pressure becomes balanced by the pressure of the confining external medium. The shock is located at a large distance from the central star—in the case of the Crab $r_{\rm shock} \sim 10^8 r_{\rm L}$. At the shock the magnetic energy of the wind is randomised and the particles carried by the wind become efficiently accelerated via the Fermi I-order process. They are further injected into the nebula downstream of the shock, where they radiate in a synchrotron process, observed as diffuse emission from radio to X-rays; in gamma-rays a dominant process is the inverse-Compton (IC) scattering on the cosmic microwave background radiation and on the synchrotron photons produced in the nebula itself (synchrotron self-Compton (SSC) emission).

Both the diffuse radiation from the nebula, as well as the high-energy pulses of emission observed from the central point-like source, provide the observational evidence that efficient mechanisms of particle acceleration are at work, challenging our understanding of the physics of relativistic flows. In the following we will concentrate on one particular underlying question: how and where the enormous magnetic energy carried by the wind is dissipated and transferred to the energised population of radiating particles?

7.1.2 Problems

The energy of the central rotator is transported over long distances, and, therefore, the wind zone proves to be an important ingredient in the global description of the system. On the other hand, very little direct information is provided by observations, in which the wind region appears underluminous (Weisskopf et al. 2000). This is related to the fact that the plasma in the wind is expected to be cold and, therefore, not producing any emission. However, if due to some internal dissipation the plasma were heated, the relativistic beaming and the radial dependence of the emissivity would cause the emission to appear as point-like (Kirk et al. 2009).

The point-like emission from the relativistic, MHD pulsar wind would be seen by a distant observer as pulsed (Kirk et al. 2002), because the internal dissipation in the wind, due to e.g. reconnection, is expected to occur abruptly in a thin, corrugated layer—a current sheet (see Sect. 7.2.2). Reconnection regions are known to be powerful particle accelerators, and, therefore, this scenario provides a promising possibility to explain puzzling very high-energy γ-ray observations from the Crab (Ansoldi et al. 2016) and Vela (Rudak et al. 2015; Leung et al. 2014) pulsars, as well as the millisecond pulsar J0614–3329 (Xing and Wang 2016). This is an alternative to the magnetospheric models, which associate the γ-ray pulses with cyclotron-self Compton scattering (Lyutikov 2013), synchro-curvature radiation (Viganò and Torres 2015) and SSC emission inside the light cylinder (Harding and Kalapotharakos 2015).

The pulsed emission, however, constitutes only a fraction of the total energy budget. On the global scale, the dissipation of the magnetic energy is believed to take place somewhere close to the termination shock. The properties of a shock, in particular its ability to dissipate the energy, strongly depend on the magnetisation of the upstream plasma. This magnetisation is quantified by the parameter σ. In a cold plasma it measures the ratio of the magnetic energy flux to the kinetic energy flux in the flow:

$$\sigma = \frac{B^2}{8\pi N m c^2 \Gamma}, \tag{7.1}$$

where B is the strength of the magnetic field, N is the plasma density in the laboratory frame and Γ is the Lorentz factor of the flow. The magnetisation is thought to be large close to the light cylinder, where the wind is launched. However, Kennel and Coroniti (1984a) have shown that the observational limits on the magnetisation can be put downstream of the termination shock, in the nebula, where it must be small. It seems, therefore, that the transition between large- and small-σ regimes requires some magnetic dissipation in the wind zone or at the shock itself. The mechanism responsible for this dissipation is unknown, with the magnetic reconnection being one of the most important candidates.

In the following we review the physics of relativistic pulsar winds and the dissipation processes that can lead to emission observed from those outflows. We start by considering their launching from the magnetospheres in Sect. 7.2.1, and we discuss their "striped" structure in Sect. 7.2.2. Further, in Sect. 7.3.1, we summarise the arguments in favour of the wind model of pulsed high-energy emission. In Sect. 7.3.2 we consider a dissipation mechanism—relativistic reconnection, which is likely to set in the current sheet of the wind; a population of particles accelerated in the course of reconnection can be responsible for the pulses of emission, which would be, therefore, a direct probe of the physics in the current sheet. In Sect. 7.3.3 we discuss the implications of this model for the gamma-ray spectra of pulsars and its predictions for future observations. Finally, in Sect. 7.4, we describe the structure of shocks of striped winds and the σ-problem—a failure to describe the conversion of the Poynting flux into the particle energy flux before the wind reaches the termination shock. A solution for this problem is still not clear, but we discuss two scenarios that have been proposed, namely the driven reconnection of wind stripes at the shock and the generation of an electromagnetic dissipative precursor to the shock.

7.2 MHD Picture: Striped Winds

7.2.1 *Launching from a Magnetosphere*

Understanding of pulsar wind launching requires the knowledge of (1) the magnetic field topology, (2) the location of particle acceleration regions, and (3) the mechanisms of supplying charges to the outflow. The latter are closely linked to radiation processes, which in the strong-field regime often change their character or become dominated by exotic phenomena, resulting from energy quantization in the direction perpendicular to the field (Landau states) or the nonconservation of transverse photon momenta which can be absorbed by the field (Harding 1991).

The magnetic field is usually modelled as a rotating dipole (strongly distorted close to the light cylinder), but higher multipoles may also contribute (Barnard and Arons 1982; Gil and Mitra 2001; Pétri 2015). The strength of the magnetic field can be measured in two ways: (1) from the pulsar spin down, assuming that it results from the dipole emission, (2) directly from the energies of cyclotron absorption lines in the X-ray spectra of neutron stars (for discussion see e.g. Harding 1991). These measurements imply that the magnetic fields at the stellar surface are very strong $B_* \gtrsim 10^{12}\,$G.

A rotating neutron star, endowed with a dipole magnetic field, generates a quadrupole electric field, in a direct analogy to a unipolar inductor: like a conductor rotating in a magnetic field, charges on the pulsar's surface are redistributed by the Lorentz force, which induces an electric potential difference between the poles and the equator. The generated electric field is thought to be strong enough to rip the charges off from the stellar surface and accelerate them along the magnetic field lines, because any transverse motion would be immediately suppressed by radiation of synchrotron photons. The accelerated charges emit curvature gamma-rays, which in the strong magnetic field can materialise into secondary $e^{\pm}$ pairs, further accelerated and radiating.

The number of pairs created in these electromagnetic cascades is not clear. The critical Goldreich-Julian density $N_{\rm GJ}$ defines the number of charges that are necessary to screen out the component of the electric field parallel to the magnetic field:

$$N_{\rm GJ} = \frac{\mathbf{B} \cdot \boldsymbol{\omega}}{2\pi e c} \frac{1}{1 - (r/r_{\rm L})^2 \sin^2\theta}, \tag{7.2}$$

where ω is the angular velocity of the pulsar. The density of pairs N, created in the magnetosphere, is usually expressed in terms of the pair multiplicity

$$\kappa = N/N_{\rm GJ}. \tag{7.3}$$

If the pair injection in the magnetosphere is somewhat below $N_{\rm GJ}$, the pulsar magnetosphere is described as the "electrosphere", in which a charge-separated plasma is confined to domes above the poles and a differentially rotating equatorial disc (Krause-Polstorff and Michel 1985a,b; Pétri et al. 2002; Spitkovsky 2004; Cerutti et al. 2015). If the beam of primary particles ripped off from the stellar surface initiates prolific cascades, a pair plasma in the magnetosphere is abundant $\kappa \gg 1$. Simulations of electromagnetic cascades suggest multiplicities in the range $\kappa \sim 10^3 - 10^5$ (see Timokhin and Harding (2015) and references therein). A constraint on the multiplicity $\kappa \sim 10^6$ is implied by the models of emission from the Crab Nebula, under the assumption that all the radiating particles are being constantly supplied by the pulsar and that these particles are in energy equipartition with the magnetic field (Rees and Gunn 1974; de Jager et al. 1996; Gallant et al. 2002) (for discussion of the radio brightness see also Lyubarsky 2003; Kirk et al. 2009). Modelling of other nebulae in general suggests $\kappa \sim 10^5$ (Bucciantini et al. 2011).

Such large values of multiplicity imply that the magnetospheric plasma easily screens the parallel component of the electric field. However, to maintain the activity, a pulsar has to sustain the constant or at least intermittent replenishment of pairs that outflow with the wind. Current theories postulate that the pair creation processes take place in limited regions of an unscreened component of the electric field—so called vacuum gaps. Their location is model-dependent: they are associated with the polar caps (located near the stellar surface close to the magnetic axis of the pulsar) (Sturrock 1971; Ruderman and Sutherland 1975; Harding et al. 1978; Daugherty and Harding 1982), outer magnetosphere (Cheng et al. 1986; Romani and Yadigaroglu 1995), and the region along the last open field line (Arons and Scharlemann 1979; Muslimov and Harding 2003, 2004; Dyks and Rudak 2003). Lyubarskii (1996) suggested that the pairs can be produced in the current sheet just outside the light cylinder, where the magnetic field in the sheet is strongest. This scenario is observed in kinetic Particle-In-Cell (PIC) simulations of pulsar magnetospheres with low obliquity, whereas the pulsars with high obliquity tend to create pairs rather in polar cap regions (Philippov et al. 2015).

The existence of gaps, although important for understanding of how pulsars really work, is usually neglected in the modelling of the global plasma dynamics. In particular, the MHD, fluid description of the plasma is thought to be valid. Since the energetics is dominated by electromagnetic fields, the particle inertia is negligible in the first approximation, but the particles are still assumed to carry charge and currents. This zero-inertia limit of MHD is called force-free electrodynamics (FFE). The equations for a force-free magnetosphere of an aligned rotator were obtained by Michel (1973), as well as Scharlemann and Wagoner (1973). The plasma, rigidly corotating with the star, is assumed to have infinite conductivity, such that in the comoving frame the electric field vanishes; after transformation into the laboratory frame, this implies the frozen-in condition between the plasma and the magnetic field

$$\mathbf{E} + (1/c)\mathbf{v} \times \mathbf{B} = 0. \tag{7.4}$$

Corotation of a plasma with the star is possible only within the light cylinder. Since the particles cannot corotate with superluminal speed, the magnetic field lines that would cross the light cylinder are bent backwards and they open beyond $r_{\rm L}$, even in the case of an aligned rotator (and in contrast with an aligned rotator in vacuum). As explained by Michel (1973), in vacuum, without the plasma which ensures the frozen-in condition, the magnetic field of such an aligned rotator is static, whereas in the presence of a plasma, the magnetic field is "rotating". As a consequence, there is a nonvanishing Poynting vector $\mathbf{S} = \mathbf{E} \times \mathbf{B}$, which describes the transport of energy; the closer to the light cylinder, the more the corotation speed increases, and so does the magnitude of the Poynting vector. Ultimately, beyond $r_{\rm L}$, it would exceed the energy density that can be transported with the speed of light. Thus, the field lines become open by the centrifugal force exerted by the corotating energy density of the electromagnetic fields.

For an oblique rotator, with the angle χ between the magnetic and rotational axes, the presence of the plasma increases the torque when compared to a rotator in vacuum. The latter emits dipole radiation at the expense of the rotational energy, and the luminosity is given by:

$$L_{\rm vac} = L_0 \sin^2 \chi, \tag{7.5}$$

where $L_0 = B_*^2 \omega^4 r_*^6/(6c^3)$, B_* is the strength of the magnetic field at the stellar equator, r_* is the stellar radius. The plasma-filled magnetosphere, on the other hand, spin downs according to

$$L_{\rm FFE} \approx \frac{3}{2} L_0 (1 + \sin^2 \chi) \tag{7.6}$$

Spitkovsky (2006) and Pétri (2012). In comparison to the vacuum case, the topology of the magnetic field lines in the force-free description is similar inside the light cylinder, but exhibits the formation of a current sheet outside this distance (Pétri 2012).

The analytical studies of pulsar magnetospheres of oblique rotators are nowadays extended to the simulations: 3D force-free (Spitkovsky 2006; Pétri 2012), resistive force-free (Kalapotharakos et al. 2012; Li et al. 2012), relativistic MHD (Komissarov 2006; Tchekhovskoy et al. 2013), and also PIC simulations (Philippov et al. 2015). In general relativity force-free simulations have been performed recently by Pétri (2016a).

An analytical two-fluid approach to describe the pulsar magnetosphere has been proposed by Petrova (2015). In this solution, electron and positron fluids exhibit a velocity shift with respect to each other, which can underlie the instabilities and production of waves, convertible into the radio emission from pulsars. Interestingly, in this model the plasma conductivity in the magnetosphere is finite due to inertial effects, which provides a physical justification for the resistivity, considered only phenomenologically as a free parameter in the resistive force-free simulations (Kalapotharakos et al. 2012; Li et al. 2012).

7.2.2 *The Striped Wind Structure*

An exact force-free monopole solution for an outflow of an aligned rotator was found by Michel (1973) and it was generalized for an oblique rotator by Bogovalov (1999). In the pulsar models, the magnetic field is believed to change smoothly from the dipolar geometry inside the light cylinder to the monopolar, radial geometry at infinity (Michel 1973). However, this change of magnetic field topology, captured numerically, has not been described analytically. Recently though, the exact dipole solution for an aligned rotator has been reported and discussed by Petrova (2016).

In the limit of negligible particle inertia, a monopole configuration is radial in the meridional plane, and in the equatorial plane these lines develop a perfect Archimedean spiral $r_s = r_L\phi$ regardless of a field topology inside the light cylinder. Globally, more appropriate is a split-monopole configuration, where two half magnetic monopoles of opposite polarity are joined together in the equatorial plane. This change in magnetic field direction must be, according to Ampère's law, accompanied by formation of a current sheet, within which the magnetic field vanishes and the pressure, necessary to keep the equilibrium, is supported by a hot plasma. In the oblique case, a current sheet oscillates around the equatorial plane as the pulsar rotates, connecting the equator with field lines of opposite polarity every half a period (see Fig. 7.1). This corrugated current sheet far from the light cylinder can be approximated by spherical shells, separating the stripes of magnetized plasma

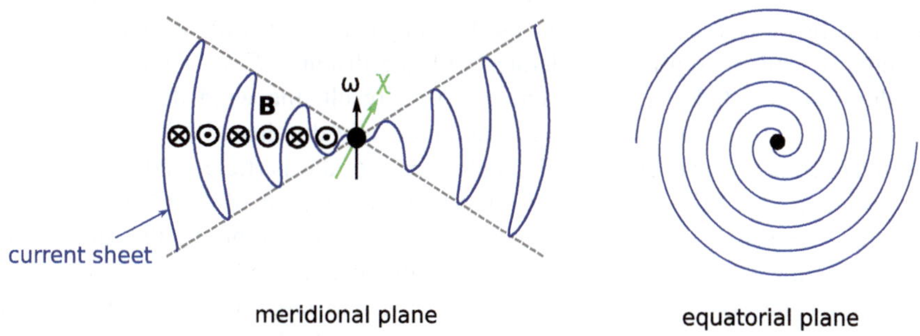

Fig. 7.1 Striped wind for an oblique rotator with the angle χ between the magnetic and the rotational axes. In the meridional plane the current sheet (*blue curve*) is corrugated and separates the stripes of opposite magnetic polarity. The dominant component of the magnetic field is toroidal. In the equatorial plane the current sheet develops an Archimedean spiral

with opposite magnetic polarity—this is the structure of a so-called striped wind (Coroniti 1990; Michel 1994). A solution for the wind with analytically prescribed current sheet and a radial velocity, has been recently reported (Pétri 2013).

In the split-monopole solution the Poynting flux has a maximum value at the equator (Michel 1973). Most of the energy, therefore, is carried within the wedge of the striped wind, defined by the spherical coordinate θ, $\pi/2 - \chi < \theta < \pi/2 + \chi$, where χ is the angle between the magnetic and rotational axes of the pulsar (obliquity). In the equatorial plane the stripes have the equal width, hence the phase-averaged magnetic field vanishes. This is not the case for the higher latitudes, for which a phase-averaged magnetic field increases with the latitude. It reaches the maximum value at the surface defined in spherical coordinates by $\theta = \pi/2 \pm \chi$. This surface passes through the cusps of the current sheet. Above it, in the polar region of the wind, the total magnetic field has only one polarity.

The radial dependence of the magnetic field components can be obtained from the magnetic flux conservation. Integrating over the volume of a sphere of a radius r, one gets from the Gauss theorem

$$B_r \propto r^{-2}. \tag{7.7}$$

Similarly, as the number of field lines in the sectional area is constant,

$$B_\phi \propto r^{-1} \tag{7.8}$$

(for a detailed discussion see e.g. Michel 1994; Kirk et al. 2009). Close to the light cylinder the radial and poloidal component of the magnetic field are comparable, but the poloidal component, which in this solution is purely radial, decreases much faster than the toroidal one. Observations of synchrotron radiation close to the termination shock, in particular its uniform linear polarization, are consistent with the toroidal structure of the magnetic field, expected in the wind at large distances from the pulsar.

The radial wind is believed to emerge from the magnetosphere as submagnetosonic. In the force-free solution the fast magnetosonic (FMS) point is located at infinity, and the outflow accelerates with the distance $\Gamma \propto r$ (Buckley 1977) at the expense of the magnetic energy. As a result, the wind described by the FFE arrives at the termination shock as kinetically dominated (Contopoulos and Kazanas 2002). However, the FFE model neglects the particle inertia, which, by definition at the FMS point become important (propagation of the FMS wave is an interplay between particle inertia and magnetic tension, implying that the former cannot be neglected). The FMS speed in a highly magnetized plasma is relativistic, defined by the Lorentz factor $\Gamma_{\rm fms} \approx \sqrt{\sigma} \gg 1$. For large-$\sigma$ flow, the FMS point should be crossed somewhere close to the light cylinder. Further out the wind becomes supermagnetosonic, described by MHD rather than FFE equations. An MHD relativistic wind is moving ballistically, with almost a constant, relativistic speed, and it stays strongly magnetically dominated up to the large distances (for a discussion of this topic see Kirk et al. 2009).

7.3 Reconnection and Pulsed Emission from the Wind

7.3.1 *Why the Wind Model?*

The striped wind is a promising site of emission of high-energy (optical to gamma-ray) nonthermal photons observed from pulsars, an idea proposed by Kirk et al. (2002). The difference between the wind and the magnetospheric scenarios lies in assumptions about the physics of production of the pulsed emission. The wind model offers an analytic description of the magnetic field structure, and the peaks in a lightcurve appear as a result of the strong beaming (since the wind is relativistic) of the radiation produced in a thin layer—a current sheet. This layer is naturally prone to dissipative processes, in particular to reconnection. On the other hand, in the magnetospheric models the peaks in a light curve are caused by the overlap of emission from different regions in the magnetosphere (caustics) (see e.g. Bai and Spitkovsky 2010); this overlap is sensitive to both the definition of gaps and the exact geometry of the magnetic field—firstly, because the geometry of the field defines how the effects of aberration of light and light travel delay add together to form the caustics; secondly, because the geometry of the field close to the light cylinder determines the shape of the polar cap on the star, which in turn controls the shape of the radiation region (Bai and Spitkovsky 2010). This requires a global modelling of the pulsar magnetospheres, and presumably can be done only via numerical simulations.

Moreover, the reconnection in the wind is expected to be a very powerful particle accelerator, whereas details of acceleration processes (and accelerating fields) that lead to magnetospheric emission are strongly model-dependent. Extreme conditions of particle acceleration are indeed required to explain recent observations of pulsars above a hundred of GeV (Ansoldi et al. 2016; Rudak et al. 2015; Xing and Wang 2016).

To understand how the pulses from the wind appear in the lightcurves, we assume that the corrugated current sheet can be described as subsequent shells, which radiate after crossing a sphere of a radius r_0 (see Fig. 7.2). A distant observer will notice a peak of emission from the shell that has crossed the radius r_0 along his line of sight, followed by a quick decline of the emissivity as the sheet propagates outwards, due to adiabatic cooling and radial dependence of magnetic field (Eqs. (7.7) and (7.8)). Depending on the viewing angle, the observer can detect up to two pulses per rotational period of a pulsar—(1) two pulses with changing separation if his line of sight lies in the wedge $\pi/2 - \chi < \theta_{\rm obs} < \pi/2 + \chi$; (2) one pulse, if the line of sight is close to $\theta_{\rm obs} \approx \pi/2 \pm \chi$; (3) no pulse if $\theta_{\rm obs} < \pi/2 - \chi$ or $\theta_{\rm obs} > \pi/2 + \chi$.

The pulses of radiation, if emitted too far from the pulsar, would be, however, smeared out. Let the observer be located is at the distance D from the pulsar. In Fig. 7.3 photons 1 and 2 are emitted at the same time from the shell passing the radius r_0 (for simplicity we assume that the shell is spherical, which is true in the first approximation far from the light cylinder). The first photon arrives at

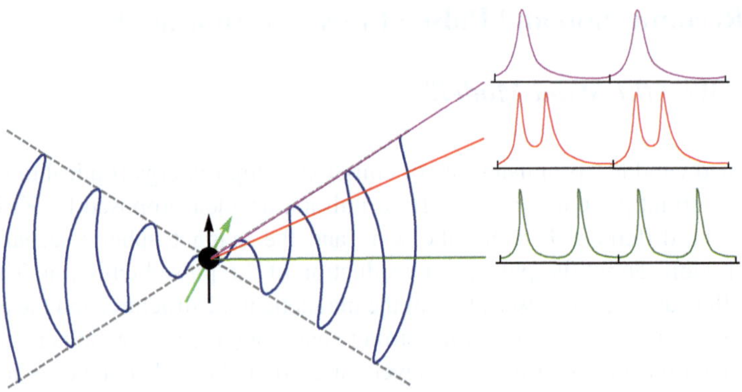

Fig. 7.2 Lightcurves from the wind. A distant observer can detect a pulse of emission when the expanding current sheet passes the radius r_0 along his line of sight. The emissivity of this sheet quickly diminishes afterwards. Depending on the viewing angle, the observer can detect up to two pulses per rotational period of a pulsar

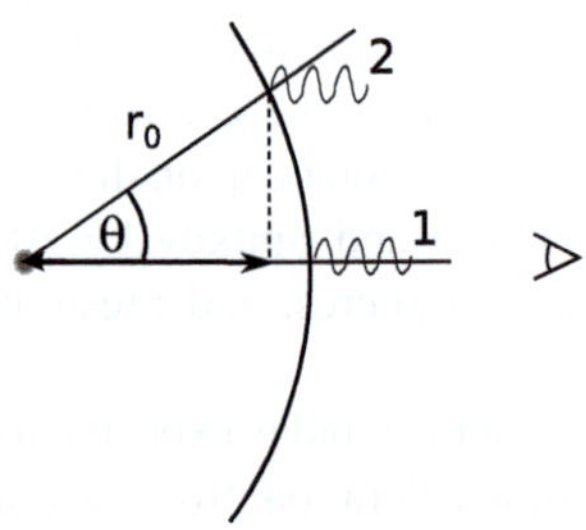

Fig. 7.3 Pulsed emission from the wind. Photons 1 and 2 are emitted at the same time from the current sheet (modelled as a thin shell) when it passes the radius r_0. A distant observer will measure a time delay between the photons. If this time difference is smaller than the time between two subsequent current sheets crossing the sphere r_0, the emission from these sheets will appear as pulsed

the observer after time: $t_1 = (D - r_0)/c$, whereas the second photon arrives after time $t_2 = (D - r_0 \cos\theta)/c$. Time time difference between arrival of the photons is therefore $\Delta t = r_0 (1 - \cos\theta)/c \approx r_0/(2\Gamma^2 c)$, where for ultrarelativistic flow with the Lorentz factor $\Gamma \gg 1$, $\theta \approx 1/\Gamma$. If this time difference is smaller than the time between two subsequent current sheets crossing the sphere r_0, the emission from these sheets will appear at the observer as pulsed. At the equator the shells are separated by half of the wavelength of the wind $\lambda/2 = \pi r_{\mathrm{L}}$. Thus, the condition for the maximal distance at which the emission emitted still will appear as pulsed can be estimated as:

$$r_0 < 2\pi r_{\mathrm{L}} \Gamma^2 \tag{7.9}$$

For a detailed discussion and the calculations taking into account an exact form of the Archimedean spiral, see Pétri (2016b).

It is worth mentioning that the pulse and off-pulse polarization features, modelled in the wind scenario for optical emission of the Crub pulsar, fit very well with this picture (Pétri and Kirk 2005). For more details see also (Kirk et al. 2009).

If the pulsed emission originates in the wind, one expects that the spectrum of the received radiation should directly probe the physics of particle acceleration in the current sheet.

7.3.2 Relativistic Reconnection and Particle Acceleration in the Wind

In highly conducting plasmas, the magnetic energy can be released by the magnetic field reconfiguration near the singular regions, where the magnetic field changes direction and the localized currents introduce some non-ideal effects, breaking the frozen-in condition (7.4). In the collisionless plasmas, these non-ideal effects include the electron kinetic/inertial terms in the Ohm's law, and microinstabilities, which on the macroscopic level play a role similar to the collisional resistivity. Rearrangement of the field topology occurs locally at so-called X-points.

In the pulsar wind, magnetic reconnection is expected to set in the current sheet, which separates the regions of oppositely polarized magnetic field. In highly magnetized plasma in the wind it proceeds in the relativistic regime, because initially the magnetic energy per particle is significantly greater than the particle energy. Thus, when the particles enter the reconnecting layer, they become accelerated to relativistic energies. While the steady-state description of relativistic reconnection has been studied analytically (Lyutikov and Uzdensky 2003; Lyubarsky 2005), numerical simulations (see Kagan et al. (2015) and references therein) demonstrate that this process is essentially time-dependent; current layers are violently unstable to the plasmoid (tearing) instability which leads to the continuous formation and ejection of multiple secondary islands—plasmoids (see Fig. 7.4).

The whole process of particle acceleration depends on how the particles penetrate and interact with plasmons, which in turn merge and grow. In the full time-dependent picture the process of particle acceleration can only be captured by fully kinetic PIC simulations (Zenitani and Hoshino 2001; Jaroschek et al. 2004; Bessho and Bhattacharjee 2007; Sironi and Spitkovsky 2014; Guo et al. 2014; Cerutti et al. 2014). The results suggest that the particles gain most of the energy when they are accelerated linearly by the electric field at X-points. When the tension force of the reconnected magnetic field advects the particles away from the X-point, they become trapped in magnetic islands, with a roughly isotropic energy distribution. The larger the size of a plasmoid, the more energetic particles it can confine. This provides a Hillas criterion for the Larmor radius of particles accelerated in the course of relativistic reconnection (Sironi et al. 2016). It has been reported (Sironi et al. 2016) that the observed highest energy particles have Larmor radius of about 3% of the system length (along the reconnection layer).

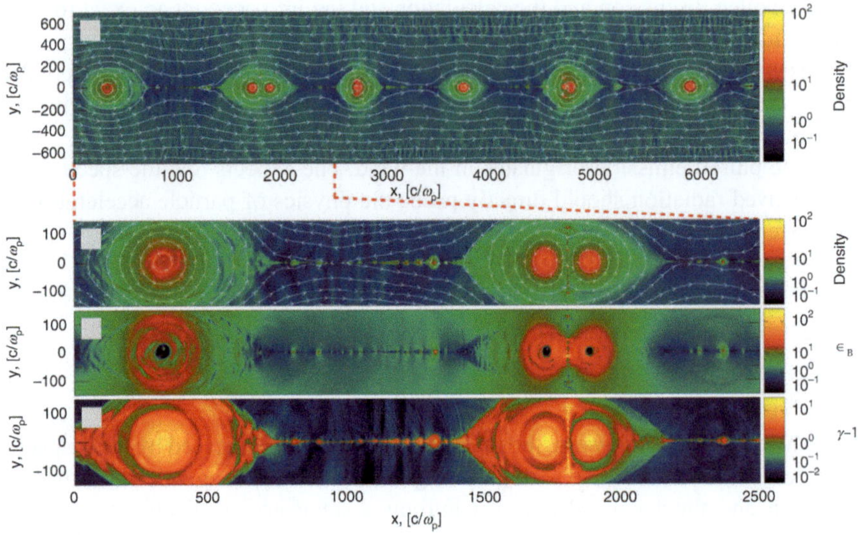

Fig. 7.4 The structure of the reconnection layer, from PIC simulations of Sironi and Spitkovsky (2014). First two panels show the particle density, the third panel shows the magnetic energy, and the bottom panel—kinetic energy per particle

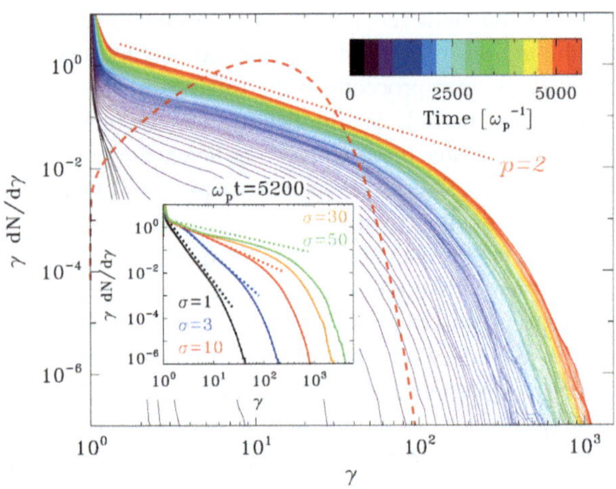

Fig. 7.5 Evolution of the particle energy spectrum during reconnection in the plasma with magnetisation $\sigma = 10$, from PIC simulations of Sironi and Spitkovsky (2014). At late times the spectrum is a power law. The dependence of the spectrum on the magnetisation is shown in the inset; the dotted lines refer to power-law slopes of $-4, -3, -2, -1.5$ from black to green

Integrated over the whole reconnection region, the energy spectrum of particles is generally observed to be a power law, with a very hard index $s < 2$ for magnetisation $\sigma > 10$ (Sironi and Spitkovsky 2014), and being the harder the larger is the magnetisation parameter (see Fig. 7.5).

The most energetic particles are the ones that interact closer to the centre of the X-point, because they are less prone to be advected away by the reconnected field and, therefore, they stay longer in the acceleration region (Sironi and Spitkovsky 2014). The maximum energy to which a power-law particle distribution extends, as well as the shape of the high-energy cutoff in this distribution, both depend on the interplay between the acceleration process and the losses that the particles experience. The particle spectra obtained from PIC simulations by Werner et al. (2016) exhibit either exponential or super-exponential behaviour. The exponential cutoff is more appropriate if the maximum energy is limited by the total available magnetic energy per particle, whereas the super-exponential shape better fits the results if the acceleration is limited by the particle escape from the acceleration region.

In pulsars there are three important timescales that play a role in shaping a particle distribution during the acceleration and radiation processes. The first one is the timescale of acceleration of a particle to a Lorentz factor γ' in the reconnection electric fields $E' = \tau B'$, where τ is the reconnection rate, and the prime denotes the quantities considered in the wind comoving frame:

$$t'_{\rm acc} = \gamma' mc/(eE'). \tag{7.10}$$

The second timescale is defined by radiative cooling, given by usual synchrotron losses of a particle with a Lorentz factor γ':

$$t'_{\rm rad} = 6\pi mc^2/(\sigma_{\rm T} c\gamma' B'^2). \tag{7.11}$$

The third timescale can be related to the particle escape from the acceleration region, when its gyroradius $r_{\rm g} = \gamma' mc/(eB')$ becomes comparable to the size of the confining region. In this case the highest energy particles are defined by the Hillas criterion (Sironi et al. 2016).

A particle is able to reach the radiation reaction limit when the timescale of the acceleration is equal to the timescale of radiative cooling in the wind comoving frame; in this case:

$$\gamma'_{\rm rad} = \left(\frac{6\pi e\tau}{\sigma_{\rm T} B'}\right)^{1/2}. \tag{7.12}$$

In fact, at the X-points the particles can reach even higher energies (Kirk 2004; Cerutti et al. 2012, 2014), because the radiation reaction force vanishes in the direct field acceleration. As a consequence, the particle distribution is not cut-off sharply at (7.12), but this energy defines where the cut-off starts in the distribution.

The escape from the acceleration region will occur when the particle gyroradius becomes a fraction ζ of a current sheet thickness $\Delta_{\rm cs} = 2\pi\beta c/\omega$ (the confining region could be similar to the size of the largest plasmoids, according to the results of simulations, it is a few percent of the system length—here $\sim r_{\rm L}$ Sironi et al.

2016), which implies the limiting energy:

$$\gamma'_{\rm esc} = \frac{\Gamma\zeta(\Delta_{\rm sc}2\pi c)eB'}{mc^2\omega}. \tag{7.13}$$

It can be expected, therefore, that in large current sheets with large plasmoids the particles will stay confined and accelerated until they reach the radiation reaction limit. On the other hand, in smaller systems, they will be mainly escaping, before reaching the radiation reaction limit.

It is interesting to consider two cases in which the limiting energy is defined by:

- the radiative cooling, and the particle distribution takes a form of a power-law with an exponential cutoff:

$$n(\gamma') \propto \gamma'^{-s} e^{-\gamma'/\gamma'_{\rm rad}}, \tag{7.14}$$

- the escape from accelerating/confining region. In this case the particle distribution is cut-off super-exponentially:

$$n(\gamma') \propto \gamma'^{-s} e^{-\gamma'^2/\gamma'^2_{\rm esc}}. \tag{7.15}$$

This approach is, in fact, phenomenological, because in reality one should solve the continuity equation with a given acceleration process and radiative losses, in order to obtain the exact form of the particle distribution function. This, however, is a difficult problem, given, in particular, the complexity of the particle acceleration at reconnection sites.

The reconnection is assumed to set in the wind at some distance $r = \hat{r} r_{\rm L}$ from a pulsar. Only a small fraction $\varepsilon_{\rm d}$ of the magnetic energy goes into acceleration of the plasma particles so that the process has some efficiency, but it does not alter the dynamics of the wind itself, and the rest of the energy stays either in the magnetic field—for instance plasmons—or in a thermal population of particles. With these assumptions the regime of the particle acceleration (radiative cooling/escape) for a given pulsar depends only on its parameters—the spin down power and the period, given that $\gamma'_{\rm rad} \sim \gamma'_{\rm esc}$ when

$$\left(\frac{\dot{E}}{10^{38}\,{\rm erg/s}}\right)^{3/2}\left(\frac{P}{0.01\,{\rm s}}\right)^{-1} \sim 0.005\left(\frac{r}{10r_{\rm L}}\right)\left(\frac{0.1}{\zeta}\right)\left(\frac{0.1}{\Delta_{\rm cs}}\right)\left(\frac{\Gamma}{100}\right)^{3/2}\left(\frac{\varepsilon_{\rm d}}{0.01}\right)^{1/2}. \tag{7.16}$$

This suggests that only the most energetic and the fastest pulsars exhibit acceleration occurring in the radiative cooling regime (Mochol and Pétri 2015). When plotted on the $P\dot{P}$-diagram (see Fig. 7.6) the dividing line separates two pulsar populations.

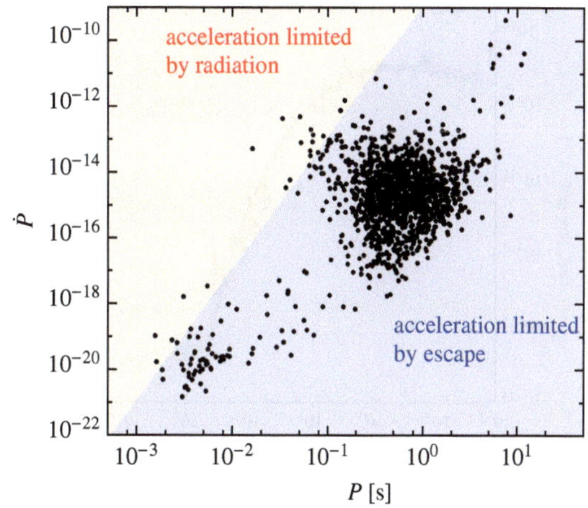

Fig. 7.6 $P\dot{P}$-diagram showing the line (7.16) which divides pulsars into two populations with different regimes of particle acceleration during reconnection in the wind

Table 7.1 Summary of two regimes of particle acceleration in the wind

Acceleration limit	Radiation	Escape
Particle distribution	$n(\gamma) = n_0\gamma^{-s}e^{-\gamma/\gamma_{\rm rad}}$	$n(\gamma) = n_0\gamma^{-s}e^{-\gamma^2/\gamma_{\rm esc}^2}$
Asymptote of synchrotron spectrum	$x^{1.3-(s+0.6)/3}e^{-1.9x^{1/3}}$	$x^{1.3-(s+0.6)/4}e^{-2x^{1/2}}$
SSC component	Detectable	Weak (not shown)

7.3.3 *Implications for Gamma-Ray Spectra of Pulsars*

Radiative signatures of each acceleration regime are determined by the power-law index of the particle distribution, as well as the shape of the cutoff at the highest energies. The most important emission process in the current sheet is the synchrotron radiation, produced by the particles moving in the leftover and/or adjacent magnetic field. In addition, the SSC spectral component may become significant in the very high energy band, because the same population of leptons that emits synchrotron photons in the sheet, can also upscatter them to higher energies in the IC mechanism. This scattering proceeds in the Thomson regime when the electron Lorentz factor γ and the photon energy normalized to the electron rest mass ϵ satisfy the condition $\gamma\epsilon \lesssim 1$. Thus, the soft photon field in the current sheet has to be sufficiently dense in order to make the scattering by the most energetic particles observable.

The behaviour of the synchrotron spectrum at the highest energies can be calculated analytically from the synchrotron flux integral using the steepest descent method (Mochol and Pétri 2015). The summary of the two considered cases is shown below in Table 7.1.

In order to model the emission from the wind the following procedure is adopted: (1) calculation of the synchrotron and SSC spectrum of an accelerated particle population (7.14) and (7.15) in the wind comoving frame, (2) transformation of the emission to the observer frame, (3) comparing the obtained spectrum of emission

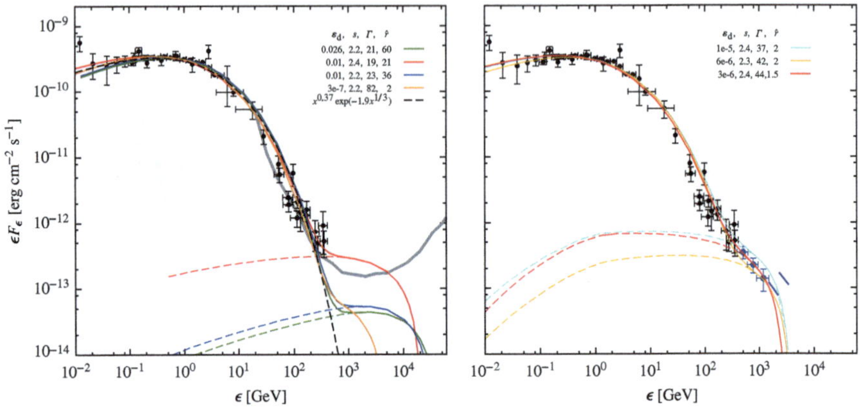

Fig. 7.7 *Left*: Several fits to the Crab spectrum up to 400 GeV. *Black points* are the data from Kuiper et al. (2001), Abdo et al. (2010) and Aleksić et al. (2014). *Thick gray line* shows the predicted CTA sensitivity (de Oña-Wilhelmi et al. 2013), *black dashed line* is the asymptote, plotted with $s = 2.2$. *Dashed lines* show SSC components, while *solid lines* are the total (synchrotron + SSC) spectrum. *Right*: Preliminary constraints on the model from the new observations from MAGIC (Ansoldi et al. 2016), shown here by the *blue points*

with the observed pulsed emission from a given pulsar, and finally (4) fitting the model to the data points in order to constrain Γ, $\hat{r}$, s and ε_d.

Two scenarios of acceleration, limited either by radiation reaction or the particle escape, are considered for an example of the Crab and Vela pulsars, respectively. The results are shown in Figs. 7.7 and 7.8.

The emphasis should be put on the fact that fitting the spectrum to the data is not unique. If the dissipation efficiency ε_d is small, the density of radiating particles is low, and, therefore, to match the data points, their emission must be generated close to the light cylinder, where the magnetic field is stronger, and, in addition, the radiation has to be boosted by a large Lorentz factor of the wind. If ϵ_d is larger, there are more accelerated particles, and, therefore, in order to not overproduce the synchrotron flux, the emission should be produced further from the light cylinder, where the magnetic field is weaker.

Interestingly, the difference between power-law indices s in the two acceleration scenarios is close to 1, as expected for the radiatively cooling population with $s \sim 2$ in comparison to the injected population with a very hard index $s \sim 1$. That would be consistent with the model, in which the cooling population is observed above the cooling break, and the escaping population is a real injection distribution.

Moreover, each of two considered regimes predicts different SSC component at TeV energies (compare Fig. 7.7 with Fig. 7.8). The maximum energy to which the synchrotron photons can be upscattered is given by

$$\epsilon_{\max} \approx 2\Gamma\gamma'_{\rm rad}mc^2 \approx 3.6\ {\rm TeV}\left(\frac{\Gamma}{10}\right)^{5/2}\left(\frac{P}{0.01{\rm s}}\right)^{1/2}\left(\frac{\varepsilon_{\rm d}}{0.01}\right)^{1/2}\left(\frac{\dot{E}}{10^{38}{\rm erg/s}}\right)^{-1/4} \tag{7.17}$$

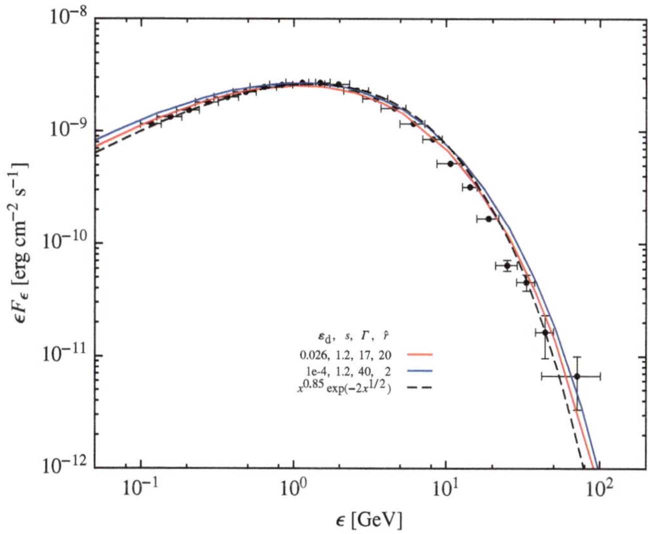

Fig. 7.8 Several best fits to the synchrotron spectrum of the Vela. For a hard particle index, the SSC component is weaker by several orders of magnitude and it is not shown in the plot. *Black dashed line* is the asymptote, plotted with $s = 1.2$. *Black points* are the data from Fermi-LAT (Leung et al. 2014)

The particle index changes the brightness of the SSC component, because it controls how many soft photons, available for upscattering (in the Thomson regime), are produced. For the steep particle index $s > 2$ (like for the Crab), there are more low energy particles that can produce low-energy synchrotron photons; in this case SSC component is brighter. From the currently available data we can constrain the upper limit on this index to be $\sim$2.4, because steeper indices would overproduce the SSC emission. On the other hand, a fit to the Vela spectrum requires a much harder index $s < 2$. In this case the SSC component is weaker by several orders of magnitude (not shown in the plot), because the energy density resides mainly in the highest energy particles, which do not produce enough low-energy synchrotron photons that can be upscattered to very high energies.

For the Crab, the key to constrain the model and the pulsar wind parameters is partly given by the recent MAGIC results (Ansoldi et al. 2016), which seems to prefer emission produced at smaller distances from the light cylinder. More accurate limits will be given by observations from the experiments HESS II and CTA.

7.4 MHD and Beyond: σ-Problem and the Structure of Termination Shocks of Striped Winds

According to MHD models, the energy transported by the wind is released in the nebula downstream of the wind termination shock, where a broad-band diffuse emission is produced. The Rankine-Hugoniot jump conditions at the shock have been solved by Kennel and Coroniti (1984a,b). With the assumption that the particles are efficiently accelerated at the shock, they calculated the synchrotron emissivity downstream of the shock as a function of the upstream wind parameters. They found that MHD shocks in highly magnetized flows are very weak, implying that the magnetisation and the flow velocity practically do not change across them. If the flow is initially highly magnetized, it stays so after passing a shock without any significant energy dissipation. Thus, in order to satisfy the observational constraints on the synchrotron emission, as well as the nonrelativistic expansion of the outer edge of the nebula, the plasma downstream, as well as upstream of the shock, has to be weakly magnetised. However, according to theoretical pulsar models, the wind is highly magnetised at its launching point and it stays so up to the termination shock. The reason is that a radial, large-σ MHD flow does not collimate as it propagates, and therefore it does not convert the energy from the electromagnetic (EM) fields to the kinetic form, arriving at the shock still Poynting dominated. Thus, it is not clear how and where the wind dissipates its EM energy to the plasma, a puzzle known as the "σ-problem". It has been shown that the phase-averaged, non-oscillating component of the fields in the wind can dissipate in the bulk of the nebula (Porth et al. 2013). Therefore the σ-problem concerns mainly the dissipation of the wave-like oscillating component of the fields.

Lyubarsky (2003) has proposed a solution, in which the striped wind, due to interaction with the shock, becomes compressed and the stripes dissipate the energy by the driven magnetic reconnection. This scenario provides not only the solution to the σ-paradox, but also explains the very hard spectrum of the nebula in the radio band as the synchrotron emission of the particles accelerated at the shock in the course of reconnection. PIC simulations (Pétri and Lyubarsky 2007; Sironi and Spitkovsky 2011) show that this mechanism operates also in 2D and 3D. The only caveat is that in order to reproduce the observed particle spectra, a high plasma density has to be assumed $\kappa \sim 10^8$, much higher than usually invoked in pulsar studies. In fact, in plasmas of the assumed densities, reconnection in the wind would start much earlier, before it arrives at the shock (Kirk and Skjæraasen 2003).

The question arises if there is a mechanism dissipating the magnetic energy in the regime of low plasma density. As pointed out by Usov (1975) and Michel (1994), according to the mass continuity $r^2 N v_r = \text{const}$, the density of current carriers N, as measured in the laboratory frame, drops as $N \propto r^{-2}$, faster than the magnetic field $B_\phi \propto r^{-1}$ that they are required to maintain. As discussed by Coroniti (1990), Michel (1994), and also Lyubarsky and Kirk (2001), the particles whose density decreases are forced to stream with higher and higher drift-speed in order to satisfy the Ampère's law in the flow, but, since the drift speed cannot reach c, the anomalous resistivity arises, which in turn will trigger the magnetic reconnection. As a result, magnetic energy would be released into heating of a plasma, which, however, will perform work on the flow, leading to its acceleration Lyubarsky and Kirk (2001). In a generic case, therefore, no significant dissipation will occur before the MHD wind arrives at the shock, which must be then itself responsible for the wave dissipation. On the other hand, if beyond a certain distance from the pulsar the flow becomes charge-starved, and therefore non-stationary, the displacement current in Ampère's law cannot be neglected anymore. Usov (1975) and Melatos and Melrose (1996) suggested that the MHD framework is not able to describe the whole physics of a diluted flow, which resembles more an EM wave in a plasma than an MHD wave. In this case one must refer to a more general, two fluid approach. Self consistent, analytic solutions of Maxwell and two-fluid equations have been discussed in numerous studies (Max and Perkins 1971; Clemmow 1974; Kennel and Pellat 1976; Asseo et al. 1984; Arka and Kirk 2012; Mochol and Kirk 2013).

The propagation of EM waves in a plasma is possible only when their frequency $\omega_{\rm w}$ exceeds the proper plasma frequency $\omega_{\rm pl}$. If the amplitude of an EM is large $a = eB/(mc\omega) \gg 1$, these waves can drive particles to relativistic Lorentz factors $\gamma \approx a \gg 1$, and the condition of propagation is less restrictive. In the vicinity of the pulsar the propagation condition of a wave with a frequency of the rotating pulsar,

$$\omega_{\rm w} = \omega_{\rm pulsar} > \omega_{\rm pl} = \left(\frac{8\pi e^2 N}{\gamma m}\right)^{1/2}, \qquad (7.18)$$

where $N = \kappa N_{\rm GJ} r_{\rm L}^2/r^2$ and $N_{\rm GJ}$ is given by Eq. (7.2), translates to

$$r > r_{\rm crit}, \qquad r_{\rm crit} = \kappa r_{\rm L} \qquad (7.19)$$

For most isolated pulsars the shock is located at a sufficiently large distance to allow the existence of the underdense region. For the Crab pulsar $\kappa \sim 10^6$, and thus the shock located at $r_{\rm shock} \sim 10^9 r_{\rm L}$, satisfies the condition (7.19). Since the EM waves become new eigenmodes in the system, they can be generated when the rarified wind interacts with the shock, forming a dissipative precursor. This precursor accelerates the particles to relativistic energies and transfers the energy from the fields into the plasma.

The two-fluid simulations (Amano and Kirk 2013) of this process have indeed demonstrated the formation of such precursors in diluted plasmas (see Fig. 7.9). The authors simulated a circularly polarized MHD wave (magnetic shear), mimicking an MHD wind, launched against a shock in a low density plasma, i.e. whose plasma frequency is smaller than the frequency of the wave. When the wave hits the shock, it is observed that EM waves are generated in front of the shock, and the system relaxes to an equilibrium in which an electromagnetic precursor is formed, and the Poynting flux of the initial wave is dissipated by the EM modes. As a consequence, a shock becomes essentially unmagnetized and thus potentially a very efficient Fermi accelerator. Recent studies (Giacche and Kirk 2017) of the particle dynamics in the precursor indeed demonstrate its ability to energise particles via wave-particle interactions. In particular, they prove that a significant fraction of such a pre-accelerated particles is reflected upstream and further picked up by the Fermi I-order mechanism.

An interesting possibility arises in pulsar binary systems, where the shocks may switch between different regimes—when the binary members are close, such that the distance between the pulsar and the shock is smaller than the critical radius

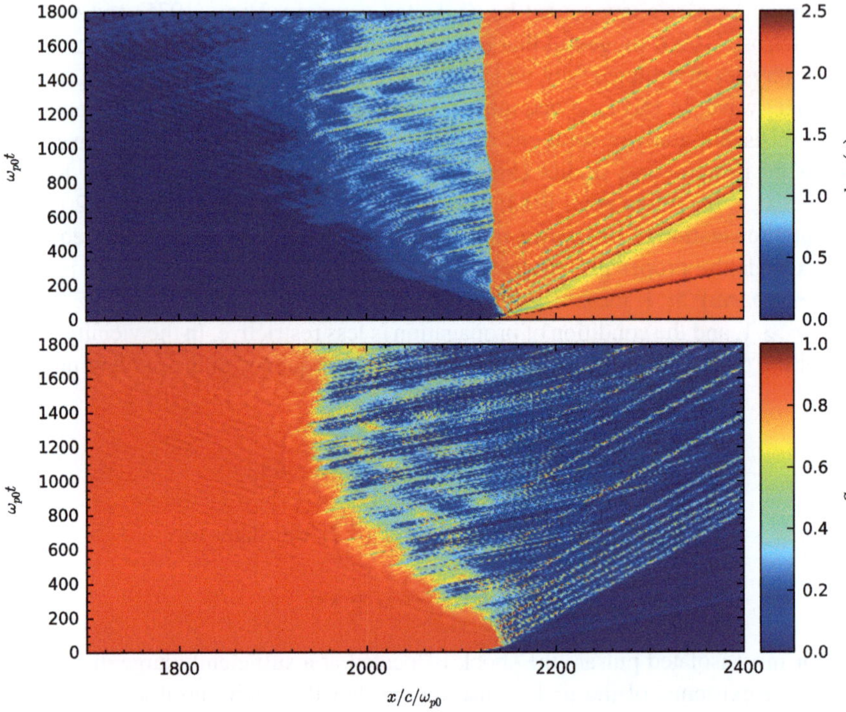

Fig. 7.9 Two-fluid simulations carried out by Amano and Kirk (2013) showing formation (space versus time) of an electromagnetic precursor, when a magnetic shear with the frequency exceeding the local plasma frequency is launched against a shock. The *top and bottom panels* show the proper density and Poynting flux, respectively. In the precursor region the Poynting flux becomes completely dissipated

$r_{\rm shock} < r_{\rm crit}$ (7.19), the shock is in the MHD regime and the energy is dissipated via driven reconnection of stripes. When the separation between binary members becomes larger, such that the shock is located at a radius $r_{\rm shock} > r_{\rm crit}$, a shock can acquire an EM precursor. At this shock regime transition, a sudden appearance of a precursor ahead of the shock should be accompanied by enhanced emission in the synchrotron and IC (on the photons from the stellar companion) processes, because the precursor emission is the most efficient close to the cut-off (Mochol and Kirk 2013). This scenario has been invoked in the eccentric gamma-ray binary B1259-63 (Mochol and Kirk 2013), where a mysterious flare, observed by the Fermi-LAT (Abdo et al. 2011; Tam et al. 2011, 2015), challenges the models (Kong et al. 2012; Khangulyan et al. 2012; Dubus and Cerutti 2013).

7.5 Summary

Pulsar winds play an important role in the coupling between the central engine—a rotating magnetized star—and the nebula, downstream of the pulsar wind shock, where the electromagnetic energy carried by the wind is released into the plasma. The mechanism leading to the energy release is not clear, however. It seems plausible that it is related to the particular structure of the equatorial component of the wind, which has a form of stripes of opposite magnetic polarity, separated by the current sheet that performs a wavy pattern in space as the pulsar rotates. This structure is believed to be prone to the relativistic magnetic reconnection.

Although it has been shown that the reconnection in the wind is not able to dissipate the electromagnetic energy before it reaches the termination shock, it provides a promising scenario for the production of the high-energy pulsed emission. The particles accelerated in the course of reconnection emit synchrotron and SSC photons, whose spectrum can reach TeV energies. Remarkably, recent observations of the Crab (Ansoldi et al. 2016), as well as the millisecond pulsar J0614–3329 (Xing and Wang 2016) confirm the detection of the very high energy spectral components, favouring the scenario, in which the gamma-rays originate from outside the light cylinder.

Most of the wind electromagnetic energy must be, however, released somewhere close to the termination shock of the wind. In this case, the stripes become compressed when the wind impacts on the shock, and the driven reconnection annihilates the alternating field component. The nonthermal spectrum of the particles accelerated in the course of this process can be observed, if the plasma carried by the wind is sufficiently dense. In the low plasma density, on the other hand, the electromagnetic fields of the wind behave more like an electromagnetic wave instead of a familiar MHD wind. Recent studies show that the shocks in diluted plasmas exhibit dissipative, electromagnetic precursors, which mediate the Poynting flux dissipation and pre-accelerate particles, which are further a subject to the Fermi process. This scenario is expected to be realised in shocks of isolated pulsars, located at large stand-off distances, $r_{\rm shock} > \kappa r_{\rm L}$.

Acknowledgements I thank Tomasz Rembiasz and Jérôme Pétri for helpful comments on the manuscript.

References

Abdo, A.A., et al.: Fermi large area telescope observations of the Crab pulsar and nebula. Astrophys. J. **708**, 1254–1267 (2010)

Abdo, A.A., et al.: Discovery of high-energy gamma-ray emission from the binary system PSR B1259-63/LS 2883 around periastron with fermi. Astrophys. J. **736**, L11 (2011)

Aleksić, J., et al.: Detection of bridge emission above 50 GeV from the Crab pulsar with the MAGIC telescopes. Astron. Astrophys. **565**, L12 (2014)

Amano, T., Kirk, J.G.: The role of superluminal electromagnetic waves in pulsar wind termination shocks. Astrophys. J. **770**, 18 (2013)
Ansoldi, S., et al.: Teraelectronvolt pulsed emission from the Crab pulsar detected by MAGIC. Astron. Astrophys. **585**, A133 (2016)
Arka, I., Kirk, J.G.: Superluminal waves in pulsar winds. Astrophys. J. **745**, 108 (2012)
Arons, J., Scharlemann, E.T.: Pair formation above pulsar polar caps – structure of the low altitude acceleration zone. Astrophys. J. **231**, 854–879 (1979)
Asseo, E., Pellat, R., Llobet, X.: Spherical propagation of large amplitude pulsar waves. Astron. Astrophys. **139**, 417–425 (1984)
Bai, X.-N., Spitkovsky, A.: Uncertainties of modeling gamma-ray pulsar light curves using vacuum dipole magnetic field. Astrophys. J. **715**, 1270–1281 (2010)
Barnard, J.J., Arons, J.: Pair production and pulsar cutoff in magnetized neutron stars with nondipolar magnetic geometry. Astrophys. J. **254**, 713–734 (1982)
Bessho, N., Bhattacharjee, A.: Fast collisionless reconnection in electron-positron plasma. Phys. Plasmas **14**(5), 056503 (2007)
Bogovalov, S.V.: On the physics of cold MHD winds from oblique rotators. Astron. Astrophys. **349**, 1017–1026 (1999)
Bucciantini, N., Arons, J., Amato, E.: Modelling spectral evolution of pulsar wind nebulae inside supernova remnants. Mon. Not. R. Astron. Soc. **410**, 381–398 (2011)
Buckley, R.: Pulsar magnetospheres with arbitrary geometry in the force-free approximation. Mon. Not. R. Astron. Soc. **180**, 125–140 (1977)
Cerutti, B., Uzdensky, D.A., Begelman, M.C.: Extreme particle acceleration in magnetic reconnection layers: application to the gamma-ray flares in the Crab nebula. Astrophys. J. **746**, 148 (2012)
Cerutti, B., Werner, G.R., Uzdensky, D.A., Begelman, M.C.: Three-dimensional relativistic pair plasma reconnection with radiative feedback in the Crab nebula. Astrophys. J. **782**, 104 (2014)
Cerutti, B., Philippov, A., Parfrey, K., Spitkovsky, A.: Particle acceleration in axisymmetric pulsar current sheets. Mon. Not. R. Astron. Soc. **448**, 606–619 (2015)
Cheng, K.S., Ho, C., Ruderman, M.: Energetic radiation from rapidly spinning pulsars. I – outer magnetosphere gaps. II – VELA and Crab. Astrophys. J. **300**, 500–539 (1986)
Clemmow, P.C.: Nonlinear waves in a cold plasma by Lorentz transformation. J. Plasma Phys. **12**, 297–317 (1974)
Contopoulos, I., Kazanas, D.: Toward resolving the Crab σ-problem: a linear accelerator? Astrophys. J. **566**, 336–342 (2002)
Coroniti, F.V.: Magnetically striped relativistic magnetohydrodynamic winds – the Crab nebula revisited. Astrophys. J. **349**, 538–545 (1990)
Daugherty, J.K., Harding, A.K.: Electromagnetic cascades in pulsars. Astrophys. J. **252**, 337–347 (1982)
de Jager, O.C., Harding, A.K., Michelson, P.F., Nel, H.I., Nolan, P.L., Sreekumar, P., Thompson, D.J.: Gamma-ray observations of the Crab nebula: a study of the synchro-compton spectrum. Astrophys. J. **457**, 253 (1996)
de Oña-Wilhelmi, E., Rudak, B., Barrio, J.A., Contreras, J.L., Gallant, Y., Hadasch, D., Hassan, T., Lopez, M., Mazin, D., Mirabal, N., Pedaletti, G., Renaud, M., de los Reyes, R., Torres, D.F., CTA Consortium: Prospects for observations of pulsars and pulsar wind nebulae with CTA. Astropart. Phys. **43**, 287–300 (2013)
Dubus, G., Cerutti, B.: What caused the GeV flare of PSR B1259-63? Astron. Astrophys. **557**, A127 (2013)
Dyks, J., Rudak, B.: Two-pole caustic model for high-energy light curves of pulsars. Astrophys. J. **598**, 1201–1206 (2003)
Gallant, Y.A., van der Swaluw, E., Kirk, J.G., Achterberg, A.: Modeling plerion spectra and their evolution. In: Slane, P.O., Gaensler, B.M. (eds.) Neutron Stars in Supernova Remnants. Astronomical Society of the Pacific Conference Series, vol. 271, p. 99. Astronomical Society of the Pacific, San Francisco (2002)

Giacche, S., Kirk, J.G.: Electron acceleration at pulsar wind termination shocks. Astrophys. J. **835**, 235 (2017)

Gil, J., Mitra, D.: Vacuum gaps in pulsars and PSR J2144-3933. Astrophys. J. **550**, 383–391 (2001)

Guo, F., Li, H., Daughton, W., Liu, Y.-H.: Formation of hard power laws in the energetic particle spectra resulting from relativistic magnetic reconnection. Phys. Rev. Lett. **113**(15), 155005 (2014)

Harding, A.K.: Physics in strong magnetic fields near neutron stars. Science **251**, 1033–1038 (1991)

Harding, A.K., Kalapotharakos, C.: Synchrotron self-compton emission from the Crab and other pulsars. Astrophys. J. **811**, 63 (2015)

Harding, A.K., Tademaru, E., Esposito, L.W.: A curvature-radiation-pair-production model for gamma-ray pulsars. Astrophys. J. **225**, 226–236 (1978)

Jaroschek, C.H., Treumann, R.A., Lesch, H., Scholer, M.: Fast reconnection in relativistic pair plasmas: analysis of particle acceleration in self-consistent full particle simulations. Phys. Plasmas **11**, 1151–1163 (2004)

Kagan, D., Sironi, L., Cerutti, B., Giannios, D.: Relativistic magnetic reconnection in pair plasmas and its astrophysical applications. Space Sci. Rev. **191**, 545–573 (2015)

Kalapotharakos, C., Kazanas, D., Harding, A., Contopoulos, I.: Toward a realistic pulsar magnetosphere. Astrophys. J. **749**, 2 (2012)

Kennel, C.F., Coroniti, F.V.: Confinement of the Crab pulsar's wind by its supernova remnant. Astrophys. J. **283**, 694–709 (1984a)

Kennel, C.F., Coroniti, F.V.: Magnetohydrodynamic model of Crab nebula radiation. Astrophys. J. **283**, 710–730 (1984b)

Kennel, C.F., Pellat, R.: Relativistic nonlinear plasma waves in a magnetic field. J. Plasma Phys. **15**, 335–355 (1976)

Khangulyan, D., Aharonian, F.A., Bogovalov, S.V., Ribó, M.: Post-periastron gamma-ray flare from PSR B1259-63/LS 2883 as a result of comptonization of the cold pulsar wind. Astrophys. J. **752**, L17 (2012)

Kirk, J.G.: Particle acceleration in relativistic current sheets. Phys. Rev. Lett. **92**(18), 181101 (2004)

Kirk, J.G., Skjæraasen, O.: Dissipation in poynting-flux-dominated flows: the σ-problem of the Crab pulsar wind. Astrophys. J. **591**, 366–379 (2003)

Kirk, J.G., Skjæraasen, O., Gallant, Y.A.: Pulsed radiation from neutron star winds. Astron. Astrophys. **388**, L29–L32 (2002)

Kirk, J.G., Lyubarsky, Y., Petri, J.: The theory of pulsar winds and nebulae. In: Becker, W. (ed.) Astrophysics and Space Science Library. Astrophysics and Space Science Library, vol. 357, p. 421. Springer, Berlin (2009)

Komissarov, S.S.: Simulations of the axisymmetric magnetospheres of neutron stars. Mon. Not. R. Astron. Soc. **367**, 19–31 (2006)

Kong, S.W., Cheng, K.S., Huang, Y.F.: Modeling the multiwavelength light curves of PSR B1259-63/LS 2883. II. The effects of anisotropic pulsar wind and doppler boosting. Astrophys. J. **753**, 127 (2012)

Krause-Polstorff, J., Michel, F.C.: Electrosphere of an aligned magnetized neutron star. Mon. Not. R. Astron. Soc. **213**, 43P–49P (1985)

Krause-Polstorff, J., Michel, F.C.: Pulsar space charging. Astron. Astrophys. **144**, 72–80 (1985)

Kuiper, L., Hermsen, W., Cusumano, G., Diehl, R., Schönfelder, V., Strong, A., Bennett, K., McConnell, M.L.: The Crab pulsar in the 0.75–30 MeV range as seen by CGRO COMPTEL. A coherent high-energy picture from soft X-rays up to high-energy gamma-rays. Astron. Astrophys. **378**, 918–935 (2001)

Leung, G.C.K., Takata, J., Ng, C.W., Kong, A.K.H., Tam, P.H.T., Hui, C.Y., Cheng, K.S.: Fermi-LAT detection of pulsed gamma-rays above 50 GeV from the vela pulsar. Astrophys. J. **797**, L13 (2014)

Li, J., Spitkovsky, A., Tchekhovskoy, A.: Resistive solutions for pulsar magnetospheres. Astrophys. J. **746**, 60 (2012)

Lyubarskii, Y.E.: A model for the energetic emission from pulsars. Astron. Astrophys. **311**, 172–178 (1996)
Lyubarsky, Y.E.: The termination shock in a striped pulsar wind. Mon. Not. R. Astron. Soc. **345**, 153–160 (2003)
Lyubarsky, Y.E.: On the relativistic magnetic reconnection. Mon. Not. R. Astron. Soc. **358**, 113–119 (2005)
Lyubarsky, Y., Kirk, J.G.: Reconnection in a striped pulsar wind. Astrophys. J. **547**, 437–448 (2001)
Lyutikov, M.: Inverse compton model of pulsar high-energy emission. Mon. Not. R. Astron. Soc. **431**, 2580–2589 (2013)
Lyutikov, M., Uzdensky, D.: Dynamics of relativistic reconnection. Astrophys. J. **589**, 893–901 (2003)
Max, C., Perkins, F.: Strong electromagnetic waves in overdense plasmas. Phys. Rev. Lett. **27**, 1342–1345 (1971)
Melatos, A., Melrose, D.B.: Energy transport in a rotation-modulated pulsar wind. Mon. Not. R. Astron. Soc. **279**, 1168–1190 (1996)
Michel, F.C.: Rotating magnetosphere: a simple relativistic model. Astrophys. J. **180**, 207–226 (1973)
Michel, F.C.: Rotating magnetospheres: an exact 3-D solution. Astrophys. J. **180**, L133 (1973)
Michel, F.C.: Magnetic structure of pulsar winds. Astrophys. J. **431**, 397–401 (1994)
Mochol, I., Kirk, J.G.: Propagation and stability of superluminal waves in pulsar winds. Astrophys. J. **771**, 53 (2013)
Mochol, I., Kirk, J.G.: Radiative damping and emission signatures of strong superluminal waves in pulsar winds. Astrophys. J. **776**, 40 (2013)
Mochol, I., Pétri, J.: Very high energy emission as a probe of relativistic magnetic reconnection in pulsar winds. Mon. Not. R. Astron. Soc. **449**, L51–L55 (2015)
Muslimov, A.G., Harding, A.K.: Extended acceleration in slot gaps and pulsar high-energy emission. Astrophys. J. **588**, 430–440 (2003)
Muslimov, A.G., Harding, A.K.: High-altitude particle acceleration and radiation in pulsar slot gaps. Astrophys. J. **606**, 1143–1153 (2004)
Pétri, J.: The pulsar force-free magnetosphere linked to its striped wind: time-dependent pseudo-spectral simulations. Mon. Not. R. Astron. Soc. **424**, 605–619 (2012)
Pétri, J.: Phase-resolved polarization properties of the pulsar striped wind synchrotron emission. Mon. Not. R. Astron. Soc. **434**, 2636–2644 (2013)
Pétri, J.: Multipolar electromagnetic fields around neutron stars: exact vacuum solutions and related properties. Mon. Not. R. Astron. Soc. **450**, 714–742 (2015)
Pétri, J.: General-relativistic force-free pulsar magnetospheres. Mon. Not. R. Astron. Soc. **455**, 3779–3805 (2016)
Pétri, J.: Theory of pulsar magnetosphere and wind. ArXiv e-prints, Aug 2016
Pétri, J., Kirk, J.G.: The polarization of high-energy pulsar radiation in the striped wind model. Astrophys. J. **627**, L37–L40 (2005)
Pétri, J., Lyubarsky, Y.: Magnetic reconnection at the termination shock in a striped pulsar wind. Astron. Astrophys. **473**, 683–700 (2007)
Pétri, J., Heyvaerts, J., Bonazzola, S.: Global static electrospheres of charged pulsars. Astron. Astrophys. **384**, 414–432 (2002)
Petrova, S.A.: Axisymmetric force-free magnetosphere of a pulsar – II. Transition from the self-consistent two-fluid model. Mon. Not. R. Astron. Soc. **446**, 2243–2250 (2015)
Petrova, S.A.: A novel look at the pulsar force-free magnetosphere. ArXiv e-prints, Aug 2016
Philippov, A.A., Spitkovsky, A., Cerutti, B.: Ab initio pulsar magnetosphere: three-dimensional particle-in-cell simulations of oblique pulsars. Astrophys. J. **801**, L19 (2015)
Porth, O., Komissarov, S.S., Keppens, R.: Solution to the sigma problem of pulsar wind nebulae. Mon. Not. R. Astron. Soc. **431**, L48–L52 (2013)
Rees, M.J., Gunn, J.E.: The origin of the magnetic field and relativistic particles in the Crab nebula. Mon. Not. R. Astron. Soc. **167**, 1–12 (1974)

Romani, R.W., Yadigaroglu, I.-A.: Gamma-ray pulsars: emission zones and viewing geometries. Astrophys. J. **438**, 314–321 (1995)

Rudak, B., et al.: Pulsations from the Vela Pulsar Down to 20 GeV with H.E.S.S. II Talk at 6th International Fermi Symposium, Arlington (2015)

Ruderman, M.A., Sutherland, P.G.: Theory of pulsars – polar caps, sparks, and coherent microwave radiation. Astrophys. J. **196**, 51–72 (1975)

Scharlemann, E.T., Wagoner, R.V.: Aligned rotating magnetospheres. General analysis. Astrophys. J. **182**, 951–960 (1973)

Sironi, L., Spitkovsky, A.: Acceleration of particles at the termination shock of a relativistic striped wind. Astrophys. J. **741**, 39 (2011)

Sironi, L., Spitkovsky, A.: Relativistic reconnection: an efficient source of non-thermal particles. Astrophys. J. **783**, L21 (2014)

Sironi, L., Giannios, D., Petropoulou, M.: Plasmoids in relativistic reconnection, from birth to adulthood: first they grow, then they go. Mon. Not. R. Astron. Soc. **462**, 48–74 (2016)

Spitkovsky, A.: Electrodynamics of pulsar magnetospheres. In: Camilo, F., Gaensler, B.M. (eds.) Young Neutron Stars and Their Environments. IAU Symposium, vol. 218, p. 357. Astronomical Society of the Pacific, San Francisco, CA (2004)

Spitkovsky, A.: Time-dependent force-free pulsar magnetospheres: axisymmetric and oblique rotators. Astrophys. J. **648**, L51–L54 (2006)

Sturrock, P.A.: A model of pulsars. Astrophys. J. **164**, 529 (1971)

Tam, P.H.T., Huang, R.H.H., Takata, J., Hui, C.Y., Kong, A.K.H., Cheng, K.S.: Discovery of GeV γ-ray Emission from PSR B1259-63/LS 2883. Astrophys. J. **736**, L10 (2011)

Tam, P.H.T., Li, K.L., Takata, J., Okazaki, A.T., Hui, C.Y., Kong, A.K.H.: High-energy observations of PSR B1259-63/LS 2883 through the 2014 periastron passage: connecting X-rays to the GeV flare. Astrophys. J. **798**, L26 (2015)

Tchekhovskoy, A., Spitkovsky, A., Li, J.G.: Time-dependent 3D magnetohydrodynamic pulsar magnetospheres: oblique rotators. Mon. Not. R. Astron. Soc. **435**, L1–L5 (2013)

Timokhin, A.N., Harding, A.K.: On the polar cap cascade pair multiplicity of young pulsars. Astrophys. J. **810**, 144 (2015)

Usov, V.V.: Wave zone structure of NP 0532 and infrared radiation excess of Crab nebula. Astrophys. Space Sci. **32**, 375–377 (1975)

Viganò, D., Torres, D.F.: Modelling of the γ-ray pulsed spectra of Geminga, Crab, and Vela with synchro-curvature radiation. Mon. Not. R. Astron. Soc. **449**, 3755–3765 (2015)

Weisskopf, M.C., Hester, J.J., Tennant, A.F., Elsner, R.F., Schulz, N.S., Marshall, H.L., Karovska, M., Nichols, J.S., Swartz, D.A., Kolodziejczak, J.J., O'Dell, S.L.: Discovery of spatial and spectral structure in the X-ray emission from the Crab nebula. Astrophys. J. **536**, L81–L84 (2000)

Werner, G.R., Uzdensky, D.A., Cerutti, B., Nalewajko, K., Begelman, M.C.: The extent of power-law energy spectra in collisionless relativistic magnetic reconnection in pair plasmas. Astrophys. J. **816**, L8 (2016)

Xing, Y., Wang, Z.: Fermi study of gamma-ray millisecond pulsars: the spectral shape and pulsed 25–200 GeV emission from J0614-3329. ArXiv e-prints, Apr 2016

Zenitani, S., Hoshino, M.: The generation of nonthermal particles in the relativistic magnetic reconnection of pair plasmas. Astrophys. J. **562**, L63–L66 (2001)

Chapter 8
Radiative Models of Pulsar Wind Nebulae

Joseph D. Gelfand

Abstract Since pulsar wind nebulae (PWNe) are the product of the interaction between the magnetized, highly relativistic particle outflow powered by a neutron star's rotational energy and its surroundings, their emission is sensitive to the properties of the central neutron star, its pulsar wind, and environment. Therefore, a PWN's broadband spectral energy distribution (SED) can be used to determine the physical characteristics of all the components of this system—including quantities difficult, if not impossible, to measure using other means. This is best done by fitting the observed SED of a PWN with a model for both the radiative and dynamical emission of the system. In this chapter, I will describe the physics behind the single-zone ("0D") models most often used to interpret the properties of PWNe, discuss their advantages and shortcomings, as well as speculate on the future of this field.

8.1 Introduction

Even before the discovery of pulsars (Hewish et al. 1968), it was suggested that some bright astronomical objects could be powered by the rotational energy of the strongly magnetized neutron star created in a supernova explosion (Pacini 1968). Shortly after the discovery of pulsars, it was realized the bulk of such a neutron star's rotational energy is carried away by highly-relativistic particles (primarily electrons and positrons) escaping along "open" magnetic field lines (Goldreich and Julian 1969). The interaction between this "pulsar wind" and its surroundings creates a pulsar wind nebula (PWN) which radiates light across the entire electromagnetic spectrum, from the lowest frequency radio waves to the highest energy γ-rays (Fig. 8.1). In fact, such sources not only dominate the population of identified Galactic TeV sources (e.g., Carrigan et al. 2013), but the Crab Nebula—the PWN powered by the pulsar created in SN 1054—lends its name to a unit

J.D. Gelfand (✉)
NYU Abu Dhabi, PO Box 129188, Abu Dhabi, United Arab Emirates
e-mail: jg168@nyu.edu

D.F. Torres (ed.), *Modelling Pulsar Wind Nebulae*, Astrophysics and Space Science Library 446, DOI 10.1007/978-3-319-63031-1_8

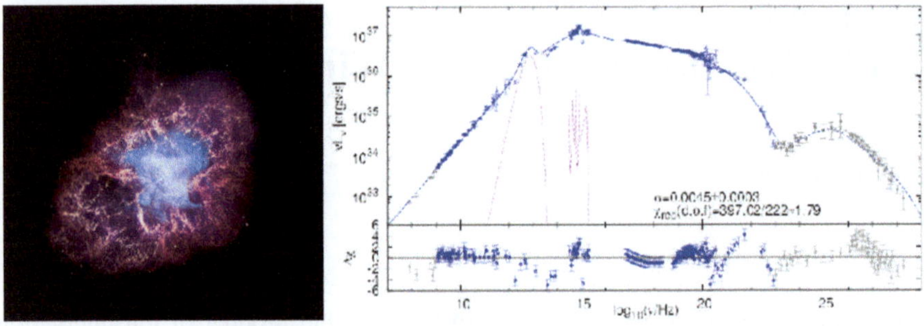

Fig. 8.1 *Left*: Broadband image of the Crab Nebula—X-ray (*blue*): NASA/CXC/SAO/F.Seward; Optical (*red and yellow*): NASA/ESA/ASU/J.Hester and A. Loll; Infrared (*purple*): NASA/JPL-Caltech/Univ. Minn./R.Gehrz. *Right*: Broadband SED of the Crab Nebula (Fig. 1 from Meyer et al. 2010)

of intensity,[1] and was long used to calibrate X-ray and γ-ray observatories (e.g., Toor and Seward 1974; Meyer et al. 2010; Weisskopf et al. 2010; Bird and for the VERITAS Collaboration 2015) until the discovery of the long and short-term variability (Wilson-Hodge et al. 2011).

As long realized and expertly discussed in recent reviews (e.g., Gaensler and Slane 2006; Bucciantini 2014; Kargaltsev et al. 2015; Reynolds et al. 2017; Slane 2017), the origin of PWNe suggests their study are important for addressing numerous open questions in high-energy astrophysics, including:

- How are neutron stars created in core-collapse supernovae?
- How are particles created in the magnetospheres of neutron stars?
- How are these particles then accelerated to the PeV energies inside the PWN?

Their answers lie in measuring several physical quantities of these systems, e.g., the initial spin-down luminosity of the pulsar, the total number of particles inside the PWN, the energy spectrum of particles injected into the PWN at the termination shock. Currently, the best way of doing so is to fit the observed dynamical properties and spectral energy distribution (SED) of a PWN inside a supernova remnant (SNR) with the properties predicted by an evolutionary models for such systems (e.g., Torres et al. (2014), Gelfand et al. (2015) and both references therein and to these works). In Sect. 8.2, I will describe the physics underlying the various "one-zone" ("0-dimensional") models used for this purpose. These models, which by their very nature, significantly simplify the true nature of PWNe, are currently the only tool that self-consistently predicts both their dynamical and radiative evolution. In Sect. 8.3, I discuss the results derived from applying such models. Lastly, in Sect. 8.4, I summarize the current state of the field and discuss possible future directions.

[1] "1 Crab" corresponds to a $2-10\,\mathrm{keV}$ flux of 2×10^{-8} ergs cm^{-2} s^{-1}.

8.2 One-Zone Models for the Evolution of a PWN Inside a SNR

The development of "one-zone" models for the evolution of a PWN inside a SNR began shortly after the discovery of pulsars (e.g., Pacini and Salvati 1973; Reynolds and Chevalier 1984; Reynolds and Chanan 1984), and were contemporaneous to the first multi-dimensional models. While these multi-dimensional models, by necessity, assumed steady-state conditions (e.g., Kennel and Coroniti 1984a,b), the one zone models incorporated both the evolution of the pulsar and the SNR. As subsequent work expanded on these earlier models, the basic assumptions underlying this approach have largely remained the same (e.g., Reynolds and Chevalier 1984):

1. the PWN is powered by the rotational energy of the pulsar,
2. the bulk of the pulsar's rotational energy is injected into the PWN in the form of relativistic particles and magnetic fields, and
3. the PWN is a spherical and homogeneous bubble located inside the SNR produced by the progenitor explosion.

It is the last feature which differentiates the one-zone models described here from more complicated, multi-dimensional treatments of PWN evolution. One implication of this assumption is that one-zone models cannot reproduce the fascinating morphological structures (e.g., Weisskopf et al. 2000) nor the spectral variations observed particularly at higher (X-ray, e.g. Mori et al. (2004) and γ-ray, e.g. Aharonian et al. 2006a) photon energies observed in some PWN. However, as pointed out by Bucciantini (2014), the quality of data for most PWNe is such that one-zone models are the only justifiable approach, and these models constitute a good starting point for more detailed studies of individual systems.

While there are differences between the various one-zone models discussed in the literature, they all include:

- the injection of particles and magnetic fields into the PWN (Sect. 8.2.1),
- the evolution of the particles and magnetic field inside the PWN (Sect. 8.2.2),
- and the interaction between the PWN and surrounding SNR (Sect. 8.2.3).

8.2.1 *Injection of Relativistic Particle and Magnetic Fields*

PWNe have long been assumed to be primarily powered by rotational energy of the associated neutron star. Therefore, most models assume the rate the pulsar injects energy into the PWN follows the classic spin-down equation (Goldreich and Julian 1969):

$$\dot{E}_{\rm psr}(t) = \dot{E}_0 \left(1 + \frac{t}{\tau_{\rm sd}}\right)^{-\frac{p+1}{p-1}} \tag{8.1}$$

where $\dot{E}_0$ is the initial spin-down luminosity of the pulsar, t is its age, τ_{sd} is the spin-down timescale, and p is the braking index (often denoted as n in other work). These quantities are related to the true age of the pulsar t_{age}, since (e.g., Pacini and Salvati 1973; Gaensler and Slane 2006):

$$\tau_{sd} = \frac{2\tau_c}{p-1} - t_{age}, \tag{8.2}$$

where τ_c is the "characteristic age" of the pulsar is defined to be:

$$\tau_c \equiv \frac{P}{2\dot{P}}, \tag{8.3}$$

where P and $\dot{P}$ are, respectively, the measured period and period-derivative of the central pulsar. These same parameters, p and τ_{sd}, can be used to determine the neutron star's initial spin period P_0, since (Gaensler and Slane 2006):

$$P_0 = P\left(1 + \frac{t}{\tau_{sd}}\right)^{-\frac{1}{p-1}}, \tag{8.4}$$

where P is the current period of the pulsar. Furthermore, the braking index p is related to the configuration and evolution of the pulsar's surface magnetic field, where $p \equiv 3$ corresponds to magnetic dipole radiation from a constant-strength, purely dipolar field (e.g., Goldreich and Julian 1969).

In most models, the full spin-down luminosity of the pulsar $\dot{E}_{psr}$ is injected into the PWN either in the form of relativistic particles $\dot{E}_p$ or magnetic fields $\dot{E}_B$. This ignores the possibility that the pulsed γ-ray emission, which can have a luminosity $L_\gamma \sim 0.1\dot{E}_{psr}$ (e.g., Abdo et al. 2013), decreases the total energy injected into the PWN. Most models (e.g. Gelfand et al. (2009), Tanaka and Takahara (2010), Martín et al. (2012) and reference both therein and to these works) furthermore assume that magnetic fields comprise a constant fraction η_B of the energy injected into the PWN, such that:

$$\dot{E}_B = \eta_B \dot{E}_{psr} \tag{8.5}$$

$$\dot{E}_p = (1-\eta_B)\dot{E}_{psr}. \tag{8.6}$$

While ions may be present in the pulsar wind (e.g. Spitkovsky and Arons 2004; Li et al. 2010), most models assume that the relativistic electrons and positrons (both energetically and in number) dominate the particles injected into the PWN. Furthermore, one-zone models typically assume these particles are primarily accelerated at or before ("upstream") the termination shock formed where the ram pressure of the unshocked pulsar wind equals the pressure inside the PWN. This does not allow for re-acceleration of particles after ("downstream" of) the termination shock (e.g., Olmi et al. 2014), possibly resulting from magnetic dissipation as observed in recent three-dimensional magneto-hydrodynamical (MHD) simulations of PWNe (e.g., Porth et al. 2014b). This is also incompatible with the radio-emitting electrons in some PWNe, e.g. the Crab Nebula, being a relic of their formation (e.g., Kennel and Coroniti 1984b; Olmi et al. 2014).

The subsequent evolution of the PWN also depends on the spectrum of particles injected at the termination shock. While a single power-law is the simplest spectrum to use (e.g., Gelfand et al. 2009), it fails to reproduce both the radio and X-ray synchrotron emission of the PWN (e.g., Reynolds and Chevalier 1984). Typically, models use one of two functional forms for this spectrum: a Maxwellian plus a power-law (MPL; e.g. Fang and Zhang 2010; Slane et al. 2010), or—more commonly—a broken power-law (BPL, e.g., Bucciantini et al. 2011; Tanaka and Takahara 2011; Torres et al. 2014; Gelfand et al. 2015). It is worth mentioning that other functional forms, such as two discontinuous power-law, have been considered in the literature (de Jager et al. 2008; Vorster et al. 2013).

The main motivation for choosing a MPL injection spectrum is that this shape is expected from "standard" Fermi acceleration processes and reproduced in "particle-in-cell" simulations of shock acceleration in a relativistic electron-positron outflow (e.g., Spitkovsky 2008). Such a choice requires four parameters: the temperature of the Maxwellian $kT_{\rm max}$, and the minimum $E_{\rm min}$, and maximum $E_{\rm max}$ energy, and particle index p (defined such that $dN/dE \propto E^{-p}$) of the power-law component. However, this choice of injection spectrum makes it difficult to reproduced the observed radio spectral index of PWNe ($\alpha \sim \pm 0.1$ where flux density $S_\nu \propto \nu^\alpha$; see recent review by Reynolds et al. 2017). This is because the radio-emitting electrons typically fall on the "Rayleigh-Jeans" part of the Maxwellian spectrum, which naturally results in a spectral index $\alpha > 0$. While this can be potentially alleviated by having the temperature of the Maxwellian change with time, physically motivated prescriptions of such an evolution are unable to do so (e.g, Atoyan (1999)).

Models which assume a BPL injection spectrum are typically more successful at reproducing the broadband SED of a PWN. While some numerical simulations have succeeded in generating such a particle spectrum at the termination shock of a PWN (e.g., Sironi and Spitkovsky 2011, 2014), they require physical conditions unlikely to be satisfied within a PWN (e.g., Gelfand et al. 2015). Typically, five parameters are used to describe this injection spectrum: the minimum $E_{\rm min}$, break $E_{\rm break}$, and maximum $E_{\rm max}$ energy of the injected particles, as well as the particle indices below p_1 and above p_2 the break energy. In general, p_1 and p_2 are assumed to be constant, while $E_{\rm min}$ and $E_{\rm break}$ are either assumed to be constant with time (e.g., Tanaka and Takahara 2010; Torres et al. 2014; Gelfand et al. 2015) or a constant fraction of the maximum energy $E_{\rm max}$ (e.g., Bucciantini et al. 2011). However, there exists several different treatments of $E_{\rm max}$:

- $E_{\rm max}$ is set to a constant value (e.g., Gelfand et al. 2009; Tanaka and Takahara 2010, 2011; Gelfand et al. 2015). While there is no strong physical motivation, it represents the "simplest" choice and possibly avoids systematic errors resulting from the assumptions inherent in the other choices below.

- $E_{\max}$ is the electric potential of the neutron star magnetosphere (e.g., Bucciantini et al. 2011; Granot et al. 2017):

$$E_{\max} \equiv e\Phi = e\sqrt{\frac{\dot{E}_{\rm psr}}{c}}, \tag{8.7}$$

  where e is the charge of the electron. The physical motivation for this choice is that the electric potential of the neutron star is the only "natural" energy scale of this system. This choice is *not* specific to a particular physics mechanism for producing high-energy particles inside the PWN, but represents a limit based on the total energy available for such acceleration.
- $E_{\max}$ is the energy such that the particle's Larmor radius is a factor ϵ (a free parameter constrained by $\epsilon < 1$) of the size of the pulsar wind's termination shock (e.g., Venter and de Jager 2007; Zhang et al. 2008; Martín et al. 2012):

$$E_{\max} \equiv \epsilon e \kappa \sqrt{\eta_{\rm B} \frac{\dot{E}_{\rm psr}}{c}}, \tag{8.8}$$

  where e is the charge of the electron and κ is the "magnetic compression ratio ($\kappa \simeq 3$ for strong shocks; e.g., Venter and de Jager 2007). This prescription assumes the highest energy particles in the PWN result from Fermi acceleration at the termination shock, a probable but not certain scenario. Furthermore, the radius of the termination shock $R_{\rm TS}$ depends on both the filling factor of the unshocked pulsar wind emanating from the neutron star magnetosphere and the pressure inside the PWN (e.g., Slane (2017) and references therein). As a result, the derivation of Eq. (8.8) makes assumptions concerning both the nature of the pulsar wind and the evolution of the PWN which may be incorrect.

All of these prescriptions are almost certainly simplifications of reality, since the maximum energy of particles accelerated at the termination shock is likely sensitive to a myriad of different processes (e.g., Sironi et al. 2013). Practically speaking, the chosen evolution of $E_{\max}$ has little impact on a model's ability to reproduce the observed properties of a SED (e.g., the comparison of model fits to G54.1+0.3 presented by Gelfand et al. 2015), but does (somewhat) impact the total number of particles inside the PWN—which has consequences for evaluating models of particle creation in the neutron star magnetosphere.

8.2.2 Evolution of Particles and Magnetic Field Inside a PWN

The evolution of the PWN is driven by the behavior of the particles and magnetic field injected into the nebula at the termination shock. The assumption of uniform pressure P, magnetic field $\mathbf{B}$, and particle distribution (density ρ and velocity $\mathbf{v}$)

throughout the nebula (e.g., Reynolds and Chevalier 1984; Gelfand et al. 2009) inherent in the one-zone approach results in the following constraints:

$$\nabla \rho \equiv 0 \tag{8.9}$$

$$\nabla B \equiv 0 \tag{8.10}$$

$$\nabla P \equiv 0. \tag{8.11}$$

These assumptions, as simple as they may be, are physically motivated. Both the sound and Alfvén speeds inside the PWN are expected to be $\approx c/\sqrt{3}$, suggesting that any significant inhomogeneities in pressure of magnetic field strength would be short lived. In fact, they are supported by a recent three-dimensional (3D) MHD simulation of a PWN, which found the pressure inside the PWN is nearly constant, and that—away from the termination shock—the magnetic field inside the PWN has a primarily random orientation and strength which varies by factors of a few, as shown in Fig. 8.2 (Porth et al. 2014b).

However, as discussed below, there is still room for considerable variations in the treatment of particles and magnetic fields in this framework. Currently, the evolution of the PWN's magnetic field strength is calculated using one of the following approaches:

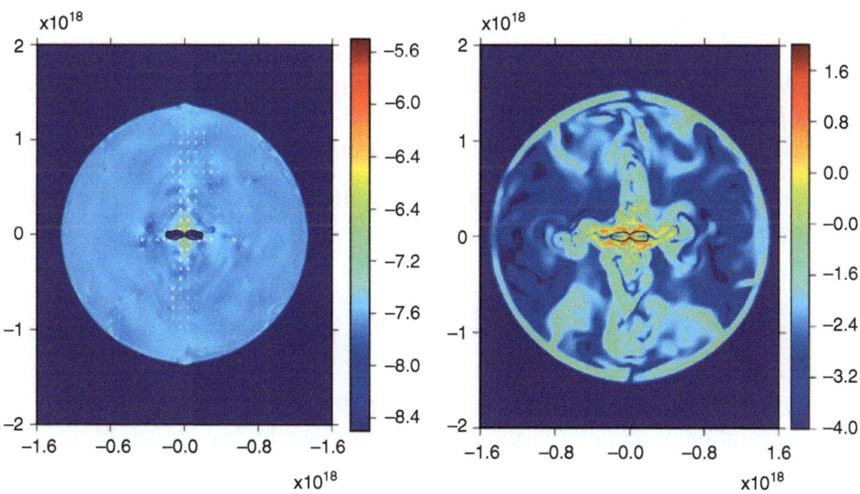

Fig. 8.2 *Left*: The total pressure distribution ($\log_{10} P_{\rm pwn}$) in a 3D MHD simulation of a PWN (Adapted from Fig. 3 of Porth et al. 2014b.) *Right*: The distribution of $\log_{10} E_B/E_{\rm pwn}$ in a 3D MHD simulation of a PWN (Adapted from Fig. 8 of Porth et al. 2014b)

- a conservation of magnetic energy inside the PWN, such that (e.g., Tanaka and Takahara 2010; Martín et al. 2012; Vorster et al. 2013):

$$\int_0^t \eta_B \dot{E}_{psr} dt' \equiv \frac{4\pi}{3} R_{pwn}^3 \times \frac{B^2}{8\pi} \tag{8.12}$$

where R_{pwn} and B are, respectively, the radius of and magnetic field strength inside the PWN at a given time t. In this prescription, at late times (defined to be $t \gg \tau_{sd}$) the magnetic field strength $B_{pwn} \propto t^{-1.3} - t^{-1.5}$ (Tanaka and Takahara 2010; Vorster et al. 2013), in agreement with earlier works (e.g., Reynolds and Chevalier 1984), or
- a conservation of magnetic flux $\Phi_{B,pwn}$ inside the PWN (e.g., Gelfand et al. 2009; Bucciantini et al. 2011)

$$\Phi_{B,pwn} \sim B_{pwn} R_{pwn}^2, \tag{8.13}$$

or $B_{pwn} \propto R_{pwn}^{-2}$. Magnetic energy is *not* conserved in this case, but is $\propto R_{pwn}^{-1}$ when additional input from the pulsar is ignored (Gelfand et al. 2009). If magnetic flux is conserved, the evolution of a PWN's magnetic field is the same as a $\gamma = 4/3$ ideal gas—i.e., the same as the relativistic particles in the PWN. As a result, the ratio of magnetic to particle energy inside the PWN remains constant (e.g. Gelfand et al. 2009; Bucciantini et al. 2011) when adiabatic (as opposed to radiative) processes dominate the evolution of particles inside the PWN—likely to be the case when $t > \tau_{sd}$ (e.g., Gelfand et al. 2009). This is consistent with the MHD steady-state solution for a PWN (Kennel and Coroniti 1984a,b). Under this prescription, $B_{pwn} \propto t^{-1.7}$ when $t \gg \tau_{sd}$—a steeper decline than when the total magnetic energy inside the PWN is conserved.

Both prescriptions result in a magnetic field evolution $B_{pwn}(t)$ very different than the functional form

$$B_{pwn}(t) = \frac{B_0}{1 + \left(\frac{t}{\tau_B}\right)^k} \tag{8.14}$$

used by some authors (e.g., Venter and de Jager 2007; Zhang et al. 2008).

The evolution of particles inside the PWN is driven by two classes of physical mechanisms: radiative processes, which involve the emission of or interaction with a photon, and dynamical processes, which are connected to the individual and bulk motion of particles within the nebula. The relative significance of the radiative and dynamical processes changes significantly with time and particle energy: at earlier times ($t < \tau_{sd}$) and higher particle energies, radiative losses dominate, while at later times and lower particle energies, dynamical processes play a more important role (e.g., Gelfand et al. 2009; Tanaka and Takahara 2010; Martín et al. 2012). However, both must be simultaneously calculated to self-consistently determine both the radiative and dynamical evolution of a PWN.

There is considerable agreement in the literature that the dominant radiative processes are synchrotron and inverse Compton (IC) emission. Synchrotron radiation, which results from the interaction between relativistic electrons and positrons and the PWN's magnetic field, is primarily responsible for the PWN's radio to hard X-ray (photon energy $E_\gamma \lesssim 0.1-1$ MeV) emission. Therefore, calculating the emission from the PWN is these wavebands is somewhat sensitive to the chosen prescription for the nebula's magnetic field as described above.

IC emission, resulting from the interaction between photons and the relativistic electrons and positrons in the PWN, dominate its highest energy (MeV–TeV) photon emission. There are likely three origins for the "target" photons needed for IC emission:

1. the cosmic microwave background (CMB),
2. the interstellar radiation field (ISRF), and
3. the synchrotron radiation from the PWN itself (i.e., synchrotron self-Compton (SSC) emission).

Most models treat the CMB as a blackbody with temperature $T_{\rm CMB} \approx 2.7528$ K, in accordance with latest measurements. The ISRF is thought to have two separate components—far-infrared (FIR), primarily thermal, emission from nearby dust, and near-infrared / optical light from stars. The FIR background is likely the dominant contributor due to Klein-Nishima suppression of scattering of optical photons (e.g., Tanaka and Takahara 2010). Typically, the ISRF is modeled as one or more blackbodies of temperature and energy density (or normalization) chosen to reproduce the observed SED (e.g., Bucciantini et al. 2011; Gotthelf et al. 2014; Slane et al. 2010, 2012; Tanaka and Takahara 2011; Torres et al. 2014). Determining the spectrum of SSC emission is computationally intensive because it requires first calculating the synchrotron emission (e.g., Tanaka and Takahara 2010). Thankfully, the SSC luminosity of a PWN is always less than its synchrotron luminosity, suggesting it is dynamically unimportant, and SSC dominates a PWN's γ-ray emission for a very limited range of parameter space and time (e.g., Torres et al. 2013).

If there are ions in the PWN (whether they enter the nebula through the pulsar wind or through its interaction with the surrounding SNR, other emission mechanisms are possible, even if they are unlikely to dominate a particular waveband (e.g., Bucciantini et al. 2011). For example, proton-proton collisions inside the PWN could produce neutral pions which then decay into two γ-rays ($\pi^0 \rightarrow \gamma + \gamma$; e.g., Bucciantini et al. 2011). Furthermore, the decay of charged pions ($\pi^\pm$) to high-energy leptons would modify the PWN's particle spectrum, affecting their subsequent radiation. Lastly, leptons scattering off any ions in the PWN can result in bremsstrahlung emission (e.g., Martín et al. 2012).

The evolution of particles inside the PWN is also affected by dynamical processes. The most important is adiabatic "losses," which results in (e.g., Martín et al. 2012):

$$\dot{E}_{\rm ad} = -\frac{1}{3}(\nabla \cdot \mathbf{v})E, \tag{8.15}$$

where **v** is the velocity of particles inside the PWN. The spherical symmetry inherent in the one-zone model approximation suggests (e.g., Martín et al. 2012):

$$\mathbf{v}(r,t) = v_{\rm pwn}(t)\left(\frac{r}{R_{\rm pwn}(t)}\right)\hat{r}, \tag{8.16}$$

which results in (e.g., Martín et al. 2012):

$$\dot{E}_{\rm ad}(t) = -\frac{v_{\rm pwn}(t)}{R_{\rm pwn}(t)}E. \tag{8.17}$$

(See Gelfand et al. (2009) for an alternative derivation of this formula from conservation of PV^{γ}, where $\gamma = 4/3$ is the adiabatic index of relativistic particles.)

While several models (e.g., Gelfand et al. (2009), Tanaka and Takahara (2010) and Bucciantini et al. (2011)) only account for adiabatic processes when calculating the dynamical evolution of particles in the PWN, it is possible that diffusion and escape can both substantially affect the evolution of the PWN (e.g. Martín et al. 2012). Modeling diffusion requires determining the (an assumed isotropic) diffusion coefficient D inside the nebula (e.g., Tang and Chevalier 2012; Vorster and Moraal 2013). One way of determining D is to fit ting the change in X-ray photon index Γ as a function of distance from the pulsar inside the PWN. This approach has reproduced the observed spectral variations of a few PWN (e.g., Tang and Chevalier 2012), resulting in diffusion coefficients of $D \sim 10^{26} - 10^{27}\,\mathrm{cm^2\,s^{-1}}$.

An alternate method is calculate D for a particular magnetic field configuration (e.g., Porth et al. 2016). Since the PWN is magnetized, charged particles gyrate around a particular magnetic field line, allowing them to transfer to other field lines typically $\lesssim 1\ r_g$ (where r_g is the gyro-radius) away—a process known as "Bohm diffusion." This typically results in a diffusion coefficient $D^B \sim 10^{26}\,\mathrm{cm^2\,s^{-1}}$ with a diffusion timescale $\tau^B \sim 10^3 - 10^4$ years, suggesting this process is important for older PWNe. However, turbulence inside the nebula can greatly enhance diffusion. Recent 3D simulations of PWNe suggests that such turbulence is dominated by eddies with sizes similar to that of the termination shock, resulting in a diffusion coefficient $D^T \sim 10^{27}\,\mathrm{cm^2\,s^{-1}}$ and timescales of ~300 years (e.g., Porth et al. 2016)—much smaller than the Bohm diffusion timescale above. Such models are also able to reproduce the observed the observed changes in X-ray flux and photon index inside a PWN (e.g., Porth et al. 2016).

Lastly, particles can and do escape from the PWN. The timescale for particle escape $\tau_{\rm esc}$ is often estimated by determining how long it would take particles to travel from the center to the edge of the PWN assuming Bohm diffusion (e.g., Zhang et al. 2008; de Jager and Djannati-Ataï 2009; Tang and Chevalier 2012), resulting in $\tau_{\rm esc} \gtrsim 10^4$ years. This is comparable to the age of older PWNe, and has been used to explain extended emission around Vela X (Hinton et al. 2011). Escape may be enhanced at the edge of the nebula, where particles within a Larmor radius of the outer boundary can gyrate outside the PWN or escape as result of mixing between the PWN and its surroundings driven by Rayleigh-Taylor instabilities (e.g.,

Bucciantini et al. 2004; Porth et al. 2014a). These latter effects have yet to be fully incorporated into one-zone models.

All of these processes, both the radiative and dynamical ones, affect the particle energy inside the PWN. As such, it affects the total pressure (particle + magnetic) inside PWN, which plays a significant role in its interaction with the surrounding SNR described in Sect. 8.2.3.

8.2.3 Interaction Between the PWN and SNR

As mentioned in Sect. 8.1, neutron stars are produced in core-collapse supernovae, the powerful explosions which mark the death of the most massive stars. Therefore, young neutron stars and their PWNe are embedded inside the SNR (Fig. 8.3) resulting from the interaction between the supernova ejecta and the surrounding

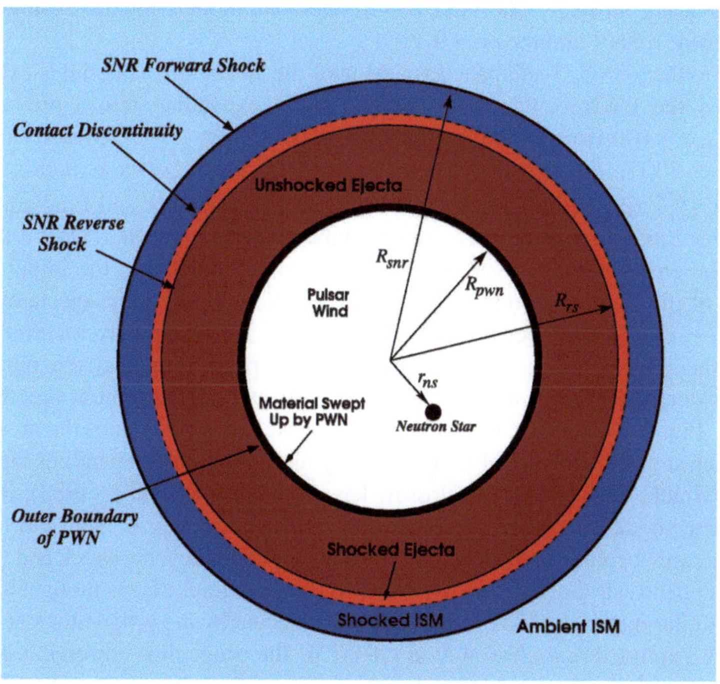

Fig. 8.3 Diagram of a PWN inside a SNR in the free expansion stage of its evolution. In this image, the ratio between the thickness of the mass shell surrounding the PWN and the radius of the PWN is 1/24, as determined by van der Swaluw et al. (2001), and the ratio of the SNR forward shock, contact discontinuity, and reverse shock radii are equal to the values given in Chevalier (1982) for his n = 9, s = 0 case. The *colors* denote the nature of the material within each region. (Reproduction of Fig. 6 in Gelfand et al. (2005), reproduced by permission of the AAS)

interstellar medium (ISM). Therefore, the evolution of the PWN is tightly coupled with that of the surrounding SNR.

When the PWN is young, it is expanding into the inner-most, and therefore slowest moving supernova ejecta. In this case, the density of the surrounding ejecta $\rho_{\rm sn}$ is typically given as:

$$\rho_{\rm sn}(r,t) = At^{-3}\left(\frac{r}{t}\right)^{-m} \tag{8.18}$$

where r is the distance from the explosion center, t is the time, and A is a constant which depends on the properties of the progenitor star and explosion (Blondin et al. (2001), Chevalier (2005) and references therein). The value of m also depends on the properties of the progenitor star: while most PWN evolutionary models assume $m \approx 0$ (e.g., Blondin et al. 2001; Gelfand et al. 2007, 2009), suitable for stars that explode as red super-giants, $m = -1.06$ may be more appropriate for stars which have a radiative envelope when they explode (Chevalier 2005). Most (e.g., Gelfand et al. 2007, 2009; Tanaka and Takahara 2010; Bucciantini et al. 2011; Martín et al. 2012), but not all (e.g., Chevalier 2005), one-zone models for the evolution of a PWN inside a SNR assume $m = 0$.

Due to the $\gamma = 5/3$ adiabatic expansion of the ejecta, this material is extremely cold and the PWN is therefore assumed to be expanding into a pressure-less $P_{\rm snr}(R_{\rm pwn}) = 0$ environment (e.g. Reynolds and Chevalier 1984; Gelfand et al. 2009; Chevalier 2005). During this period, the PWN's radius grows as approximately $R_{\rm pwn} \propto t^{\frac{6-m}{5-m}}$ (e.g., Reynolds and Chevalier 1984; Chevalier and Fransson 1992; Chevalier 2005), though radiative losses might slow this somewhat. Only if the total rotational energy of the pulsar, $E_{\rm psr} \approx \dot{E}_0\tau_{\rm sd}$, is comparable to the initial kinetic energy of the supernova ejecta $E_{\rm sn}$ will the PWN reach the outer ejecta envelope $\rho \propto r^{-9} - r^{-11}$ (Blondin et al. 2001; Chevalier 2005)—an unlikely occurrence.

As the PWN expands into the inner-most supernova ejecta, the outer-most supernova ejecta sweeps up and shocks the surrounding material at the "forward shock" (Fig. 8.3; Chevalier 1982; Truelove and McKee 1999). Since the pressure of the shocked external material is significantly higher than the expanding supernova ejecta (which, due to adiabatic losses, has $P_{\rm snr} \approx 0$; Chevalier 1982), it drives a "reverse shock" into the remnant (Fig. 8.3). The evolution of both the location and pressure "behind" (i.e., downstream, or between the reverse shock and forward shock) depends strongly on its environment. At early times, this is thought to be the circumstellar material (CSM) shaped by mass ejected by the progenitor star shortly before it exploded (e.g., Yaron et al. 2017). If the progenitor powered a constant stellar wind, this will result in a density $\rho_{\rm csm} \propto r^{-2}$ environment (e.g., Chevalier 1982) though reality is likely more complicated (e.g., van Marle et al. 2012). At later times, the SNR is likely expanding into the surrounding ISM, which is assumed to have a uniform density ($\rho_{\rm ism} \propto r^0$) on the relevant size scales. Where this transition in external medium occurs, and the effect it has on the structure and evolution of the SNR, is poorly understood (e.g., Truelove and McKee 1999) and likely sensitive to the particulars of the progenitor and its immediate surroundings—and therefore ignored in most one-zone models.

In any case, eventually the PWN and reverse shock will meet. At this point, the PWN is no longer inside an essentially pressure-less environment, but now surrounded by hot gas. The much higher pressure outside the PWN than inside, not surprisingly, significantly impacts the dynamical evolution of the PWN (e.g., Reynolds and Chevalier 1984). In fact, simulations suggest that the reverse-shock heated material will crush the PWN, causing a rapid decrease in radius which results in a significant increase in the pressure and magnetic field strength inside the PWN (e.g., Blondin et al. 2001; van der Swaluw et al. 2001, 2003; Bucciantini et al. 2003). Eventually, the pressure in the PWN will be higher than that of the surrounding SNR, causing it to re-expand (e.g., Blondin et al. 2001; van der Swaluw et al. 2001; Bucciantini et al. 2003). Due to the considerable difference in mass density inside ($\rho_{\rm pwn} \approx 0$) and outside ($\rho_{\rm snr}(R_{\rm pwn}) \gg 0$) the PWN, at this point the shell of material surrounding the PWN is highly susceptible to Raleigh-Taylor instabilities, which may disrupt the PWN (e.g., Bucciantini et al. 2004; Gelfand et al. 2009).

Currently, one-zone models are the *only* technique which self-consistently determine the radiative and dynamical properties of the PWN before and after it collides with the SNR reverse shock (e.g., Gelfand et al. 2007, 2009; Martín et al. 2016). The PWN's dynamical evolution is generally determined by calculating the motion of the surrounding shell of swept-up material, neglecting the dynamical effects of any plasma instabilities on this shell (e.g., Bucciantini et al. 2004; Porth et al. 2014a). The equations of motion for this shell are (e.g., Gelfand et al. 2007, 2009; Martín et al. 2016[2]):

$$\frac{dR_{\rm sw}}{dt} = v_{\rm pwn} \tag{8.19}$$

$$\frac{dM_{\rm sw}}{dt} = \begin{cases} 4\pi R_{\rm pwn}^2 \rho_{\rm snr}(R_{\rm pwn}) \left[v_{\rm pwn} - v_{\rm snr}(R_{\rm pwn})\right] & v_{\rm pwn} > v_{\rm snr}(R_{\rm pwn}) \\ 0 & v_{\rm pwn} \leq v_{\rm snr}(R_{\rm pwn}) \end{cases} \tag{8.20}$$

$$\frac{d(M_{\rm sw} v_{\rm pwn})}{dt} = 4\pi R_{\rm pwn}^2 [P_{\rm pwn} - P_{\rm snr}(R_{\rm pwn})]. \tag{8.21}$$

where $R_{\rm sw}(\equiv R_{\rm pwn})$ is the radius of the shell (assumed to be the radius of the PWN $R_{\rm pwn}$), $v_{\rm pwn}$ is the velocity of the shell (assumed to be expansion velocity of the PWN), $M_{\rm sw}$ is the mass of the swept-up material, $P_{\rm pwn}$ is the total pressure inside the PWN, equal to (e.g., Gelfand et al. 2009):

$$P_{\rm pwn} = \frac{B_{\rm pwn}^2}{8\pi} + \frac{E_{\rm p}}{4\pi R_{\rm pwn}^3} \tag{8.22}$$

[2]There are subtle differences between the equations used in these works and the ones below, which do not significantly impact the final results.

where E_p is the particle energy inside the PWN (calculated using the procedures described in Sects. 8.2.1 and 8.2.2), and $\rho_{snr}(R_{pwn})$, $v_{snr}(R_{pwn})$, and $P_{snr}(R_{pwn})$ are respectively the density, velocity, and pressure of material inside the SNR just outside the PWN.

These last quantities are typically calculated using analytical models for the structure and evolution of a SNR (e.g., Chevalier 1982; Bandiera 1984; Truelove and McKee 1999), ignoring any impact the PWN might have on the evolution of the SNR—a safe assumption, given that the initial rotational energy of the pulsar ($E_{psr} \lesssim 10^{49} - 10^{50}$ ergs in most cases) is likely much smaller than the initial kinetic energy of the supernova ejecta ($E_{sn} \sim 10^{51}$ ergs). These descriptions for the evolution of the SNR depend on the mass M_{ej} and initial kinetic energy E_{sn} of the supernova ejecta, as well as the initial ejecta density profile (in particular, the value of the m quantity defined in Eq. (8.18)) and the structure of the surrounding medium which together determine when the PWN collides with the reverse shock. As stated earlier, most (e.g., Gelfand et al. 2009; Martín et al. 2016) PWN models assume a constant density ($m = 0$) ejecta core and a $\rho \propto r^{-9}$ ejecta envelope and the SNR is expanding into a constant density ISM, with (Chevalier 2005) a notable exception. This prescription, while simplistic, reproduces the results of more complicated simulations as shown in Fig. 8.4.

Not incorporated into one-zone models are asymmetries in the interaction between the PWN and reverse shock resulting from the spatial ("kick") velocity of the neutron star (Fig. 8.3; van der Swaluw et al. 2004) and gradients in the surrounding ISM (Temim et al. 2015). In fact, such asymmetries are required to explain the radio and X-ray morphologies of some systems (e.g. Temim et al. 2015). Furthermore, to date, one-zone models do not account for the eventual passage of the pulsar out of the SNR. As the neutron (and presumably its PWN) approach the SNR shell, the pulsar will begin traveling supersonically relative to its local environment

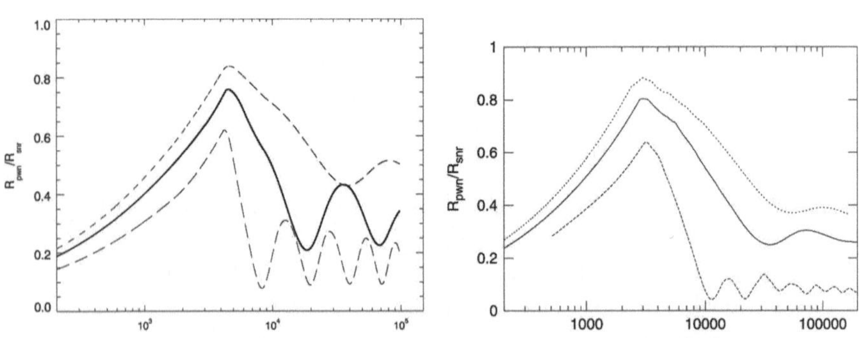

Fig. 8.4 R_{pwn}/R_{snr} for models A (*solid line*), B (*long-dashed line*), and C (*short-dashed line*) in Blondin et al. (2001). The left panel shows the result of the model presented in Gelfand et al. (2007), while the bottom panel is a reproduction of Fig. 3 in Blondin et al. (2001) reproduced by permission of the AAS. This figure is a reproduction of Fig. 7 in Gelfand et al. (2007), reproduced by permission of the AAS

(van der Swaluw et al. 2004). In this case, the pulsar wind will not be confined by the pressure of the surrounding gas, but by the ram pressure associated with this motion, resulting in a bow-shock nebula (van der Swaluw et al. 2004).

8.3 Results and Applications of One-Zone Models

As described in Sects. 8.2.1–8.2.3, current one-zone models self-consistently predict the dynamical (e.g., size and expansion rate) and radiative (e.g., broadband SED) properties of a PWN inside a SNR accounting for:

- the time-variable injection of energy into the PWN by the neutron star, which depends on its initial spin-down luminosity $\dot{E}_0$, spin-down timescale τ_{sd}, and braking index p,
- the magnetization of the pulsar wind η_B,
- the spectrum of particles injected into the PWN at the termination shock (which typically involves five parameters: E_{min}, E_{break}, E_{max}, p_1, and p_2 as described in Sect. 8.2.1),
- and the interaction between the PWN and surrounding SNR, which is sensitive to the mass M_{ej} and initial kinetic energy E_{sn} of the supernova ejecta as well as the density of the surrounding ISM n_{ism}.

Such models have been applied by various authors to $\sim$15 PWNe, with different analyses of the same system generally giving similar results (e.g., G54.1+0.3; Gelfand et al. 2015). For pretty much all systems modeled to date, the best fits require $\eta_B \ll 1$, or a particle dominated wind injected into the PWN at the termination shock. Not only is this hard to reconcile with the magnetic-dominated wind expected to originate from the neutron star magnetosphere (the so-called "sigma" problem; see Porth et al. (2013) for a discussion and possible solution), this requires a rate of particle creation in the neutron star difficult to explain (Timokhin and Harding 2015).

The (relative) simplicity of one-zone models and the power of modern computers makes it feasible to identify different combinations of the $\gtrsim$10 model parameters which reproduce the observed properties of a particular system. Not surprisingly, such searches of the multi-dimensional parameter space find considerable degeneracies between some model parameters. Some of these are not surprising: the self-similarity of SNR evolution models (e.g., Chevalier 1982) effectively couples the values of E_{sn}, M_{ej}, and n_{ism} (Fig. 8.5; e.g., Gelfand et al. 2014), while the spin-down equation for pulsars (Eq. (8.1)) results in a dependence between p and τ_{sd} (e.g., Gelfand et al. 2014, 2015; Gotthelf et al. 2014). However, perhaps less obvious is the correlation between the distance d to the PWN and wind magnetization η_B seen in analyses of two young PWNe: G54.1+0.3 (Gelfand et al. 2014) and Kes 75 (Fig. 8.5; Gelfand et al. 2015). Such searches can also provide statistically significant constraints on the values of model parameters. For example, the analysis of G54.1+0.3 mentioned above constrained the supernova ejecta mass

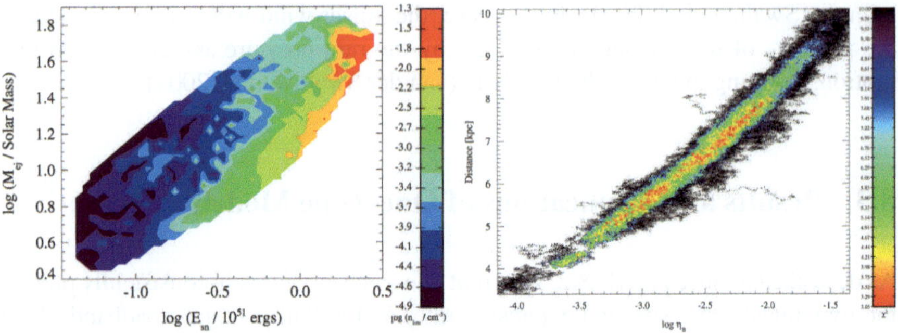

Fig. 8.5 *Left*: ISM density $n_{\rm ism}$ (*color scale*) for different values of the initial kinetic energy $E_{\rm sn}$ and mass $M_{\rm ej}$ of the supernova ejecta for parameter combinations which reproduce the observed properties of G54.1+0.3 (Fig. 2 from Gelfand et al. (2015), reproduced with permission from AAS). *Right*: Trial values of the pulsar wind magnetization $\eta_{\rm B}$ and distance d to Kes 75. The colors indicate the χ^2 of the trial, with *red indicating a better fit* and *black indicating a worse fit* (Fig. 4 from Gelfand et al. 2014)

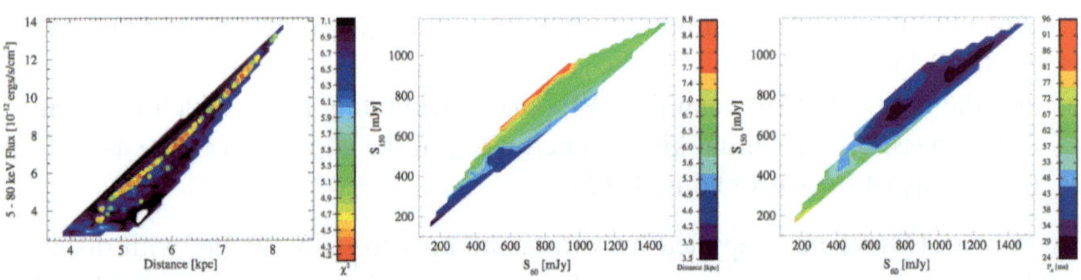

Fig. 8.6 *Left*: χ^2 for trials with different distances d and predicted $5-80$ keV flux of G54.1+0.3, with red signifying a lower χ^2 (better fit) and black a higher χ^2 (worse fit). The *dots indicate trials* with $\chi^2 \leq 5.10$ and are included to better demonstrate the degeneracy between these two parameters. (Reproduction of Fig. 12 in Gelfand et al. (2015), reproduced by permission of the AAS.) *Middle and Right*: Distance d to G54.1+0.3 (*middle*) and initial spin period P_0 of PSR J1930+1852 (*right*) for the values of the 60 MHz S_{60} and 150 MHz S_{150} flux densities. (Reproduction of Fig. 12 in Gelfand et al. (2015), reproduced by permission of the AAS)

$M_{\rm ej} \gtrsim 10\,M_\odot$—which implies a $\sim 15-20\,M_\odot$ progenitor star (Gelfand et al. 2015), conclusion supported by a recent, independent analysis of the IR shell around this PWN (Temim et al. 2017).

Furthermore, such parameter exploration not only predicts the values of additional observables, but determine the importance of their measurement. For example, a recent study of G54.1+0.3 not only predicted the low frequency radio flux density and hard X-ray flux and photon index of this source, but indicated that measuring these quantities will significantly constrain its distance and the initial spin period of its associated pulsar PSR J1930+1852 (Fig. 8.6; Gelfand et al. 2015). This style of analysis is especially useful in trying to determine the advances which will result from new observing facilities such as the Square Kilometer Array (Gelfand et al. 2015), *ATHENA*, and the Cerenkov Telescope Array (Antoniadis et al. 2015) among others.

Additionally, as mentioned in Sect. 8.2.3, one-zone models apply both before and after the PWN collides with the SNR reverse shock (e.g., Reynolds and Chevalier 1984; Gelfand et al. 2007, 2009; Martín et al. 2016). Therefore, they can be used to consistently analyze a large number of systems, allowing one to search for connections between the properties of the neutron star, pulsar wind, and progenitor star and supernova explosion. For example, the analysis by Chevalier (2005) of $\sim$10 PWNe found no connection between the properties of the central neutron star (e.g., initial spin period P_0) and the type of supernova explosion. The modeling of four PWNe by Tanaka and Takahara (2011) identified a possible anti-correlation between the initial rotational energy of the pulsar E_{psr} and (spin-down inferred) dipole surface magnetic field strength. The study of six PWNe by Bucciantini et al. (2011) found a very similar break energy E_{break} in particle spectrum injected into the PWN at the termination shock despite a wide range in characteristic age τ_c and spin-down luminosity $\dot{E}$ of the associated pulsars—suggesting a similar acceleration mechanism despite the physical differences between these systems. The largest such study to date was by Torres et al. (2014), who analyzed nine PWNe both securely associated with a "young" ($\tau_c \lesssim 10^4$ years) radio pulsar and detected at TeV energies. This analysis also found a small spread in E_{break} between the different systems, supporting the notion of a common acceleration mechanism, and that no statistically significant correlation between most parameters—except for the maximum particle energy E_{max} and spin-down luminosity $\dot{E}_{psr}$, though this may be a consequence of using Eq. (8.8) to determine the evolution of E_{max} in their model (Fig. 8.7; Torres et al. 2014). As new observing facilities come online, the number of PWNe suited for such an analysis will only grow, allowing for more concrete conclusions for the connection—or lack thereof—between the properties of a neutron star and both its progenitor star and explosion and pulsar wind.

Last, but not least, one-zone models are also useful in interpreting the nature of extended (primarily X-ray) emission associated with magnetars. A class of isolated neutron stars with photon luminosity greater than their spin-down luminosity ($L_\gamma \gg \dot{E}_{psr}$), their emission is primarily powered by the decay of an extremely strong surface magnetic field (Thompson and Duncan 1996). This class of neutron stars can generate an outflow of relativistic particles several different ways: through a rotation-powered mechanism similar to that of "normal" pulsars, as suggested by the detection of pulsed radio emission from some members (e.g., Rea et al. 2012), the decay of the surface magnetic field (e.g., Thompson and Blaes 1998), or as a consequence of giant flares (e.g., Frail et al. 1999; Gaensler et al. 2005; Gelfand et al. 2005). As such, it is possible to use one zone models to predict the properties of the "magnetar wind nebula" (MWN) powered by such a neutron star (Tanaka 2016), modifying the rate the magnetar injects energy into its MWN to account for the non-rotation powered processes listed above.

Currently, the most convincing candidate for a MWN is the diffuse X-ray emission detected around magnetar SWIFT J1834.9−0846 (Younes et al. 2016). Unlike other extended X-ray sources observed around magnetars, its hard X-ray spectrum and constant X-ray flux despite significant variability in the magnetar's X-ray flux suggests it is not the result of X-rays scattering off dust along the line

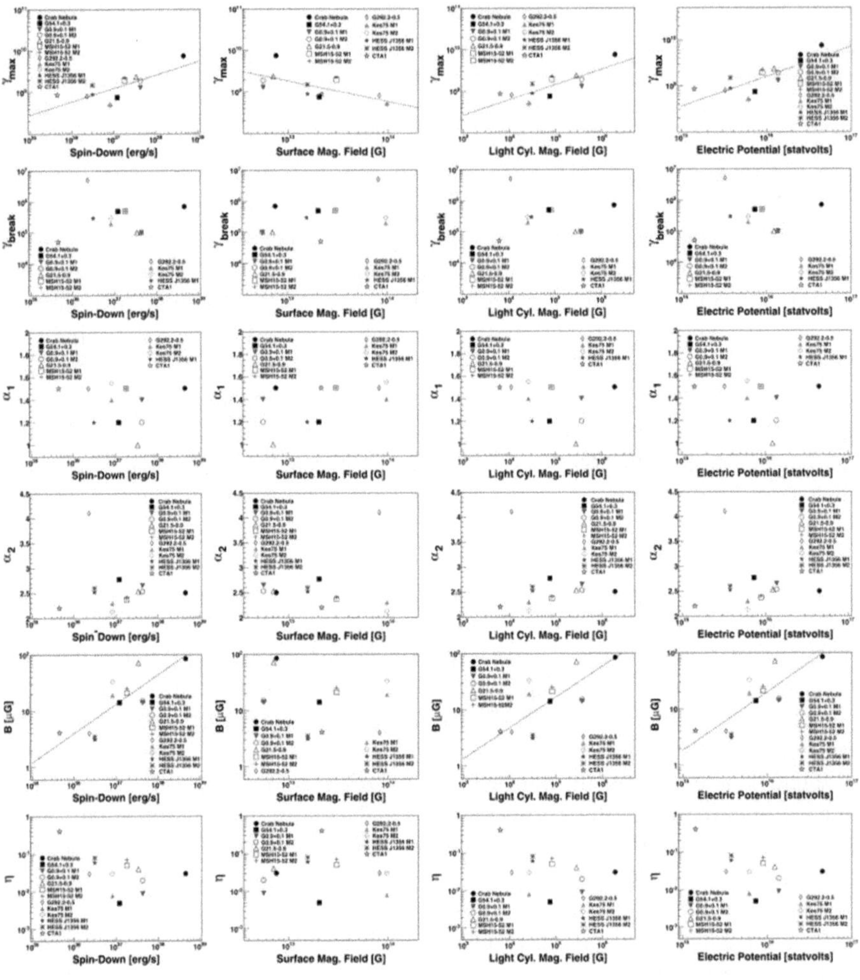

Fig. 8.7 PWNe properties in the y-axis of all plots as a function of pulsar properties in the x-axis of all plots. The values of all magnitudes refer to the current time. From *left to right*, we plot the obtained values of $\gamma_{\rm max}$ ($E_{\rm max} = m\gamma_{\rm max}c^2$), $\gamma_{\rm b}$ ($E_{\rm break} = m\gamma_{\rm b}c^2$), α_1 (p_1), α_2 (p_2), $B_{\rm pwn}(t_{\rm age})$, and $\eta_{\rm B}$, as a function of (*from top to bottom*) spin-down, surface magnetic field, light cylinder magnetic field, and potential. (Figure 20 from Torres et al. 2014)

of sight (Younes et al. 2016). This magnetar is also located within SNR W41, and coincident with an extended TeV (Aharonian et al. 2006b) and GeV (Nolan et al. 2012) γ-ray source of dimensions comparable to the remnant.

Recently, there have been three separate analyses of this source, all trying to determine if the X-ray (and γ-ray) emission can be powered by the rotational energy of this magnetar. Two of these analyses conclude another energy source is required—either wind-braking (Tong 2016) or by other outflows possibly associated with the magnetar's bursting activity (Granot et al. 2017). The analysis by Tong

(2016) comes to this conclusion, in large part, by assuming equipartition between the magnetic field and particles inside the PWN—a condition not yet observed from in any PWN (e.g., Tanaka and Takahara 2011; Bucciantini et al. 2011; Gelfand et al. 2014; Torres et al. 2014; Gelfand et al. 2015). Granot et al. (2017) reaches this conclusion assuming the highest energy particles injected by the magnetar into its MWN is set by the electric potential of its magnetosphere ($E_{max} \equiv e\Phi$ as defined in Eq. (8.7)). Under this assumption, the magnetar can not currently produce particles responsible for its hard X-ray emission, and radiative and adiabatic losses have eliminated any such particles injected earlier in its history (Granot et al. 2017). However, Torres (2017) is able to reproduce both the X-ray and γ-ray emission from this assuming that are emitted by a "normal" PWN which is currently being compressed by the reverse shock in SNR W41. While additional data and modeling are needed to determine which interpretation, the fact that all three analyses essentially use one-zone models similar (or, in the case of Torres (2017), practically identical) to those described in Sects. 8.2.1 and 8.2.2 points to the power of this technique.

8.4 Summary and Possible Future Developments in Modeling PWNe

As hopefully demonstrated above, one-zone models for the evolution of a PWN inside a SNR are an extremely powerful tool for understanding these sources, in particular for measuring the birth properties of the neutron star—which relates to its formation during a core-collapse supernova, the magnetization and particle spectrum of its pulsar wind—which relates to the generation and acceleration of particles in this outflow, and the properties of its associated supernova and environment—which provides important information concerning its progenitor star.

As described in Sects. 8.2.1 and 8.2.2, while the various one-zone models are built upon the same general principles, there can differ significantly in the details—especially in their treatment of E_{max} and the escape of particles from the PWN. While Martín et al. (2012) details the difference in PWN evolution which can arise from differing assumptions (Fig. 8.8), a better understanding of the systematic errors associated with this approach is sorely needed. This, in part, can be accomplished by comparing the evolution predicted by one-zone models and more complicated 2D and 3D MHD simulations. Unfortunately, such a comparison is not currently possible since it is unfeasible to run these multi-dimensional simulations over the entire lifetime of known PWNe. However, it may be possible to incorporate into one-zone models more "realistic" prescriptions for particle injection into and escape from the PWN as well as the magnetic and pressure structure of the nebula derived from multi-dimensional simulations (e.g., Olmi et al. 2014; Porth et al. 2014a,b; Sironi and Spitkovsky 2014). Additionally, another important extension of one-zone models is to somehow incorporate the almost certainly asymmetric interaction

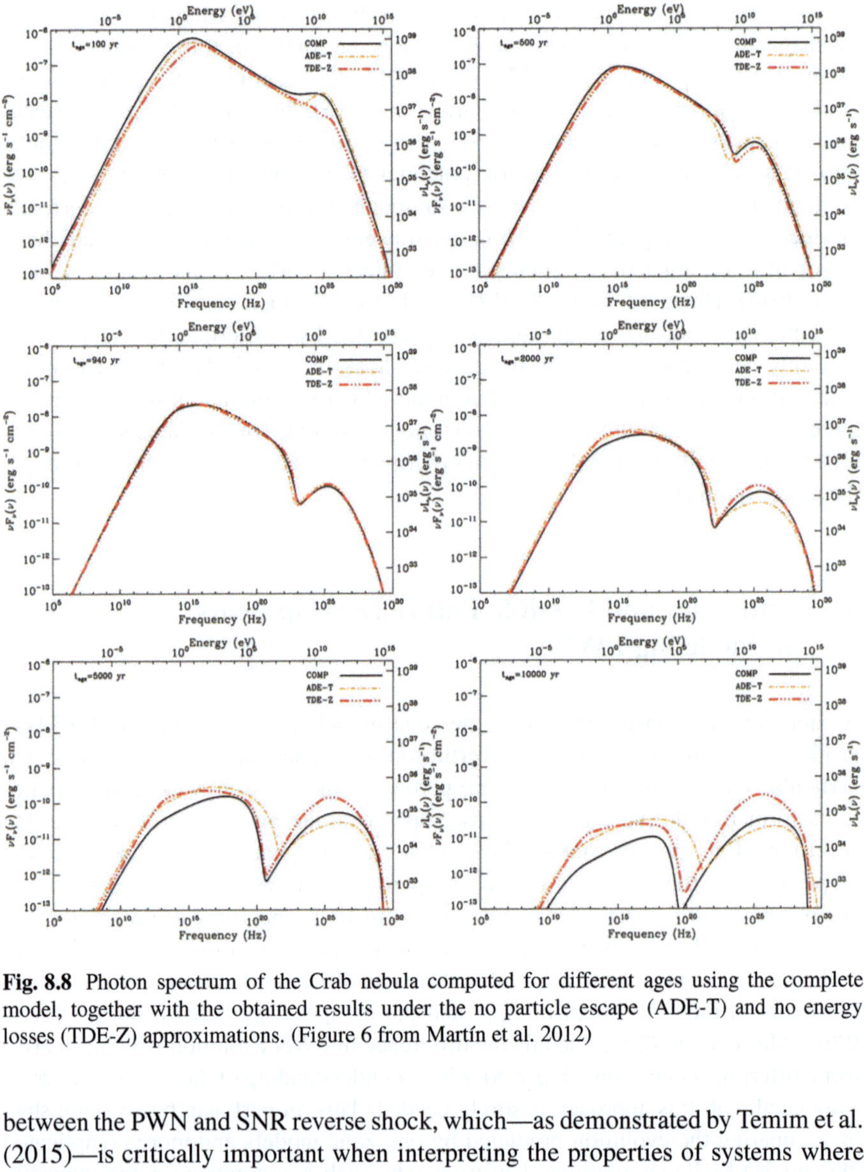

Fig. 8.8 Photon spectrum of the Crab nebula computed for different ages using the complete model, together with the obtained results under the no particle escape (ADE-T) and no energy losses (TDE-Z) approximations. (Figure 6 from Martín et al. 2012)

between the PWN and SNR reverse shock, which—as demonstrated by Temim et al. (2015)—is critically important when interpreting the properties of systems where this collision has already occurred.

Lastly, new observing facilities have the potential of revolutionizing this field. Currently, there are only ~10–20 PWNe suitable for the detailed analyses described in Sect. 8.3. As a result, it is very difficult to draw strong conclusions concerning the relationship, or lack thereof, between the properties of the progenitor star and supernova, neutron star, and pulsar wind. Not only will the improved sensitivities promised by the Square Kilometer Array, *ATHENA*, and the Cerenkov Telescope Array likely lead to a significant number of PWN discoveries, the improved measurements of the radio, X-ray, and γ-ray properties of known PWNe will significantly increase the number of systems suitable for the detailed analyses (including a full exploration of the possible parameter space) described in Sect. 8.3.

In summary, while the design and implementation of one-zone models for the evolution of a PWN inside a SNR still requires (possibly unrealistically simple) assumptions about the underlying physics, they remain the most powerful tool for interpreting the observed properties of known systems. Furthermore, their ability to predict the results of new observations make them the only verifiable method of analyzing these sources, and provides a powerful test of their validity. Future improvements in our theoretical understanding of these sources will only help make one-zone models more realistic, while the discovery and detailed study of additional systems will only make the various conclusions drawn by fitting the observed properties of large numbers of PWNe with one-zone models more secure. Despite the age ($\gtrsim$30 years; Reynolds and Chevalier 1984) of this technique, one-zone models still have much to offer.

Acknowledgements I would like to thank Diego Torres for both the invitation to write this chapter, and his patience with its submission. I would also like to thank my collaborators for their patience and many useful discussions on this subject. JDG acknowledges the support of the NYU Abu Dhabi Research Enhancement Fund under grant RE022, and the NASA Astrophysical Data Analysis Program under grant 16-ADAP6-0095.

References

Abdo, A.A., Ajello, M., Allafort, A., Baldini, L., Ballet, J., Barbiellini, G., Baring, M.G., Bastieri, D., Belfiore, A., Bellazzini, R., et al.: The second Fermi large area telescope catalog of gamma-ray pulsars. Astrophys. J. Suppl. Ser. **208**, 17 (2013). doi:10.1088/0067-0049/208/2/17

Aharonian, F., Akhperjanian, A.G., Bazer-Bachi, A.R., Beilicke, M., Benbow, W., Berge, D., Bernlöhr, K., Boisson, C., Bolz, O., Borrel, V., Braun, I., Brown, A.M., Bühler, R., Büsching, I., Carrigan, S., Chadwick, P.M., Chounet, L.-M., Cornils, R., Costamante, L., Degrange, B., Dickinson, H.J., Djannati-Ataï, A., O'C. Drury, L., Dubus, G., Egberts, K., Emmanoulopoulos, D., Espigat, P., Feinstein, F., Ferrero, E., Fiasson, A., Fontaine, G., Funk, S., Funk, S., Füßling, M., Gallant, Y.A., Giebels, B., Glicenstein, J.F., Goret, P., Hadjichristidis, C., Hauser, D., Hauser, M., Heinzelmann, G., Henri, G., Hermann, G., Hinton, J.A., Hoffmann, A., Hofmann, W., Holleran, M., Horns, D., Jacholkowska, A., de Jager, O.C., Kendziorra, E., Khélifi, B., Komin, N., Konopelko, A., Kosack, K., Latham, I.J., Le Gallou, R., Lemière, A., Lemoine-Goumard, M., Lohse, T., Martin, J.M., Martineau-Huynh, O., Marcowith, A., Masterson, C., Maurin, G., McComb, T.J.L., Moulin, E., de Naurois, M., Nedbal, D., Nolan, S.J., Noutsos, A., Orford, K.J., Osborne, J.L., Ouchrif, M., Panter, M., Pelletier, G., Pita, S., Pühlhofer, G., Punch, M., Raubenheimer, B.C., Raue, M., Rayner, S.M., Reimer, A., Reimer, O., Ripken, J., Rob, L., Rolland, L., Rowell, G., Sahakian, V., Santangelo, A., Saugé, L., Schlenker, S., Schlickeiser, R., Schröder, R., Schwanke, U., Schwarzburg, S., Shalchi, A., Sol, H., Spangler, D., Spanier, F., Steenkamp, R., Stegmann, C., Superina, G., Tavernet, J.-P., Terrier, R., Théoret, C.G., Tluczykont, M., van Eldik, C., Vasileiadis, G., Venter, C., Vincent, P., Völk, H.J., Wagner, S.J., Ward, M.: Energy dependent γ-ray morphology in the pulsar wind nebula HESS J1825-137. Astron. Astrophys. **460**, 365–374 (2006a). doi:10.1051/0004-6361:20065546

Aharonian, F., Akhperjanian, A.G., Bazer-Bachi, A.R., Beilicke, M., Benbow, W., Berge, D., Bernlöhr, K., Boisson, C., Bolz, O., Borrel, V., Braun, I., Breitling, F., Brown, A.M., Chadwick, P.M., Chounet, L.-M., Cornils, R., Costamante, L., Degrange, B., Dickinson, H.J., Djannati-Ataï, A., Drury, L.O., Dubus, G., Emmanoulopoulos, D., Espigat, P., Feinstein, F., Fontaine, G., Fuchs, Y., Funk, S., Gallant, Y.A., Giebels, B., Gillessen, S., Glicenstein, J.F., Goret, P., Hadjichristidis, C., Hauser, M., Heinzelmann, G., Henri, G., Hermann, G., Hinton, J.A., Hofmann, W., Holleran, M., Horns, D., Jacholkowska, A., de Jager, O.C., Khélifi, B., Komin, N., Konopelko, A., Latham, I.J., Le Gallou, R., Lemière, A., Lemoine-Goumard, M., Leroy, N., Lohse, T., Martin, J.M., Martineau-Huynh, O., Marcowith, A., Masterson, C., McComb, T.J.L., de Naurois, M., Nolan, S.J., Noutsos, A., Orford, K.J., Osborne, J.L., Ouchrif, M., Panter, M., Pelletier, G., Pita, S., Pühlhofer, G., Punch, M., Raubenheimer, B.C., Raue, M., Raux, J., Rayner, S.M., Reimer, A., Reimer, O., Ripken, J., Rob, L., Rolland, L., Rowell, G., Sahakian, V., Saugé, L., Schlenker, S., Schlickeiser, R., Schuster, C., Schwanke, U., Siewert, M., Sol, H., Spangler, D., Steenkamp, R., Stegmann, C., Tavernet, J.-P., Terrier, R., Théoret, C.G., Tluczykont, M., Vasileiadis, G., Venter, C., Vincent, P., Völk, H.J., Wagner, S.J.: The H.E.S.S. survey of the inner galaxy in very high energy gamma rays. Astrophys. J. **636**, 777–797 (2006b). doi:10.1086/498013

Antoniadis, J., Guillemot, L., Possenti, A., Bogdanov, S., Gelfand, J., Kramer, M., Mignani, R., Stappers, B., Torne, P.: Multi-wavelength, multi-messenger pulsar science in the SKA Era. In: Advancing Astrophysics with the Square Kilometre Array (AASKA14), p. 157 (2015)

Atoyan, A.M.: Radio spectrum of the Crab nebula as an evidence for fast initial spin of its pulsar. Astron. Astrophys. **346**, 49–52 (1999)

Bandiera, R.: Convective supernovae. Astron. Astrophys. **139**, 368–374 (1984)

Bird, R., for the VERITAS Collaboration: Raster scanning the Crab nebula to produce an extended VHE calibration source. ArXiv e-prints (2015)

Blondin, J.M., Chevalier, R.A., Frierson, D.M.: Pulsar wind nebulae in evolved supernova remnants. Astrophys. J. **563**, 806–815 (2001). doi:10.1086/324042

Bucciantini, N.: Review of the theory of pulsar-wind nebulae. Astron. Nachr. **335**, 234–239 (2014). doi:10.1002/asna.201312024

Bucciantini, N., Arons, J., Amato, E.: Modelling spectral evolution of pulsar wind nebulae inside supernova remnants. Mon. Not. R. Astron. Soc. **410**, 381–398 (2011). doi:10.1111/j.1365-2966.2010.17449.x

Bucciantini, N., Blondin, J.M., Del Zanna, L., Amato, E.: Spherically symmetric relativistic MHD simulations of pulsar wind nebulae in supernova remnants. Astron. Astrophys. **405**, 617–626 (2003). doi:10.1051/0004-6361:20030624

Bucciantini, N., Amato, E., Bandiera, R., Blondin, J.M., Del Zanna, L.: Magnetic Rayleigh-Taylor instability for pulsar wind nebulae in expanding supernova remnants. Astron. Astrophys. **423**, 253–265 (2004). doi:10.1051/0004-6361:20040360

Carrigan, S., Brun, F., Chaves, R.C.G., Deil, C., Gast, H., Marandon, V., for the H. E. S. S. Collaboration: Charting the TeV Milky way: H.E.S.S. galactic plane survey maps, catalog and source populations. ArXiv e-prints (2013)

Chevalier, R.A.: Self-similar solutions for the interaction of stellar ejecta with an external medium. Astrophys. J. **258**, 790–797 (1982). doi:10.1086/160126

Chevalier, R.A.: Young core-collapse supernova remnants and their supernovae. Astrophys. J. **619**, 839–855 (2005). doi:10.1086/426584

Chevalier, R.A., Fransson, C.: Pulsar nebulae in supernovae. Astrophys. J. **395**, 540–552 (1992). doi:10.1086/171674

de Jager, O.C., Djannati-Ataï, A.: Implications of HESS observations of pulsar wind nebulae. In: Becker, W. (ed.) Astrophysics and Space Science Library, vol. 357, p. 451. Springer, Berlin (2009). doi:10.1007/978-3-540-76965-1_17

de Jager, O.C., Slane, P.O., LaMassa, S.: Probing the radio to X-ray connection of the vela X pulsar wind nebula with fermi LAT and H.E.S.S. Astrophys. J. **689**, 125 (2008). doi:10.1086/595959

Fang, J., Zhang, L.: Multiband emission from pulsar wind nebulae: a possible injection spectrum. Astron. Astrophys. **515**, 20 (2010). doi:10.1051/0004-6361/200913615

Frail, D.A., Kulkarni, S.R., Bloom, J.S.: An outburst of relativistic particles from the soft γ-ray repeater SGR1900+14. Nature **398**, 127–129 (1999). doi:10.1038/18163

Gaensler, B.M., Slane, P.O.: The evolution and structure of pulsar wind nebulae. Annu. Rev. Astron. Astrophys. **44**, 17–47 (2006). doi:10.1146/annurev.astro.44.051905.092528

Gaensler, B.M., Kouveliotou, C., Gelfand, J.D., Taylor, G.B., Eichler, D., Wijers, R.A.M.J., Granot, J., Ramirez-Ruiz, E., Lyubarsky, Y.E., Hunstead, R.W., Campbell-Wilson, D., van der Horst, A.J., McLaughlin, M.A., Fender, R.P., Garrett, M.A., Newton-McGee, K.J., Palmer, D.M., Gehrels, N., Woods, P.M.: An expanding radio nebula produced by a giant flare from the magnetar SGR 1806-20. Nature **434**, 1104–1106 (2005). doi:10.1038/nature03498

Gelfand, J.D., Slane, P.O., Temim, T.: The properties of the progenitor, neutron star, and pulsar wind in the supernova remnant Kes 75. Astron. Nachr. **335**, 318–323 (2014). doi:10.1002/asna.201312039

Gelfand, J.D., Slane, P.O., Temim, T.: The properties of the progenitor supernova, pulsar wind, and neutron star inside PWN G54.1+0.3. Astrophys. J. **807**, 30 (2015). doi:10.1088/0004-637X/807/1/30

Gelfand, J.D., Slane, P.O., Zhang, W.: A dynamical model for the evolution of a pulsar wind nebula inside a nonradiative supernova remnant. Astrophys. J. **703**, 2051–2067 (2009). doi:10.1088/0004-637X/703/2/2051

Gelfand, J.D., Lyubarsky, Y.E., Eichler, D., Gaensler, B.M., Taylor, G.B., Granot, J., Newton-McGee, K.J., Ramirez-Ruiz, E., Kouveliotou, C., Wijers, R.A.M.J.: A rebrightening of the radio nebula associated with the 2004 December 27 giant flare from SGR 1806-20. Astrophys. J. **634**, 89–92 (2005). doi:10.1086/498643

Gelfand, J.D., Gaensler, B.M., Slane, P.O., Patnaude, D.J., Hughes, J.P., Camilo, F.: The radio emission, X-ray emission, and hydrodynamics of G328.4+0.2: a comprehensive analysis of a luminous pulsar wind nebula, its neutron star, and the progenitor supernova explosion. Astrophys. J. **663**, 468–486 (2007). doi:10.1086/518498

Gelfand, J., Breton, R., Ng, C.-Y., Hessels, J., Stappers, B., Roberts, M., Possenti, A.: Pulsar wind nebulae in the SKA era. In: Advancing Astrophysics with the Square Kilometre Array (AASKA14), p. 46 (2015)

Goldreich, P., Julian, W.H.: Pulsar electrodynamics. Astrophys. J. **157**, 869 (1969). doi:10.1086/150119

Gotthelf, E.V., Tomsick, J.A., Halpern, J.P., Gelfand, J.D., Harrison, F.A., Boggs, S.E., Christensen, F.E., Craig, W.W., Hailey, J.C., Kaspi, V.M., Stern, D.K., Zhang, W.W.: NuSTAR discovery of a young, energetic pulsar associated with the luminous gamma-ray source HESS J1640-465. Astrophys. J. **788**, 155 (2014). doi:10.1088/0004-637X/788/2/155

Granot, J., Gill, R., Younes, G., Gelfand, J., Harding, A., Kouveliotou, C., Baring, M.G.: Learning about the magnetar Swift J1834.9-0846 from its wind nebula. Mon. Not. R. Astron. Soc. **464**, 4895–4926 (2017). doi:10.1093/mnras/stw2554

Hewish, A., Bell, S.J., Pilkington, J.D.H., Scott, P.F., Collins, R.A.: Observation of a rapidly pulsating radio source. Nature **217**, 709–713 (1968). doi:10.1038/217709a0

Hinton, J.A., Funk, S., Parsons, R.D., Ohm, S.: Escape from vela X. Astrophys. J. **743**, 7 (2011). doi:10.1088/2041-8205/743/1/L7

Kargaltsev, O., Cerutti, B., Lyubarsky, Y., Striani, E.: Pulsar-wind nebulae. Recent progress in observations and theory. Space Sci. Rev. **191**, 391–439 (2015). doi:10.1007/s11214-015-0171-x

Kennel, C.F., Coroniti, F.V.: Confinement of the Crab pulsar's wind by its supernova remnant. Astrophys. J. **283**, 694–709 (1984a). doi:10.1086/162356

Kennel, C.F., Coroniti, F.V.: Magnetohydrodynamic model of Crab nebula radiation. Astrophys. J. **283**, 710–730 (1984b). doi:10.1086/162357

Li, H., Chen, Y., Zhang, L.: Lepto-hadronic origin of γ-rays from the G54.1+0.3 pulsar wind nebula. Mon. Not. R. Astron. Soc. **408**, 80–84 (2010). doi:10.1111/j.1745-3933.2010.00934.x

Martín, J., Torres, D.F., Pedaletti, G.: Molecular environment, reverberation, and radiation from the pulsar wind nebula in CTA 1. Mon. Not. R. Astron. Soc. **459**, 3868–3879 (2016). doi:10.1093/mnras/stw684

Martín, J., Torres, D.F., Rea, N.: Time-dependent modelling of pulsar wind nebulae: study on the impact of the diffusion-loss approximations. Mon. Not. R. Astron. Soc. **427**, 415–427 (2012). doi:10.1111/j.1365-2966.2012.22014.x

Meyer, M., Horns, D., Zechlin, H.-S.: The Crab nebula as a standard candle in very high-energy astrophysics. Astron. Astrophys. **523**, 2 (2010). doi:10.1051/0004-6361/201014108

Mori, K., Burrows, D.N., Hester, J.J., Pavlov, G.G., Shibata, S., Tsunemi, H.: Spatial variation of the X-ray spectrum of the Crab nebula. Astrophys. J. **609**, 186–193 (2004). doi:10.1086/421011

Nolan, P.L., Abdo, A.A., Ackermann, M., Ajello, M., Allafort, A., Antolini, E., Atwood, W.B., Axelsson, M., Baldini, L., Ballet, J., et al.: Fermi large area telescope second source catalog. Astrophys. J. Suppl. Ser. **199**, 31 (2012). doi:10.1088/0067-0049/199/2/31

Olmi, B., Del Zanna, L., Amato, E., Bandiera, R., Bucciantini, N.: On the magnetohydrodynamic modelling of the Crab nebula radio emission. Mon. Not. R. Astron. Soc. **438**, 1518–1525 (2014). doi:10.1093/mnras/stt2308

Pacini, F.: Rotating neutron stars, pulsars and supernova remnants. Nature **219**, 145–146 (1968). doi:10.1038/219145a0

Pacini, F., Salvati, M.: On the evolution of supernova remnants. Evolution of the magnetic field, particles, content, and luminosity. Astrophys. J. **186**, 249–266 (1973). doi:10.1086/152495

Porth, O., Komissarov, S.S., Keppens, R.: Solution to the sigma problem of pulsar wind nebulae. Mon. Not. R. Astron. Soc. **431**, 48–52 (2013). doi:10.1093/mnrasl/slt006

Porth, O., Komissarov, S.S., Keppens, R.: Rayleigh-Taylor instability in magnetohydrodynamic simulations of the Crab nebula. Mon. Not. R. Astron. Soc. **443**, 547–558 (2014a). doi:10.1093/mnras/stu1082

Porth, O., Komissarov, S.S., Keppens, R.: Three-dimensional magnetohydrodynamic simulations of the Crab nebula. Mon. Not. R. Astron. Soc. **438**, 278–306 (2014b). doi:10.1093/mnras/stt2176

Porth, O., Vorster, M.J., Lyutikov, M., Engelbrecht, N.E.: Diffusion in pulsar wind nebulae: an investigation using magnetohydrodynamic and particle transport models. Mon. Not. R. Astron. Soc. **460**, 4135–4149 (2016). doi:10.1093/mnras/stw1152

Rea, N., Pons, J.A., Torres, D.F., Turolla, R.: The fundamental plane for radio magnetars. Astrophys. J. **748**, 12 (2012). doi:10.1088/2041-8205/748/1/L12

Reynolds, S.P., Chanan, G.A.: On the X-ray emission from Crab-like supernova remnants. Astrophys. J. **281**, 673–681 (1984). doi:10.1086/162144

Reynolds, S.P., Chevalier, R.A.: Evolution of pulsar-driven supernova remnants. Astrophys. J. **278**, 630–648 (1984). doi:10.1086/161831

Reynolds, S.P., Pavlov, G.G., Kargaltsev, O., Klingler, N., Renaud, M., Mereghetti, S.: Pulsar-wind nebulae and magnetar outflows: observations at radio, X-ray, and gamma-ray wavelengths. Space Sci. Rev. (2017). doi:10.1007/s11214-017-0356-6

Sironi, L., Spitkovsky, A.: Acceleration of particles at the termination shock of a relativistic striped wind. Astrophys. J. **741**, 39 (2011). doi:10.1088/0004-637X/741/1/39

Sironi, L., Spitkovsky, A.: Relativistic reconnection: an efficient source of non-thermal particles. Astrophys. J. **783**, 21 (2014). doi:10.1088/2041-8205/783/1/L21

Sironi, L., Spitkovsky, A., Arons, J.: The maximum energy of accelerated particles in relativistic collisionless shocks. Astrophys. J. **771**, 54 (2013). doi:10.1088/0004-637X/771/1/54

Slane, P.: Pulsar wind nebulae. ArXiv e-prints (2017)

Slane, P., Castro, D., Funk, S., Uchiyama, Y., Lemiere, A., Gelfand, J.D., Lemoine-Goumard, M.: Fermi detection of the pulsar wind nebula HESS J1640-465. Astrophys. J. **720**, 266–271 (2010). doi:10.1088/0004-637X/720/1/266

Slane, P., Hughes, J.P., Temim, T., Rousseau, R., Castro, D., Foight, D., Gaensler, B.M., Funk, S., Lemoine-Goumard, M., Gelfand, J.D., Moffett, D.A., Dodson, R.G., Bernstein, J.P.: A broadband study of the emission from the composite supernova remnant MSH 11-62. Astrophys. J. **749**, 131 (2012). doi:10.1088/0004-637X/749/2/131

Spitkovsky, A.: Particle acceleration in relativistic collisionless shocks: Fermi process at last? Astrophys. J. **682**, 5 (2008). doi:10.1086/590248

Spitkovsky, A., Arons, J.: Time dependence in relativistic collisionless shocks: theory of the variable "Wisps" in the Crab nebula. Astrophys. J. **603**, 669–681 (2004). doi:10.1086/381568

Tanaka, S.J.: A broadband emission model of magnetar wind nebulae. Astrophys. J. **827**, 135 (2016). doi:10.3847/0004-637X/827/2/135

Tanaka, S.J., Takahara, F.: A model of the spectral evolution of pulsar wind nebulae. Astrophys. J. **715**, 1248–1257 (2010). doi:10.1088/0004-637X/715/2/1248

Tanaka, S.J., Takahara, F.: Study of four young TeV pulsar wind nebulae with a spectral evolution model. Astrophys. J. **741**, 40 (2011). doi:10.1088/0004-637X/741/1/40

Tang, X., Chevalier, R.A.: Particle transport in young pulsar wind nebulae. Astrophys. J. **752**, 83 (2012). doi:10.1088/0004-637X/752/2/83

Temim, T., Slane, P., Kolb, C., Blondin, J., Hughes, J.P., Bucciantini, N.: Late-time evolution of composite supernova remnants: deep chandra observations and hydrodynamical modeling of a crushed pulsar wind nebula in SNR G327.1-1.1. Astrophys. J. **808**, 100 (2015). doi:10.1088/0004-637X/808/1/100

Temim, T., Dwek, E., Arendt, R.G., Borkowski, K.J., Reynolds, S.P., Slane, P., Gelfand, J.D., Raymond, J.C.: A massive shell of supernova-formed dust in SNR G54.1+0.3. Astrophys. J. **836**, 129 (2017). doi:10.3847/1538-4357/836/1/129

Thompson, C., Blaes, O.: Magnetohydrodynamics in the extreme relativistic limit. Phys. Rev. D **57**, 3219–3234 (1998). doi:10.1103/PhysRevD.57.3219

Thompson, C., Duncan, R.C.: The soft gamma repeaters as very strongly magnetized neutron stars. II. Quiescent neutrino, X-ray, and alfven wave emission. Astrophys. J. **473**, 322 (1996). doi:10.1086/178147

Timokhin, A.N., Harding, A.K.: On the polar cap cascade pair multiplicity of young pulsars. Astrophys. J. **810**, 144 (2015). doi:10.1088/0004-637X/810/2/144

Tong, H.: On the possible wind nebula of magnetar Swift J1834.9-0846: a magnetism-powered synchrotron nebula. Res. Astron. Astrophys. **16**, 143 (2016). doi:10.1088/1674-4527/16/9/143

Toor, A., Seward, F.D.: The Crab nebula as a calibration source for X-ray astronomy. Astron. J. **79**, 995–999 (1974). doi:10.1086/111643

Torres, D.F.: Rotationally powered magnetar nebula around swift J1834.9-0846. Astrophys. J. **835**, 54 (2017). doi:10.3847/1538-4357/835/1/54

Torres, D.F., Martín, J., de Oña Wilhelmi, E., Cillis, A.: The effects of magnetic field, age and intrinsic luminosity on Crab-like pulsar wind nebulae. Mon. Not. R. Astron. Soc. **436**, 3112–3127 (2013). doi:10.1093/mnras/stt1793

Torres, D.F., Cillis, A., Martín, J., de Oña Wilhelmi, E.: Time-dependent modeling of TeV-detected, young pulsar wind nebulae. J. High Energy Astrophys. **1**, 31–62 (2014). doi:10.1016/j.jheap.2014.02.001

Truelove, J.K., McKee, C.F.: Evolution of nonradiative supernova remnants. Astrophys. J. Suppl. Ser. **120**, 299–326 (1999). doi:10.1086/313176

van der Swaluw, E., Downes, T.P., Keegan, R.: An evolutionary model for pulsar-driven supernova remnants. A hydrodynamical model. Astron. Astrophys. **420**, 937–944 (2004). doi:10.1051/0004-6361:20035700

van der Swaluw, E., Achterberg, A., Gallant, Y.A., Tóth, G.: Pulsar wind nebulae in supernova remnants. Spherically symmetric hydrodynamical simulations. Astron. Astrophys. **380**, 309–317 (2001). doi:10.1051/0004-6361:20011437

van der Swaluw, E., Achterberg, A., Gallant, Y.A., Downes, T.P., Keppens, R.: Interaction of high-velocity pulsars with supernova remnant shells. Astron. Astrophys. **397**, 913–920 (2003). doi:10.1051/0004-6361:20021488

van Marle, A.J., Keppens, R., Yoon, S.-C., Langer, N.: On the circumstellar medium of massive stars and how it may appear in GRB observations. Memorie della Societa Astronomica Italiana Supplementi **21**, 40 (2012)

Venter, C., de Jager, O.C.: Constraints on the parameters of the unseen pulsar in the PWN G0.9+0.1 from radio, X-ray, and VHE gamma-ray observations. In: Becker, W., Huang, H.H. (eds.) WE-Heraeus Seminar on Neutron Stars and Pulsars 40 years after the Discovery, p. 40. Max-Planck-Institut für Extraterrestrische Physik, Garching (2007)

Vorster, M.J., Moraal, H.: The effect of diffusion on the particle spectra in pulsar wind nebulae. Astrophys. J. **765**, 30 (2013). doi:10.1088/0004-637X/765/1/30

Vorster, M.J., Tibolla, O., Ferreira, S.E.S., Kaufmann, S.: Time-dependent modeling of pulsar wind nebulae. Astrophys. J. **773**, 139 (2013). doi:10.1088/0004-637X/773/2/139

Weisskopf, M.C., Hester, J.J., Tennant, A.F., Elsner, R.F., Schulz, N.S., Marshall, H.L., Karovska, M., Nichols, J.S., Swartz, D.A., Kolodziejczak, J.J., O'Dell, S.L.: Discovery of spatial and spectral structure in the X-ray emission from the Crab nebula. Astrophys. J. **536**, 81–84 (2000). doi:10.1086/312733

Weisskopf, M.C., Guainazzi, M., Jahoda, K., Shaposhnikov, N., O'Dell, S.L., Zavlin, V.E., Wilson-Hodge, C., Elsner, R.F.: On calibrations using the Crab nebula and models of the nebular X-ray emission. Astrophys. J. **713**, 912–919 (2010). doi:10.1088/0004-637X/713/2/912

Wilson-Hodge, C.A., Cherry, M.L., Case, G.L., Baumgartner, W.H., Beklen, E., Narayana Bhat, P., Briggs, M.S., Camero-Arranz, A., Chaplin, V., Connaughton, V., Finger, M.H., Gehrels, N., Greiner, J., Jahoda, K., Jenke, P., Kippen, R.M., Kouveliotou, C., Krimm, H.A., Kuulkers, E., Lund, N., Meegan, C.A., Natalucci, L., Paciesas, W.S., Preece, R., Rodi, J.C., Shaposhnikov, N., Skinner, G.K., Swartz, D., von Kienlin, A., Diehl, R., Zhang, X.-L.: When a standard candle flickers. Astrophys. J. **727**, 40 (2011). doi:10.1088/2041-8205/727/2/L40

Yaron, O., Perley, D.A., Gal-Yam, A., Groh, J.H., Horesh, A., Ofek, E.O., Kulkarni, S.R., Sollerman, J., Fransson, C., Rubin, A., Szabo, P., Sapir, N., Taddia, F., Cenko, S.B., Valenti, S., Arcavi, I., Howell, D.A., Kasliwal, M.M., Vreeswijk, P.M., Khazov, D., Fox, O.D., Cao, Y., Gnat, O., Kelly, P.L., Nugent, P.E., Filippenko, A.V., Laher, R.R., Wozniak, P.R., Lee, W.H., Rebbapragada, U.D., Maguire, K., Sullivan, M., Soumagnac, M.T.: Confined dense circumstellar material surrounding a regular type II supernova: the unique flash-spectroscopy event of SN 2013fs. ArXiv e-prints (2017)

Younes, G., Kouveliotou, C., Kargaltsev, O., Gill, R., Granot, J., Watts, A.L., Gelfand, J., Baring, M.G., Harding, A., Pavlov, G.G., van der Horst, A.J., Huppenkothen, D., Göğüş, E., Lin, L., Roberts, O.J.: The wind nebula around magnetar swift J1834.9-0846. Astrophys. J. **824**, 138 (2016). doi:10.3847/0004-637X/824/2/138

Zhang, L., Chen, S.B., Fang, J.: Nonthermal radiation from pulsar wind nebulae. Astrophys. J. **676**, 1210–1217 (2008). doi:10.1086/527466

Chapter 9
MHD Modeling: Aims, Usage, Scales Assessed, Caveats, Codes

A. Mignone

Abstract This chapter is intended to provide a review of the current state of the art modeling in the field of computational magnetohydrodynamics (MHD). The equations of MHD are first derived starting from a kinetic description and its applicability in terms of temporal and spatial scales are discussed. The chapter then focuses on Godunov-type methods which have became widespread over last decades, owing to their ability to describe high-Mach number flows and discontinuities. These schemes lean on a conservative formulation of the equations and on the solution of the Riemann problem at cell interfaces. Such flows are often encountered in astrophysical environments such as supernovae remnants, pulsar wind nebulae and relativistic jets. Finally, advanced modeling using adaptive mesh refinement techniques is also described together with an example application.

9.1 Overview

Matter in the visible universe exists for the most part in the form of a plasma, that is, a collection of charge particles and the electromagnetic fields generated from them. An exact description of a plasma that requires knowledge of every particle position at all times is unattainable, except for extremely simple situations. Instead, different levels of approximation can be adopted that are sufficiently accurate to capture the essential dynamics of the system at the scale of interest.

As it is often the case, the same phenomenon may be described using different viewpoints, each modeling a limited range of behavior (Bellan 2006). The price to pay for choosing one method or the other is measured by the amount of information that one is willing to sacrifice in order to provide an acceptable description of the system at the desired spatial and temporal scales and for which the chosen model is appropriate. Stated roughly, the larger the spatial scale one of the system, the less the wealth of information one is able to cope with. Since, as we shall see, a

A. Mignone (✉)
Physics Department, Turin University, Torino, Italy
e-mail: mignone@ph.unito.it

D.F. Torres (ed.), *Modelling Pulsar Wind Nebulae*, Astrophysics and Space Science Library 446, DOI 10.1007/978-3-319-63031-1_9

numerical approach is often unavoidable in modeling complex plasma interactions, the available computational power must also be weighed in the balance.

At small scales, a plasma is best described using the Fokker Plank equation, governing the time evolution of the plasma distribution function $f_s \equiv f_s(\mathbf{x}, \mathbf{v}, t)$ in phase space, where s is used to denote the specie (ions or electrons in a simple two-component plasma),

$$\frac{\partial f_s}{\partial t} + \mathbf{v} \cdot \nabla_{\mathbf{x}} f_s + \frac{e_s}{m_s c}\left(\mathbf{E} + \frac{\mathbf{v}}{c} \times \mathbf{B}\right) \cdot \nabla_{\mathbf{v}} f_s = \left(\frac{\partial f_s}{\partial t}\right)_{\rm coll} . \tag{9.1}$$

Here m_s and e_s are the particle mass and charge, $\mathbf{v}$ and $\mathbf{x}$ are the (independent) velocity and spatial coordinates, $\mathbf{E}$ and $\mathbf{B}$ are the electric and magnetic field vectors while $\nabla_{\mathbf{x}}$ and $\nabla_{\mathbf{v}}$ are gradient operators acting on space and velocity coordinates, respectively.

The term on the right hand side of Eq. (9.1) represents, in compact form, the effect of collisions: charged particles are deflected as they experience the weak Coulomb potential generated by all surrounding nearby charged particles. The effect of collisions is to *thermalize* the plasma, by developing a distribution of the velocities for each species. In astrophysics, however, the inter-particle distance is usually very large and the net effect of collisions can be neglected. In such cases, Eq. (9.1) without the collisional term on the right hand side becomes the Vlasov equation.

The distribution function gives the number of particles contained in an infinitesimal volume of phase space and encloses all of the relevant information of the system. The actual particle number density and charge density in ordinary three-dimensional space, for example, are computed from

$$n_s(\mathbf{x}, t) \equiv \int f_s(\mathbf{x}, \mathbf{v}, t)\, d^3 v\,, \qquad q_s(\mathbf{x}, t) \equiv \int e_s f_s(\mathbf{x}, \mathbf{v}, t)\, d^3 v\,. \tag{9.2}$$

Likewise, the average fluid velocity can be computed from

$$\mathbf{u}_s(\mathbf{x}, t) \equiv \int \mathbf{v} f_s(\mathbf{x}, \mathbf{v}, t)\, d^3 v\,, \tag{9.3}$$

The Vlasov equation must be coupled to the time-dependent Maxwell's equations of electromagnetism for a complete description:

$$\nabla \times \mathbf{E} = -\frac{1}{c}\frac{\partial \mathbf{B}}{\partial t} \tag{9.4}$$

$$\nabla \times \mathbf{B} = \frac{1}{c}\frac{\partial \mathbf{E}}{\partial t} + \frac{4\pi}{c}\mathbf{J} \tag{9.5}$$

with initial conditions satisfying the time-independent conditions

$$\nabla \cdot \mathbf{E} = 4\pi q \tag{9.6}$$

$$\nabla \cdot \mathbf{B} = 0\,. \tag{9.7}$$

In the previous equations $q \equiv q(\mathbf{x}, t)$ and $\mathbf{J} \equiv \mathbf{J}(\mathbf{x}, t)$ should be expressed in terms of the distribution functions:

$$q(\mathbf{x}, t) = \sum_s e_s \int f_s(\mathbf{x}, \mathbf{v}, t) d^3 v; \qquad \mathbf{J}(\mathbf{x}, t) = \sum_s e_s \int \mathbf{v} f_s(\mathbf{x}, \mathbf{v}, t) d^3 v \tag{9.8}$$

Equations (9.1), (9.4)–(9.5) must be solved together in the six-dimensional phase space and this operation is computationally extremely intensive. In addition, the distribution function contains an excess of information and does not directly provide macroscopic observables with the exception of very special cases (e.g. the solar wind, see Chiuderi and Velli 2015).

Fluid models can be derived from Eq. (9.1) if one is willing to give up the information contained in the velocity distribution and consider the behavior of the system in the ordinary three-dimensional space. In particular, if $\psi_s(\mathbf{v})$ is any quantity associated with each particle, we can define its weighted average value by integrating ψf over velocity coordinates:

$$\langle \psi_s \rangle = \frac{1}{n_s(\mathbf{x}, t)} \int \psi_s(\mathbf{v}) f_s(\mathbf{x}, \mathbf{v}, t)\, d^3 v \tag{9.9}$$

where $n_s(\mathbf{x}, t)$ is the particle number density defined in Eq. (9.2). Moment equations are obtained when $\psi_s(\mathbf{v})$ is a multilinear product of the velocity, in the form $\psi_s \propto v_i v_j \ldots v_k$. In this respect, the particle number density and charge density defined, respectively, by Eq. (9.2) and the first of Eq. (9.8) can be considered zero-th order moments. Moments obey evolution equations obtained by multiplying Eq. (9.1) by $\psi_s(\mathbf{v})$ and integrate over all velocities. The results is the general moment equation,

$$\frac{\partial}{\partial t}\Big(n_s \langle \psi_s \rangle\Big) + \nabla \cdot \Big(n_s \langle \mathbf{v} \psi_s \rangle\Big) - \frac{n_s e_s}{m_s}\Big[\mathbf{E} \cdot \langle \nabla_\mathbf{v} \psi_s \rangle + \Big\langle \Big(\frac{\mathbf{v}}{c} \times \mathbf{B}\Big) \cdot \nabla_\mathbf{v} \psi_s \Big\rangle\Big] = 0 \tag{9.10}$$

where, from now on, we will neglect the collisional term on the right hand side of Eq. (9.1) as appropriate for dilute collisionless plasmas in astrophysics.

In the fluid approach the system is treated as a continuous medium describing the dynamical evolution of small volumes of fluid containing a large number of particles. The continuity assumption is meaningful to model physical processes that take place on a length scale that largely exceeds the average inter-particle distance and in which the identity of a single particle is lost. In such a description, a fluid element must be considered small enough so that any macroscopic quantity has negligible variation across its dimension but large enough to contain many particles and so to be insensitive to particle fluctuations, Boyd and Sanderson (2003). The goal of the fluid approach is to develop a macroscopic model that does not depend on the detail of the microphysics and, to the largest possible extent, is independent of the knowledge of the distribution function. In doing so, however, one must be willing to accept a loss of information related to the behavior of the particle in velocity space. This loss is acceptable when individual particle velocities are not substantially different from the average speed and when particle-wave resonant effects can be neglected (Chiuderi and Velli 2015).

Finally, we note that the moment Eq. (9.10) is an *infinite* chain of equations in which the equation for the n−th moment requires knowledge of the next higher order $(n+1)$ moment. The chain is usually truncated for $n > 3$ but it requires some sort of closure in order to have a mathematically consistent set of equations. At its simplest level, for instance, a closure scheme assumes that higher order moments vanish or can be expressed in terms of lower moments.

Since a plasma is made at least of two oppositely charged fluids, ions and electrons, a two-fluid model can be readily derived by considering two equations of the form (9.10), one for the electrons and one for the ions: $s = e, i$. These, together with Maxwell's equations (9.4)–(9.5) lead to the two-fluid model. Assuming no heat flow and a scalar fluid pressure, straightforward manipulations of Eq. (9.10) lead to (see, e.g., Bellan 2006; Chiuderi and Velli 2015):

$$\begin{aligned}
\frac{\partial \rho_s}{\partial t} + \nabla\cdot(\rho_s\mathbf{u}_s) &= 0\\
\frac{\partial \rho_s\mathbf{u}_s}{\partial t} + \nabla\cdot(\rho_s\mathbf{u}_s\mathbf{u}_s + p_s\mathsf{I}) &= q_s\left(\mathbf{E} + \frac{\mathbf{u}_s}{c}\times\mathbf{B}\right)\\
\frac{\partial \mathcal{E}_s}{\partial t} + \nabla\cdot(\mathcal{E}_s\mathbf{u}_s + p_s\mathbf{u}_s) &= q_s\mathbf{E}\cdot\mathbf{u}_s\\
\frac{\partial \mathbf{B}}{\partial t} + c\nabla\times\mathbf{E} &= 0\\
\frac{\partial \mathbf{E}}{\partial t} - c\nabla\times\mathbf{B} &= -4\pi(q_i\mathbf{u}_i + q_e\mathbf{u}_e)
\end{aligned} \tag{9.11}$$

where $\rho_s = m_s n_s$ is the mass density of each specie, $\mathbf{u}_s$ is the bulk (fluid) velocity, p_s is the scalar pressure, while $\mathcal{E}_s = \rho_s e_s + \rho_s\mathbf{u}_s^2/2$ is the energy density (in what follows $\rho_s e_s$ denotes the internal energy density). The two-fluid Eq. (9.11) constitute a system of 16 hyperbolic partial differential equations in the 16 unknowns (ρ_s, $\mathbf{u}_s$, p_s, $\mathbf{E}$, $\mathbf{B}$) for $s = i, e$. Numerical attempts to solve (9.11) have been done by, e.g., Shumlak and Loverich (2003). However, the fact that the two species have very unequal masses leads to a large disparity in temporal and spatial scales. This can pose serious difficulties in their numerical treatment, even for experienced researchers.

An additional averaging processes between ions and electrons equations leads to a single-fluid description which, in many situations, can be regarded as a plausible and effective model to describe a plasma. Finally, if one assumes that electromagnetic fields and fluid fluctuate on the same time and length scales, which are necessarily determined by the slower rates of the ions (Boyd and Sanderson 2003), the single-fluid model can be further simplified to obtain the equations of magnetohydrodynamics (MHD). In this approximation rapid wave fluctuations, the faster response of electrons and high frequency processes are neglected. This will be the subject of the rest of this chapter.

9.2 Magnetohydrodynamics

As already mentioned before, the equations of MHD can be derived from the single-fluid equations under the assumptions that all characteristic time and length scales are much larger than the ion Larmor period and gyroradius and that the plasma is not relativistic. This limit the applicability of MHD to the low-frequency regime. Under these conditions, the displacement current in Ampere's law (last equation in (9.11)) and the electric field term in the momentum equations can be neglected as they are $(\mathcal{U}/c)^2$ smaller than the remaining terms, where $\mathcal{U}$ is the typical fluid velocity. Charge variations are also negligible in this regime and the system is essentially described as an electrically conducting fluid:

$$\frac{\partial \rho}{\partial t} + \nabla \cdot (\rho \mathbf{u}) = 0 \tag{9.12}$$

$$\rho \left(\frac{\partial \mathbf{u}}{\partial t} + \mathbf{u} \cdot \nabla \cdot \mathbf{u} \right) = -\nabla p + \frac{1}{c} \mathbf{J} \times \mathbf{B} \tag{9.13}$$

$$\frac{\partial (\rho e)}{\partial t} + \nabla \cdot (\rho e \mathbf{u}) = -p \nabla \cdot \mathbf{u} + \left(\frac{\mathbf{J}}{\sigma} \right)^2 \tag{9.14}$$

$$\frac{\partial \mathbf{B}}{\partial t} + c \nabla \times \mathbf{E} = 0\,. \tag{9.15}$$

Equation (9.12) is the (mass) continuity equation, Eq. (9.13) is the momentum equation (or the equation of motion), Eq. (9.14) is the first law of thermodynamics, Eq. (9.15) is the induction equation (Faraday's law). Here $\mathbf{u}$ is the fluid velocity and represents the bulk speed of the center of mass of ions and electrons while ρ and p are the sum of their densities and pressures, respectively. In the low-frequency limit, Ampere's law reduces to $\mathbf{J} = (c/4\pi)\nabla \times \mathbf{B}$ and it is not an evolutionary equation. The solenoidal condition $\nabla \cdot \mathbf{B} = 0$ will be satisfied at any time if it is initially (although this might not be true from the computational point of view).

The MHD equations provide a system of 8 partial differential equations in the unknowns (ρ, $\mathbf{u}$, ρe, $\mathbf{B}$). An equation of state (EoS) $\rho e \equiv \rho e(\rho, p)$ must be specified in order to close the system. For the rest of this chapter we shall use an ideal equation of state, $\rho e = p/(\Gamma - 1)$ where Γ is the specific heat ratio. For a more general EoS, see the recent work by Vaidya et al. (2015).

Fluid and electromagnetic fields are coupled through Ohm's law which, within the MHD regime, takes the form

$$\mathbf{E} + \frac{\mathbf{u}}{c} \times \mathbf{B} = \frac{\mathbf{J}}{\sigma} \tag{9.16}$$

where σ is the plasma electric conductivity. Note that the left hand side of Eq. (9.16) represents the electric field as seen in the fluid co-moving frame. The presence of a non-zero resistivity makes the MHD equations a mixed hyperbolic/parabolic system of partial differential equations. In the limit of infinite conductivity ($\sigma \to \infty$) we obtain the ideal MHD equations which are hyperbolic in nature and support the propagation of fast, slow, Alfvén wave modes.

For computational purposes, the ideal MHD equations are usually written in conservative form to highlight the fundamental principles of mass, momentum and (total) energy conservation. Substituting the expression for $\mathbf{J}$ into Eq. (9.13) and using straightforward vector identities (Chiuderi and Velli 2015), it is possible to show that Eq. (9.12)–(9.15) can be cast as a system of conservation laws:

$$\frac{\partial \rho}{\partial t} + \nabla \cdot (\rho \mathbf{u}) = 0 \tag{9.17}$$

$$\frac{\partial (\rho \mathbf{u})}{\partial t} + \nabla \cdot \left[\rho \mathbf{u}\mathbf{u} - \frac{\mathbf{B}\mathbf{B}}{4\pi} + \left(p + \frac{\mathbf{B}^2}{8\pi}\mathsf{I}\right)\right] = 0 \tag{9.18}$$

$$\frac{\partial \mathcal{E}}{\partial t} + \nabla \cdot \left[\left(\mathcal{E} + p + \frac{\mathbf{B}^2}{8\pi}\right)\mathbf{u} - \frac{(\mathbf{u}\cdot\mathbf{B})}{4\pi}\mathbf{B}\right] = 0 \tag{9.19}$$

$$\frac{\partial \mathbf{B}}{\partial t} + \nabla \cdot (\mathbf{u}\mathbf{B} - \mathbf{B}\mathbf{u}) = \nabla \times (\eta \nabla \times \mathbf{B}) = 0 \tag{9.20}$$

where $\mathcal{E} = \rho e + \rho \mathbf{u}^2/2 + \mathbf{B}^2/(8\pi)$ is the total energy density.

As mentioned before, MHD is a non-relativistic macroscopic theory suitable to describe a plasma at large scales, much greater than the Debye length and particles (electron/ion) gyroradii. It is valid on time scales longer than the inverses of the plasma and Larmor frequencies. MHD is therefore not able to capture kinetic effects (such as Landau damping) or the behavior of the plasma at the gyroradius or skin depth scale. Nevertheless, due to its reduced complexity, it is often regarded as a first approximation to much of the physics even when some of the conditions are not met as it allows to draw some intuitive conclusions on the system without the need of solving more complex set of equations.

9.2.1 Relativistic MHD

The equations of relativistic magnetohydrodynamic (RMHD) are derived under the physical assumptions of constant magnetic permeability and infinite conductivity, appropriate for a perfectly conducting fluid (Anile 1989). They follow from the covariant form of particle number conservation, energy-momentum conservation and Maxwell equations:

$$\frac{\partial(\rho\gamma)}{\partial t} + \nabla\cdot(\rho\gamma\mathbf{u}) = 0 \qquad (9.21)$$

$$\frac{\partial}{\partial t}\left(w\gamma^2\frac{\mathbf{u}}{c^2} + \frac{\mathbf{S}}{c^2}\right) + \nabla\cdot\left(w\gamma^2\frac{\mathbf{uu}}{c^2} - \frac{\mathbf{EE}+\mathbf{BB}}{4\pi} + p_t\mathsf{I}\right) = 0 \qquad (9.22)$$

$$\frac{\partial\mathcal{E}}{\partial t} + \nabla\cdot\left(w\gamma^2\mathbf{u} + \mathbf{S}\right) = 0 \qquad (9.23)$$

$$\frac{\partial\mathbf{B}}{\partial t} + c\nabla\times\mathbf{E} = 0 \qquad (9.24)$$

$$\frac{\partial\mathbf{E}}{\partial t} - c\nabla\times\mathbf{E} = -4\pi\mathbf{J} \qquad (9.25)$$

where $w = \rho e + p$ is the enthalpy density, $\mathbf{S} = c\mathbf{E}\times\mathbf{B}/(4\pi)$ is the Poynting flux, $p_t = p + (\mathbf{E}^2 + \mathbf{B}^2)/(8\pi)$ is the total (gas + electromagnetic) pressure while

$$\mathcal{E} = w\gamma^2 - p + \frac{\mathbf{E}^2+\mathbf{B}^2}{8\pi} \qquad (9.26)$$

is the total energy density. Note that Ampere's law (9.5) retains its original form and the displacement current is *not* neglected in the derivation. However, in the approximation of infinite conductivity, Eq. (9.25) is not necessary as the electric field is computed directly from the covariant form of Ohm's law yielding $c\mathbf{E} = -\mathbf{u}\times\mathbf{B}$. This is the ideal relativistic MHD limit.

Relativistic effects come into play whenever the plasma bulk velocity approaches the speed of light ($u \sim c$), or at extremely high temperature so that pressure becomes comparable to the rest mass energy ($p \sim \rho c^2$) or for strongly magnetized plasma so that the classical definition of the Alfvén velocity becomes superluminal ($\mathbf{B}^2 \sim \rho c^2$).

9.3 Numerical Approach: Finite Volume

Finite volume schemes have a long standing tradition for their ability to capture shocks and discontinuities in general without the need to introduce artificial dissipation. They rely on the conservative form of the equations,

$$\frac{\partial\mathbf{U}}{\partial t} + \nabla\cdot\mathsf{F}(\mathbf{U}) = 0 \qquad (9.27)$$

which stems from physical principles stating conservation of fundamental quantities such as mass, momentum and energy. Equation (9.27) is a system of partial differential equations in which $\mathbf{U}$ is an array of conserved quantities (such as density, momentum and energy) while F is the flux tensor.

For the sake of exposition, we now restrict our attention to one-dimensional Cartesian coordinates so that Eq. (9.27) simplifies to

$$\frac{\partial \mathbf{U}}{\partial t} + \frac{\partial \mathbf{F}}{\partial x} = 0\,. \tag{9.28}$$

Let's assume a computational domain defined by $x \in [x_b, x_e]$ which is discretized into N_x cells (or zones) of equal length Δx centered around their position center, x_i. Cell interfaces are located at $x_{i\pm\frac{1}{2}} = x_i \pm \Delta x/2$. Integrating over a control volume Δx and over a time interval $\Delta t = t^{n+1} - t^n$ one obtains

$$\langle \mathbf{U} \rangle_i^{n+1} = \mathbf{U}_i^n - \frac{\Delta t}{\Delta x}\left[\tilde{\mathbf{F}}_{i+\frac{1}{2}}^{n+\frac{1}{2}} - \tilde{\mathbf{F}}_{i-\frac{1}{2}}^{n+\frac{1}{2}}\right] \tag{9.29}$$

where $\langle \mathbf{U} \rangle$ and $\tilde{\mathbf{F}}$ are, respectively, the *volume* average of the solution array and the *temporal* average of the flux at the cell boundary:

$$\langle \mathbf{U} \rangle^n \equiv \frac{1}{\Delta x}\int_{x_{i-\frac{1}{2}}}^{x_{i+\frac{1}{2}}} \mathbf{U}(x, t^n)\,dx\,, \qquad \tilde{\mathbf{F}}_{i\pm\frac{1}{2}}^{n+\frac{1}{2}} \equiv \frac{1}{\Delta t}\int_{t^n}^{t^{n+1}} \mathbf{F}(x_{i\pm\frac{1}{2}}, t)dt\,. \tag{9.30}$$

Equation (9.29) is an exact expression relating the change of zone-averaged conserved quantities inside a cell to the net time-averaged flux through its cell boundaries, see Fig. 9.1. This formulation is essential to ensure exact conservation at the discrete level.

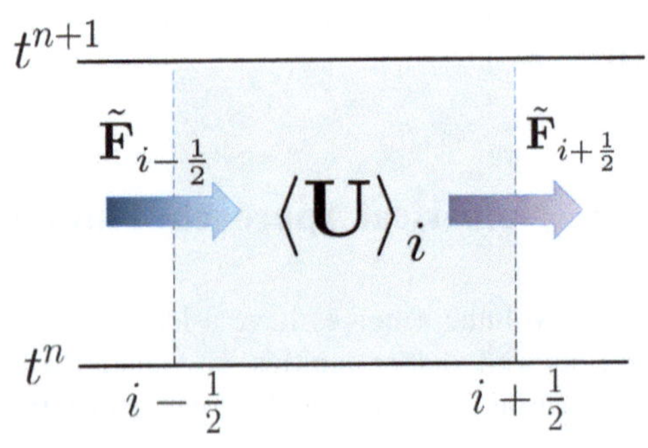

Fig. 9.1 Schematic diagram of a finite volume conservative scheme: the variation of the average value of **U** inside a computational cell is related to the net flux through its boundary

9.3.1 The Riemann Problem

A quick inspection of Eq. (9.29) shows that some approximation must be introduced in order to make this expression of any practical purpose. Indeed, the computation of the fluxes at cell interfaces requires knowledge of the solution between t^n and t^{n+1}. In shock-capturing schemes, this is achieved by determining the solution of the Riemann problem, that is, the evolution of a discontinuity separating two constant states initially adjacent to a zone interface:

$$\mathbf{U}(x, t^n) = \begin{cases} \mathbf{U}_L & \text{for } x < x_{i+\frac{1}{2}} , \\ \mathbf{U}_R & \text{for } x > x_{i+\frac{1}{2}} . \end{cases} \tag{9.31}$$

The discontinuity will break into a number of waves whose number and properties depend on the actual system of conservation laws being solved, see the schematic diagram in Fig. 9.2. Solving the Riemann problem tantamounts to computing the interface flux $\tilde{\mathbf{F}}^{n+\frac{1}{2}}_{i+\frac{1}{2}}$ for $t > t^n$. This approach dates back to the pioneering work of Godunov (1959); conservative finite volume methods in which the flux is computed by solving the Riemann problem are indeed known as *Godunov-type* methods. A Riemann solver naturally embeds the concept of upwinding: the discretization stencil should be biased towards the direction from which characteristic waves propagate.

This can be easily understood by considering the simple scalar equation with constant coefficient,

$$\frac{\partial U}{\partial t} + a\frac{\partial U}{\partial x} = 0 \tag{9.32}$$

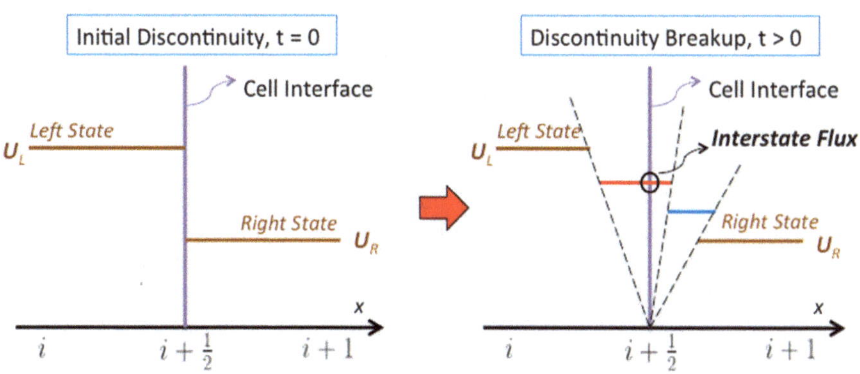

Fig. 9.2 Schematic diagram showing the breaking of two constant states initially separated by a discontinuity (*left*) into a number of intermediate waves (*right*). The flux function is computed at the cell interface, $x_{i+\frac{1}{2}}$

where a is a constant. Equation (9.32) has known analytical solution consisting of a uniform translation of the initial profile, $U(x,t) = U(x-at,0)$. The finite-volume discretization of Eq. (9.32), in the sense provided by Eq. (9.29) is therefore

$$U_i^{n+1} = U_i^n - \frac{\Delta t}{\Delta x}\left[\tilde{F}^{n+\frac{1}{2}}_{i+\frac{1}{2}} - \tilde{F}^{n+\frac{1}{2}}_{i-\frac{1}{2}}\right], \tag{9.33}$$

Note that for a first-order scheme, $\langle U\rangle \approx U$ and the interface flux is computed from the solution of the Riemann problem:

$$\tilde{F}^{n+\frac{1}{2}}_{i+\frac{1}{2}} = \begin{cases} aU_i & \text{if } a > 0 \\ aU_{i+1} & \text{if } a < 0\,. \end{cases} \tag{9.34}$$

The previous equation clarifies the concept of upwind: the information must be taken from the left (right) of the interface if the speed is positive (negative).

If $a > 0$, then Eq. (9.32) together with the interface flux (9.34) yields the so-called "forward in time, backward in space" (FTBS) scheme:

$$U_i^{n+1} = U_i^n - \frac{a\Delta t}{\Delta x}\left(U_i^n - U_{i-1}^n\right). \tag{9.35}$$

The dimensionless number $C = a\Delta t/\Delta x$ is the Courant-Friedrichs-Lewy (CFL) number and stability constraints limit its maximum value to 1. This conditions is typically used to limit the time step in Godunov-type codes. Note that the maximum allowed value for C is generally lower in more than one dimension.

Upwind methods provide a built-in mechanism for numerical dissipation which is ultimately responsible for the stability of the method. As an example, the amount of numerical diffusion introduced by the FTBS scheme can be quantified by computing the local truncation error introduced in the discretized Eq. (9.35). Expanding U_i^{n+1} and U_{i-1}^n around U_i^n and keeping terms up to second-order, inside Eq. (9.35) one finds

$$\frac{\partial U}{\partial t} + a\frac{\partial U}{\partial x} = \frac{a\Delta x}{2}\left(1 - a\frac{\Delta t}{\Delta x}\right)\frac{\partial^2 U}{\partial x^2} + O(\Delta t^2) + O(\Delta x^2)\,. \tag{9.36}$$

We thus see that the grid values U_i^n obtained with the FTBS scheme provide a first-order approximation to the true solution of the original scalar Eq. (9.32) but satisfy *exactly* a modified partial differential equation given by (9.36), see to book by LeVeque (2002). If we neglect second-order terms, then the grid values U_i^n are second-order approximation to the actual solution. For higher order schemes the first term on the right hand side of Eq. (9.36) can be eliminated and the amount of dissipation is progressively reduced.

For more complex systems of conservation laws, the solution to the Riemann problem requires the solution of nonlinear systems of equations. In MHD and relativistic MHD, for instance, the decay of the initial discontinuity gives rise to

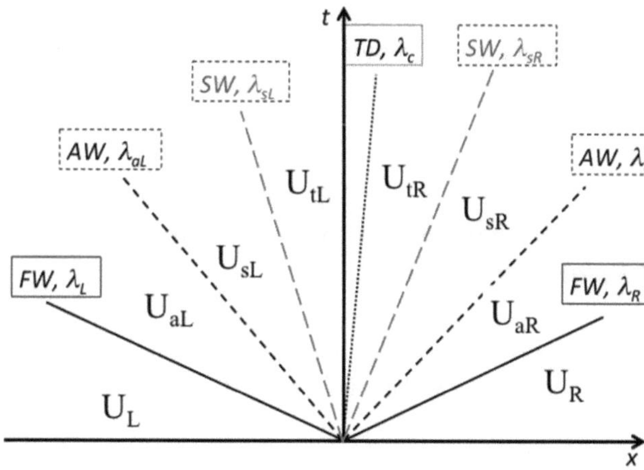

Fig. 9.3 General structure of the Riemann fan generated by two initial constant states $\mathbf{U}_L$ and $\mathbf{U}_R$. The pattern includes seven waves corresponding to a pair of fast magnetosonic waves with speed $u \pm c_f$, a pair of rotational (Alfvén waves, $u \pm v_A$), a pair of slow magnetosonic modes with speed $u \pm c_s$ and a contact (or tangential) discontinuity in the middle propagating at the fluid speed u. The seven waves bound six intermediate constant states, from left to right, U_{aL}, U_{sL}, U_{tL}, U_{tR}, U_{sR}, U_{aR}

a self-similar wave pattern in the $x - t$ plane where seven wave modes can develop. This is schematically shown in Fig. 9.3.

Fast waves (FW) are the outermost ones and enclose a pair of Alfvén waves (AW), a pair of slow waves (SW) bounding a tangential discontinuity (TD) in the middle. Fast and slow waves can be either shock or rarefaction waves where primary flow quantities (density, pressure, velocity and magnetic field) change discontinuously (for the former) or smoothly (for the latter), depending on the pressure jump and magnetic field orientation. Across the AW, thermodynamic quantities remain continuous while the tangential components of magnetic field trace circles (in classical MHD) or ellipses (in relativistic MHD), Mignone et al. (2012). Finally, through the contact mode, only density exhibits a jump while thermal pressure, velocity and magnetic field remain continuous.

Across each wave, the Rankine-Hugoniot jump conditions must be satisfied

$$\lambda^{(\kappa)} \left(\mathbf{U}_L^{(\kappa)} - \mathbf{U}_R^{(\kappa)} \right) = \mathbf{F}(\mathbf{U}_L^{(\kappa)}) - \mathbf{F}(\mathbf{U}_R^{(\kappa)}) \tag{9.37}$$

where $\kappa = 1, \ldots, 7$ lists the waves. An exact solver demands the simultaneous solution of the Rankine-Hugoniot jump conditions (9.37) across each wave through a self-consistent procedure that resolves fast, slow and Alfvén waves to the left and to the right of the contact (tangential in the degenerate case) discontinuity, always located at the center of the structure. This task is computationally expensive in classical MHD and almost impracticable in the relativistic case (see Giacomazzo and Rezzolla 2006) to be computationally efficient. For this reason, approximate methods of solution are usually preferred. The degree of approximation reflects the ability to capture and spread discontinuities over few or more computational zones and it is ultimately responsible for the amount of numerical dissipation of the numerical scheme.

9.3.2 *Approximate Riemann Solvers for Classical and Relativistic MHD*

In this section we briefly review some approximate methods of solution, that are often employed in numerical simulations of MHD flows.

9.3.2.1 The HLL Approximate Riemann Solver

The Harten Lax-van Leer solver is based on a two-wave approximation to the Riemann fan (Harten et al. 1983) (see also the book by Toro 1997). The internal structure is lumped into a single intermediate state $\mathbf{U}^{\rm hll}$ with corresponding flux $\mathbf{F}^{\rm hll}$ and only the outermost fast magnetosonic waves λ_L and λ_R are retained, see the left panel in Fig. 9.4. In analogy with Eq. (9.37), state and flux vectors inside the fan must simultaneously satisfy

$$\lambda_S \mathbf{U}^{\rm hll} - \mathbf{F}^{\rm hll} = \lambda_S \mathbf{U}_S - \mathbf{F}_S \tag{9.38}$$

with $S = L, R$. The intercell flux is then computed as

$$\tilde{\mathbf{F}} = \begin{cases} \mathbf{F}_L & \text{if} \quad \lambda_L > 0 \\ \mathbf{F}^{\rm hll} & \text{if} \quad \lambda_L \le 0 \le \lambda_R \\ \mathbf{F}_R & \text{if} \quad \lambda_R < 0 \end{cases} \tag{9.39}$$

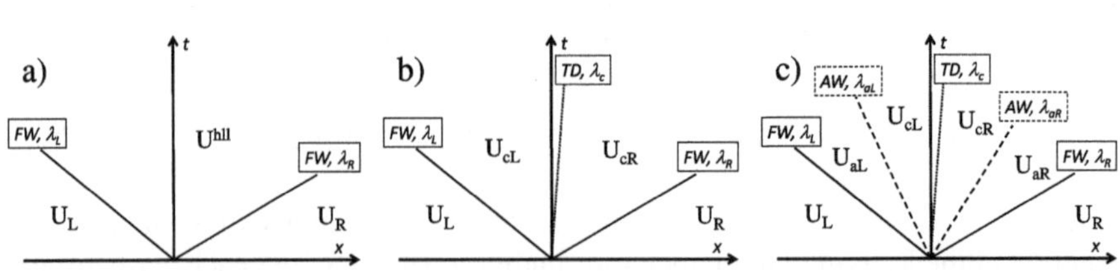

Fig. 9.4 *Left panel* (**a**): Approximate structures of the Riemann fan in the HLL solver. Fast waves (FW) have speeds λ_L and λ_R. *Middle panel* (**b**): approximate structures of the HLLC Riemann solver. The tangential discontinuity (TD) with speed λ_c separates the two intermediate states $\mathbf{U}_{cL}$ and $\mathbf{U}_{cR}$. *Right panel* (**c**): Wave structure in the HLLD approximate Riemann solver. The different waves have been labeled with FW (fast waves with speeds λ_L and λ_R), AW (Alfvén waves with speeds λ_{aL} and λ_{aR}) and TD (tangential discontinuity, with speed λ_c)

where $\mathbf{F}_L \equiv \mathbf{F}(\mathbf{U}_L)$, $\mathbf{F}_R \equiv \mathbf{F}(\mathbf{U}_R)$ while

$$\mathbf{F}^{\text{hll}} = \frac{\lambda_R \mathbf{F}_L - \lambda_L \mathbf{F}_R + \lambda_R \lambda_L (\mathbf{U}_R - \mathbf{U}_L)}{\lambda_R - \lambda_L} \tag{9.40}$$

is obtained by solving the two jump conditions given by Eq. (9.38). Solving for $\mathbf{U}^{\text{hll}}$ yields

$$\mathbf{U}^{\text{hll}} = \frac{\lambda_R \mathbf{U}_R - \lambda_L \mathbf{U}_L + \mathbf{F}_L - \mathbf{F}_R}{\lambda_R - \lambda_L} \tag{9.41}$$

Equations (9.41) and (9.40) can be regarded as the state and flux integral averages of the solution to the Riemann problem. Note also that to have a consistent system of equations, $\mathbf{F}^{\text{hll}}$ and $\mathbf{U}^{\text{hll}}$ are *independent* variables, that is, $\mathbf{F}^{\text{hll}} \neq \mathbf{F}(\mathbf{U}^{\text{hll}})$.

The outermost wave speeds can be estimated from the initial left and right input states. A very common recipe, given by Davis (1988), is

$$\begin{aligned} \lambda_L &= \min\left[\lambda_{cf-}(\mathbf{U}_L), \lambda_{cf-}(\mathbf{U}_R)\right] \\ \lambda_R &= \min\left[\lambda_{cf+}(\mathbf{U}_L), \lambda_{cf+}(\mathbf{U}_R)\right] \end{aligned} \tag{9.42}$$

where λ_{cf-} and λ_{cf+} are the leftmost and rightmost eigenmodes which, in classical MHD, correspond to $\lambda_{cf-} = |u_x| - c_f$, $\lambda_{cf+} = |u_x| + c_f$, where c_f is the fast magnetosonic speed.

The HLL approach is simple to implement, cost-effective and requires only a guess to the outermost fast speed without any particular knowledge of the solution.

9.3.2.2 The HLLC Approximate Riemann Solver

The HLLC scheme, originally introduced by Toro (1994), restores the middle contact or tangential mode in the Riemann fan which is now approximated by three waves separating four constant states, see the middle panel in Fig. 9.4. The middle wave moves at the fluid speed and separates two additional intermediate constant states $\mathbf{U}_{cL}$ and $\mathbf{U}_{cR}$ across which continuity of total pressure, velocity and tangential magnetic fields is imposed. The intermediate states $\mathbf{U}_{cL}$ and $\mathbf{U}_{cR}$ are computed from the jump conditions (9.37) with $\lambda^{(\kappa)} = \lambda_L, \lambda_c, \lambda_R$ and the interface flux can then be constructed as

$$\tilde{\mathbf{F}} = \begin{cases} \mathbf{F}_L & \text{if} \quad \lambda_L > 0 \\ \mathbf{F}_L + \lambda_L (\mathbf{U}_{cL} - \mathbf{U}_L) & \text{if} \quad \lambda_L \le 0 \le \lambda_c \\ \mathbf{F}_R + \lambda_R (\mathbf{U}_{cR} - \mathbf{U}_R) & \text{if} \quad \lambda_c \le 0 \le \lambda_R \\ \mathbf{F}_R & \text{if} \quad \lambda_R < 0 \,. \end{cases} \tag{9.43}$$

In the non-degenerate case ($B_x \neq 0$), the set of jump conditions across the three waves (including the conditions imposed across the contact mode) gives a total of 20 independent equations. Flux and state vectors for each state must then be written in terms of 10 unknowns in order to have a well-posed system of equations, see the discussion in Mignone and Bodo (2006). Although the solver involves simple algebraic manipulations in the classical case, the relativistic extension requires the solution of a quadratic equations in the case of Mignone and Bodo (2006) and an iterative scheme in Kim and Balsara (2014).

Nevertheless, the derivation of stable and mathematically consistent HLLC solvers has been the subject of investigation and a number of inconsistencies have been highlighted both in the classical derivation and in the relativistic extension. The relativistic HLLC scheme of Mignone and Bodo (2006), for instance, is well-behaved for strictly two-dimensional flows, applications to genuinely three-dimensional problems may suffer from a pathological singularity when the component of magnetic field normal to a zone interface approaches zero. This issue has been addressed in the work by Kim and Balsara (2014). We remind the reader to Batten et al. (1997), Gurski (2004), Li (2005) (for classical MHD) and to Mignone and Bodo (2006), Honkkila and Janhunen (2007), Kim and Balsara (2014) for relativistic MHD.

9.3.2.3 The HLLD Approximate Riemann Solver

The HLLD Riemann solver was originally introduced by Miyoshi and Kusano (2005) in the context of classical MHD and by Mignone et al. (2009) for the relativistic MHD equations. An isothermal version of the solver was presented by Mignone (2007). In the HLLD approach, the Riemann fan splits into 4 additional intermediate states separated by five waves, namely, two outermost fast waves (λ_R and λ_L) enclosing a pair of rotational discontinuities (λ_{aR} and λ_{aL}) separated by a tangential discontinuity λ_c, see right panel in Fig. 9.4. As before, the initial states $\mathbf{U}_L$ and $\mathbf{U}_R$ are inputs to the problem and a consistent solution must be sought in terms of a well-balanced set of equations and unknowns. Besides continuity of velocity, total pressure and magnetic field across the tangential mode, it is further required that scalar quantities carry zero-jump through the rotational waves. This leaves a well-defined problem which, in the case of classical MHD, can be solved with simple algebraic manipulation and leads to the flux function[1]:

$$\tilde{\mathbf{F}} = \begin{cases} \mathbf{F}_L & \text{if} \quad \lambda_L > 0 \\ \mathbf{F}_L + \lambda_L\,(\mathbf{U}_{aL} - \mathbf{U}_L) & \text{if} \quad \lambda_L \le 0 \le \lambda_{aL} \\ \mathbf{F}_L + \lambda_L\,(\mathbf{U}_{aL} - \mathbf{U}_L) + \lambda_{aL}(\mathbf{U}_{cL} - \mathbf{U}_{aL}) & \text{if} \quad \lambda_{aL} \le 0 \le \lambda_c \\ \mathbf{F}_R + \lambda_R\,(\mathbf{U}_{aR} - \mathbf{U}_R) + \lambda_{aR}(\mathbf{U}_{cR} - \mathbf{U}_{aR}) & \text{if} \quad \lambda_c \le 0 \le \lambda_{aR} \\ \mathbf{F}_R + \lambda_R\,(\mathbf{U}_{aR} - \mathbf{U}_R) & \text{if} \quad \lambda_{aR} \le 0 \le \lambda_R \\ \mathbf{F}_R & \text{if} \quad \lambda_R < 0 \end{cases} \tag{9.44}$$

[1]Note that Equation 16 in Mignone et al. (2009) contains a misprint.

A relativistic extension has been presented by Mignone et al. (2009) and introduces several additional complexities since the normal component of velocity is no longer constant across the rotational waves, as it is the case with classical MHD. Nevertheless, a closed-form solution can still be found by reducing the set of jump conditions to a single scalar nonlinear equation which must solved iteratively.

9.3.2.4 The Roe Approximate Riemann Solver

Roe's scheme (Roe 1981) tries to resolve the initial jump by replacing the original conservation law with a linearized system of constant coefficients. Thus one proceeds by seeking the exact solution to the following linearized Riemann problem:

$$\begin{cases} \dfrac{\partial \mathbf{U}}{\partial t} + \overline{\mathsf{A}} \cdot \dfrac{\partial \mathbf{U}}{\partial x} = \mathbf{0}\,, \\ \mathbf{U}(x,0) = \begin{cases} \mathbf{U}_L & \text{for} \quad x < 0\,, \\ \mathbf{U}_R & \text{for} \quad x > 0\,. \end{cases} \end{cases} \tag{9.45}$$

where $\overline{\mathsf{A}} \equiv \overline{\mathsf{A}}(\mathbf{U}_L, \mathbf{U}_R)$ is a constant matrix. The matrix $\overline{\mathsf{A}}$ is called a Roe matrix if the following requirements are met (Toro 1997):

- Consistency with the original conservation law: $\overline{\mathsf{A}}(\mathbf{U},\mathbf{U}) = \mathsf{A}(\mathbf{U})$, where A is simply the Jacobian of the flux, i.e., $\mathsf{A}(\mathbf{U}) = \partial \mathbf{F}(\mathbf{U})/\partial \mathbf{U}$.
- Conservation across discontinuities:

$$\mathbf{F}(\mathbf{U}_R) - \mathbf{F}(\mathbf{U}_L) = \overline{\mathsf{A}} \cdot (\mathbf{U}_R - \mathbf{U}_L)\,. \tag{9.46}$$

- Hyperbolicity: $\overline{\mathsf{A}}$ must have a complete set of real eigenvalues $\lambda_k(\mathbf{U}_L, \mathbf{U}_R)$ and associated left and right eigenvectors $\mathbf{l}_k$ and $\mathbf{r}_k$ such that

$$\overline{\mathsf{A}} \cdot \mathbf{r}_k = \lambda_k \mathbf{r}_k\,, \quad \mathbf{l}_k \cdot \overline{\mathsf{A}} = \lambda_k \mathbf{l}_k\,, \quad \mathbf{l}_k \cdot \mathbf{r}_j = \delta_{ij}\,, \tag{9.47}$$

  where the last property implies orthonormality.

Roe's scheme is thus equivalent to the solution of the Riemann problem for system of advection equation with linear constant coefficients, for which the flux function takes the form:

$$\tilde{\mathbf{F}} = \frac{\mathbf{F}_L + \mathbf{F}_R}{2} - \frac{1}{2}\sum_k (\mathbf{l}_k \cdot \Delta\mathbf{U})\,|\lambda_k|\mathbf{r}_k\,, \tag{9.48}$$

and $\Delta\mathbf{U} = \mathbf{U}_R - \mathbf{U}_L$. An extension to classical MHD, valid for arbitrary values of the specific heat ratio Γ, may be found in Cargo and Gallice (1997) while relativistic implementation can be found in Balsara (2001), Koldoba et al. (2002), Anton et al. (2010).

The Riemann solver of Roe resolves all seven waves in the Riemann fan and it is therefore more accurate than the previous solvers. However, in spite of its popularity, the Roe Riemann solver may severely fail in presence of low density flows yielding intermediate states that are nonphysical (e.g. negative pressures or densities), sharpen large protrusion in proximity of strong multi-dimensional shocks (carbuncle) or create spurious oscillations in grid aligned shocks (odd-even decoupling), see Einfeldt et al. (1991). In addition, an *entropy fix* should be used in order to avoid the creation of expansion shock instead of smooth rarefaction waves. For this reason, most popular codes provide hybrid implementations that are able to switch to a more robust (albeit more diffusive) solvers when such extreme conditions are met. An adaptive strategy is implemented in the PLUTO code and has been described in the appendix of Mignone et al. (2012).

9.3.3 Reconstruct-Solve-Update: The Basic Building Block for Finite Volume Schemes

At its simplest level, a shock-capturing finite volume scheme is based on a rather general sequence of steps allowing the solution vector to evolve to the next time level. High-order accuracy in space is ensured by properly reconstructing the solution inside each computational cell. Time-integration can be obtained using characteristic tracing or the method of lines, see the paper by Mignone et al. (2007). A simple pseudo-code is comprised of this general sequence of steps:

- *Reconstruct*: starting from the solution array $\mathbf{U}^n = \mathbf{U}(t^n)$, reconstruct interface values from zone averages using a high-order non oscillatory polynomial:

$$\begin{cases} \mathbf{U}_{i+\frac{1}{2},L} = \lim_{x\to x^-_{i+\frac{1}{2}}} \mathbf{U}_i(x) \\ \mathbf{U}_{i+\frac{1}{2},R} = \lim_{x\to x^+_{i+\frac{1}{2}}} \mathbf{U}_{i+1}(x) \end{cases} \tag{9.49}$$

  Several reconstruction schemes are available in the literature. A popular second-order scheme,

$$\mathbf{U}_{i+\frac{1}{2},L} = \mathbf{U}_i + \frac{\Delta\mathbf{U}_i}{2}, \qquad \mathbf{U}_{i+\frac{1}{2},R} = \mathbf{U}_{i+1} - \frac{\Delta\mathbf{U}_{i+1}}{2}, \tag{9.50}$$

  makes use of *slope limiters* in order to avoid the introduction of unwanted new extrema. Slope-limiter methods (see, for instance, the book by Toro (1997) and the recent paper by Mignone et al. 2014) ensure that the reconstructed values are bounded by neighboring cells by expressing the final slope as a weighted average of the forward and backward derivatives. This implies that zero gradient is assigned near local extrema.

  Extensive practice suggests that, most often, a more robust reconstruction algorithm is obtained using *primitive* variables (such as velocity and pressure)

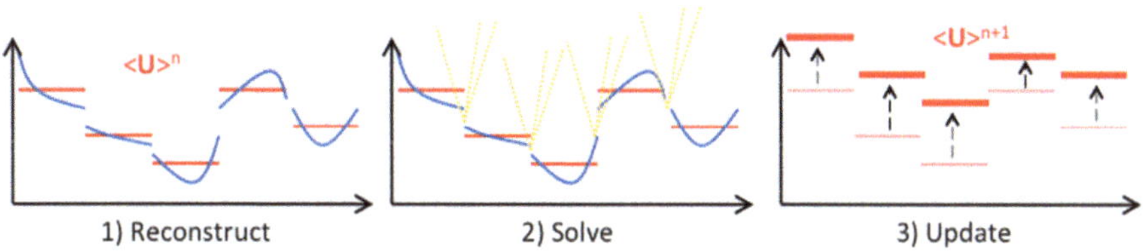

Fig. 9.5 The Reconstruct-Solve-Update step sequence typical of a Godunov-type code: (1) the solution is first reconstructed inside a computational zone, (2) a Riemann solver is used to compute the intercell flux and (3) the solution is updated to the next time level

rather than conservative variables (such as momentum and energy). Another robust choice, which requires the characteristic decomposition of the conservation law, is given by characteristic variables $\mathbf{w} = \mathsf{L}\mathbf{U}$.

Higher than second-order reconstruction schemes, although less widespread, are employed. See, for instance, Del Zanna et al. (2007), Dumbser and Zanotti (2009), McCorquodale and Colella (2011), Zanotti et al. (2015).

- *Solve*: compute the interface flux by solving the Riemann problems between adjacent discontinuous states obtained at zone interfaces during the reconstruction step. This task can be achieved by using any of the Riemann solvers described in Sect. 9.3.2.
- *Update*: update the solution array of conserved variables to the next time level using Eq. (9.29) or a higher order discretization methods.

This three-step sequence of step is schematically depicted in Fig. 9.5 and may be iterated more than once in the case of a multi-stage Runge-Kutta scheme which has become very popular among practitioners. A robust second-order time stepping method is the following predictor-corrector scheme:

$$\begin{aligned} \langle\mathbf{U}\rangle_i^* &= \mathbf{U}_i^n - \frac{\Delta t}{\Delta x}\left[\tilde{\mathbf{F}}_{i+\frac{1}{2}}^n - \tilde{\mathbf{F}}_{i-\frac{1}{2}}^n\right] \\ \langle\mathbf{U}\rangle_i^{n+1} &= \frac{\mathbf{U}_i^n + \mathbf{U}_i^*}{2} - \frac{\Delta t}{2\Delta x}\left[\tilde{\mathbf{F}}_{i+\frac{1}{2}}^* - \tilde{\mathbf{F}}_{i-\frac{1}{2}}^*\right] . \end{aligned} \tag{9.51}$$

commonly known as the second-order TVD Runge-Kutta scheme.

9.4 Comparison on Selected Benchmarks

9.4.1 Simple Advection

The effect of spatial reconstruction on the amount of numerical diffusion can be observed in a simple advection test. Figure 9.6 shows the numerical solution, after one period, of the uniform advection equation with initial condition given by a simple square wave. The reconstruction algorithm changes from flat (left) to linear (middle) and piecewise parabolic (right panel). The solutions are oscillation-free and

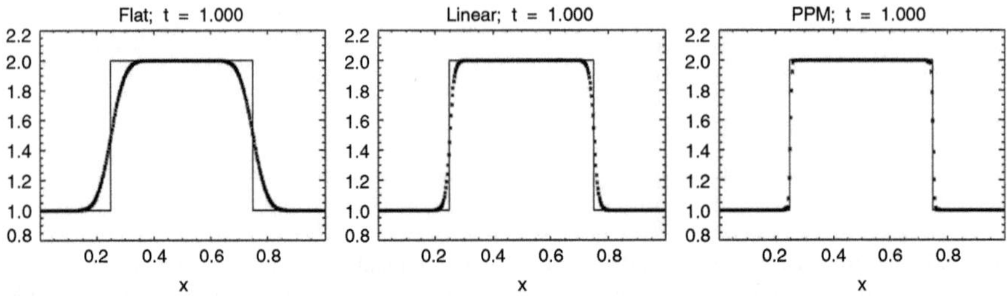

Fig. 9.6 Analytical (*solid line*) and numerical (*points*) solutions for the uniform advection of a square wave after one period. From left to right, results are shown for the first-order scheme, using piecewise linear reconstruction and piecewise parabolic reconstruction

the speed of the discontinuous wave is correctly captured, although it considerably smeares out for the first-order scheme while it is resolved on $\approx$ 6 zones with the piecewise parabolic method (PPM, in the version of Mignone et al. 2012).

9.4.2 Alfvén Wave Decay

In the next test, we consider the propagation of circularly polarized Alfvén waves by solving the equations of relativistic MHD, as described in Del Zanna et al. (2007) (see also Mignone et al. 2012). The computational domain is the unit periodic square $x, y \in [0, 1]$, initially filled with constant density and gas pressure ($\rho = p = 1$) while magnetic and velocity vectors are initialized as

$$\mathbf{B} = B_0\,(1,\, \eta\cos\phi,\, \eta\sin\phi)\,, \qquad \mathbf{u} = -v_A\left(0,\, \frac{B_y}{B_0},\, \frac{B_z}{B_0}\right)\,, \tag{9.52}$$

where $\phi = 2\pi(x+y)$ is the phase corresponding to a wave propagating along the main diagonal, $B_0 = 1$ is the (constant) magnetic field component in the direction of propagation, $\eta = 1$ is the amplitude. The wave propagate at the Alfvén velocity v_A given by

$$v_A^2 = \frac{2\alpha}{1+\sqrt{1-4\eta^2\alpha^2}}, \qquad \alpha = \frac{B_0^2}{w_g + B_0^2(1+\eta^2)}\,. \tag{9.53}$$

thus corresponding to $v_A \approx 0.382$ for the chosen parameters.

The problem has an exact solution given by Eqs. (9.52)–(9.53) with $\phi \to \phi - \omega t$ where $\omega = \sqrt{8}\pi v_A$ is the angular frequency. The ideal EoS is employed with $\Gamma = 4/3$.

The left panel in Fig. 9.7 shows the errors of u_z (in L1 norm) after one period $T = 1/(\sqrt{2}v_A)$ as a function of the mesh size and for three different Riemann solvers: HLL, HLLC and HLLD. All solvers yield 2nd-order accurate solutions although HLLC and HLLD exhibit somewhat smaller errors when compared to HLL. The

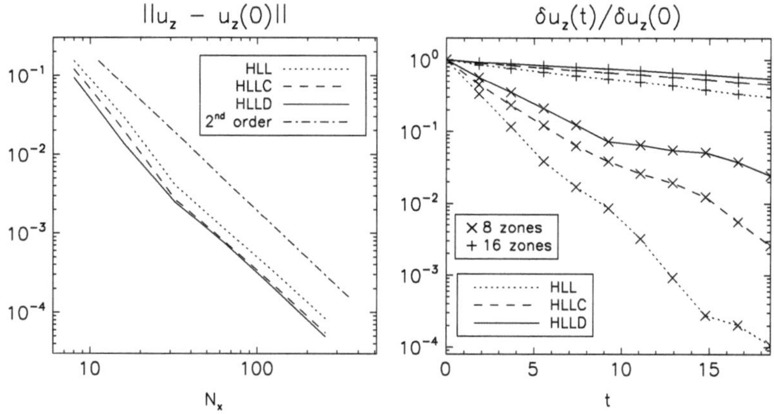

Fig. 9.7 *Left*: L1 errors of u_z after one period as a function of the grid resolution N_x and for different Riemann solvers (HLL, HLLC and HLLD, see the legend). *Right*: normalized amplitude of u_z as a function of time for the selected Riemann solvers using 8 (*plus signs*) and 16 (x *signs*) cells per wavelength

right panel of Fig. 9.7 shows the decay of the wave amplitude, defined as $\delta u_z = \max(u_z) - \min(u_z)$ (normalized to its initial value) after ten revolutions, using 8 and 16 zones per wavelength. At the end of the computation, the results obtained with HLL, HLLC and HLLD at the lowest resolution show that the wave amplitude has been reduced, respectively, to $\sim 10^{-5}$, $\sim 2.5 \times 10^{-3}$ and $\sim 2 \times 10^{-2}$ of the nominal value. At larger resolutions differences becomes less pronounced, see Mignone et al. (2012).

These results suggest that the dissipative properties of the Riemann solver may be crucial in the resolution of small-scale structures often encountered, for instance, in turbulent environments such as PWN torus, jet cocoons, accretion disks, and so forth. Here proper resolution must be carefully chosen in order to model the spectrum of fluctuations up to the desired accuracy.

9.4.3 1D MHD Shock Tube

In this test we solve the ideal MHD equations with initial condition given by the following one-dimensional Riemann problem:

$$\begin{aligned} \{\rho, u_x, B_x, B_y, p\}_L &= \left\{1, 10, \frac{5}{\sqrt{4\pi}}, \frac{5}{\sqrt{4\pi}}, 20\right\} \\ \{\rho, u_x, B_x, B_y, p\}_R &= \left\{1, -10, \frac{5}{\sqrt{4\pi}}, \frac{5}{\sqrt{4\pi}}, 1\right\} , \end{aligned} \tag{9.54}$$

consisting of two colliding flows having two different temperatures and moving at the same (opposite) speed.

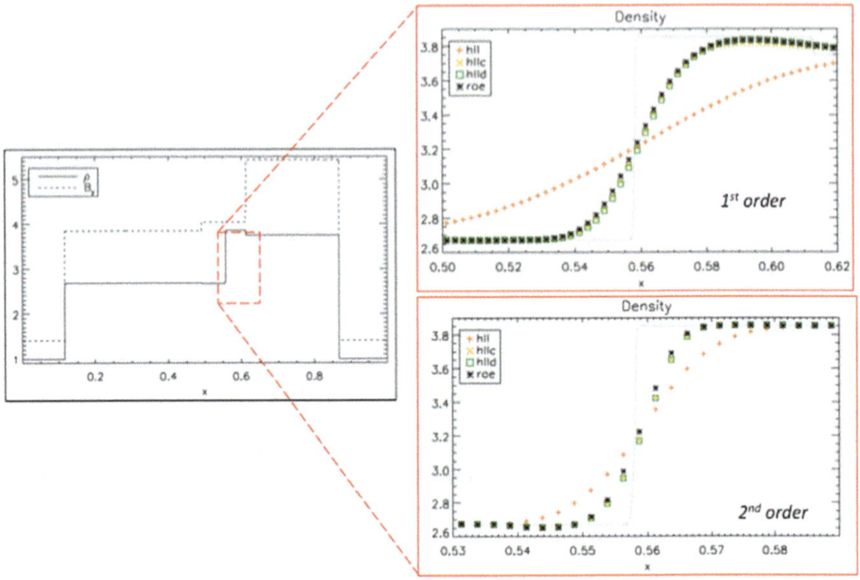

Fig. 9.8 *Left-half*: density (*solid line*) and *y*-component of magnetic field (*dashed line*) at $t = 0.08$. *Right-half*: close-up of the central region (*close to the contact discontinuity*) showing density computed with the 1st order (*top*) and 2nd order schemes and different Riemann solvers. Results obtained with HLL, HLLC, HLLD and Roe are shown with *red, orange, green and black symbols*

The solution, shown in left half of Fig. 9.8 at $t = 0.08$, consists of two outermost fast magnetosonic shocks (located at $x \approx 0.12$ and $x \approx 0.87$) enclosing two slow shocks and a contact discontinuity in the middle ($x \approx 0.75$).

Computations have been repeated for different Riemann solvers (HLL, HLLC, HLLD and Roe) using both first- and second-order schemes, with uniform grid resolution $N_x = 400$. The enlargements on the right-half of Fig. 9.8 show the numerical solutions obtained in proximity of the contact wave mode. Here the larger amount of diffusion introduced by the HLL Riemann solver is clearly visible for the first-order scheme while the other solvers behave similarly. When switching to the second-order method (bottom-right) differences between solvers are reduced while still keeping the same tendency observed before, with HLL and Roe being the most and least diffusive solvers, respectively.

9.4.4 Axisymmetric Pulsar Wind Nebulae

The choice of the Riemann solver (and the numerical scheme in general) can have a profound impact on the results of the computation. This effect, as well as the bearing of mesh resolution and interpolation methods, is often underrated and may, at least to some extent, cause a loss of confidence in the results (Mignone et al. 2004).

As an example, we consider here an application of the PLUTO code to the pulsar wind model described by Olmi et al. (2014, 2015). Computations are performed using axisymmetric spherical coordinates (r, θ) and a computational domain ranging from $r_{\min} = 0.025$ to $r_{\max} = 10$ (in light years) with a logarithmic stretching so that $\Delta r \propto r$ while $\theta \in [0, \pi]$. The numerical resolution is 512×512 zones. Our parameters are chosen so that the wind has Lorentz factor $\gamma = 10$, anisotropy parameter $\alpha = 0.1$ and magnetization $\sigma = 0.02$.

Figure 9.9 shows the density maps obtained at $t = 100, 300$ years (top panels) and at $t = 500, 700$ years (bottom panels) using the HLL (right halves) and HLLC (left halves) Riemann solvers. Note that to avoid numerical instabilities in the initial transient phase, the accuracy of the scheme has been switched to 1st order for $t < 2.5$ years.

The choice of the Riemann solver affects both the morphology and dynamics of the nebula as it is clearly evident downstream of the termination shock where the wind is slowed down and the initial toroidal magnetic field is largely amplified. The post-shock flow is rapidly randomized by the development of hydromagnetic

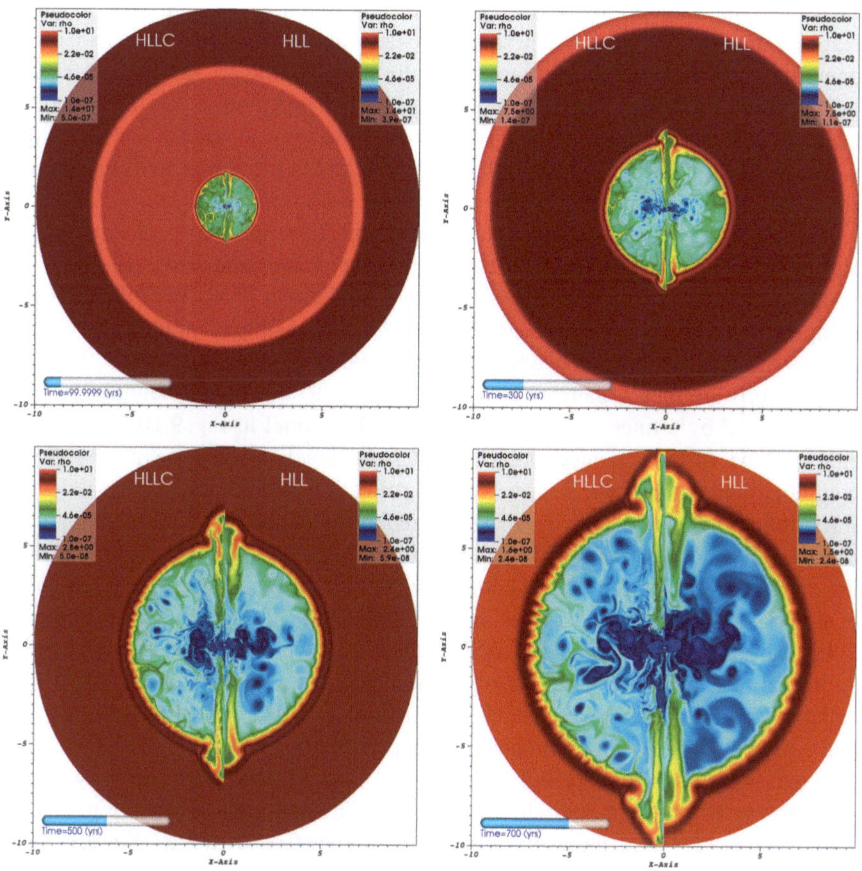

Fig. 9.9 Dynamical evolution of the pulsar wind nebula showing density (in log scale). From left to right, top to bottom: system at $t = 100$, 300, 500 and $t = 700$ years

turbulence and magnetic hoop stresses are effective in collimating the jet along the polar axis. Results obtained with HLLC give a larger wealth of detail in the inner regions (downstream of the termination shock) where the flow appear to be more turbulent as a consequence of the reduced numerical viscosity. Also, the jet appears to propagate faster with HLLC rather than HLL.

9.5 Adaptive Mesh Refinement

Proper computational modelling of astrophysical environments may become extremely challenging when great disparities in the spatial and temporal scales arise in the problem. Under these circumstances, which represent a key aspect for the next exa-scale computing era, an approach based on a static (fixed) grid can make the computation very inefficient or even prohibitive in the most extreme cases. Typically, such conditions occur when the flow dynamics exhibit very localized features that evolve on a much shorter scale when compared to the rest of the computational domain. To overcome these limitations, one possibility is to change or adapt the computational grid dynamically in space and time so that the features of interest can be adequately captured and resolved. Adaptive mesh refinement (AMR) is one such technique and can lead, for a certain class of problems, to a considerable speed-up while minimizing memory and storage requirements.

Two basic approaches, along with their variants, have gained popularity over the past decades and can be roughly classified as *patch-based* or *cell-based*.

In a patch-based approach to AMR (Berger and Colella 1989), the solution is first computed on a coarse base-level grid representing the whole computational domain and, as the computation proceeds, individual grid cells are flagged for refinement. Tagged cells are then organized into larger aggregates to form rectangular patches characterized by a finer mesh spacing, see the left panel in Fig. 9.10. The process can then be further iterated up to the desired level of refinement so that the computational domain will be covered by a collection of hierarchically nested grids. Each refinement level will be characterized by a collection of non-overlapping rectangular patches onto which the single grid algorithm can be re-used. This is the approach used by the ENZO code (Bryan et al. 2014), PLUTO code (Mignone et al. 2012), the VAC (van der Holst et al. 2008) and the FLASH codes (Fryxell et al. 2000) (although both VAC and FLASH shares ideas with the cell-based approach).

In contrast, in a cell-based (or tree-based, Khokhlov 1998) approach to AMR, cells may be individually refined and organized into a tree data structure (quadtree and octree in two and three dimensions, respectively), see right panel in Fig. 9.10. The tree structure provides full connectivity having each "oct" to easily access neighboring parent cells and children "octs". This approach is used, e.g., by the Ramses code (Teyssier 2002). For a more comprehensive review see Keppens et al. (2012).

As an example, we show in Fig. 9.11 the evolution of a double magnetized current sheet in using the patch-based approach of the PLUTO code. The initial

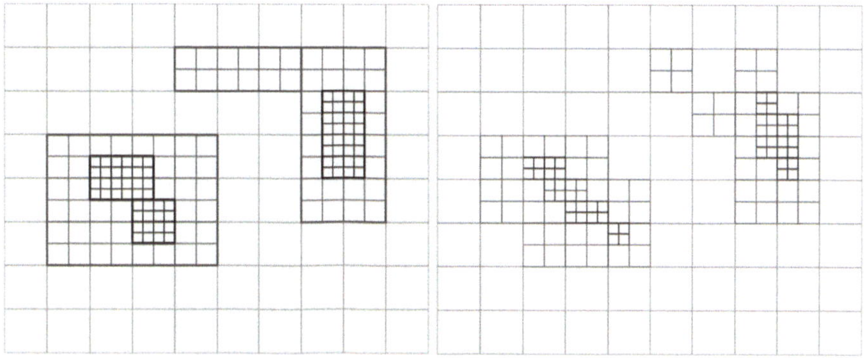

Fig. 9.10 Patch-based (*left*) versus cell-based (*right*) adaptive refinement

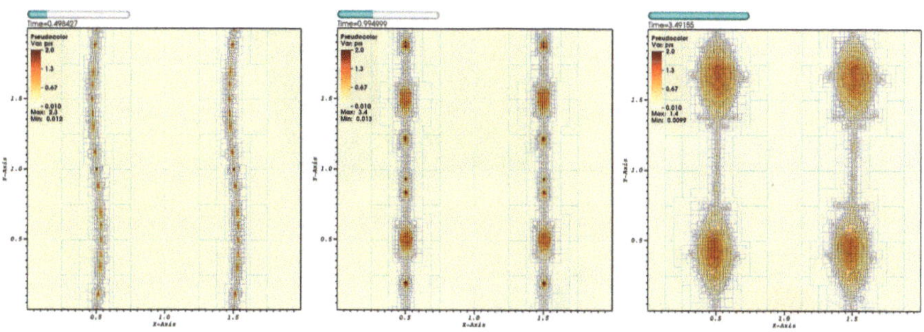

Fig. 9.11 Density maps showing the evolution of the current sheet at $t = 0.5, 1, 3.5$. Refinement levels are overplotted in *blue*

configuration is described in Mignone et al. (2012) and here we report the results using 5 levels of refinement starting from a base grid of 64×64 grid points, yielding an effective resolution of 2048×2048 zones. The current sheets are violently unstable to the tearing instability which drives the formation of islands or plasmoids with larger gas pressure.

Islands propagate parallel to the magnetic field in the y-direction and coalesce to form bigger plasmoids which are often invoked as promising candidates for particle acceleration. As reconnection occurs, the field is dissipated and its magnetic energy is transformed into thermal energy, driving Alfvén and compressional waves which further seed reconnection. The refinement is localized in proximity of the current sheets, where the current is large. This provides a computational speed-up by a factor of ~ 5 over the static grid computation at the equivalent resolution. Larger gains can be obtained when the filling factor of the refined regions becomes smaller and smaller.

Similarly, Fig. 9.12 shows the results from a 3D simulations of the crab nebula using the PLUTO code (see Olmi et al. (2016) for a more detailed description and similar computations has been presented earlier by Porth et al. 2014). The computation has been performed using Cartesian coordinates starting from a base

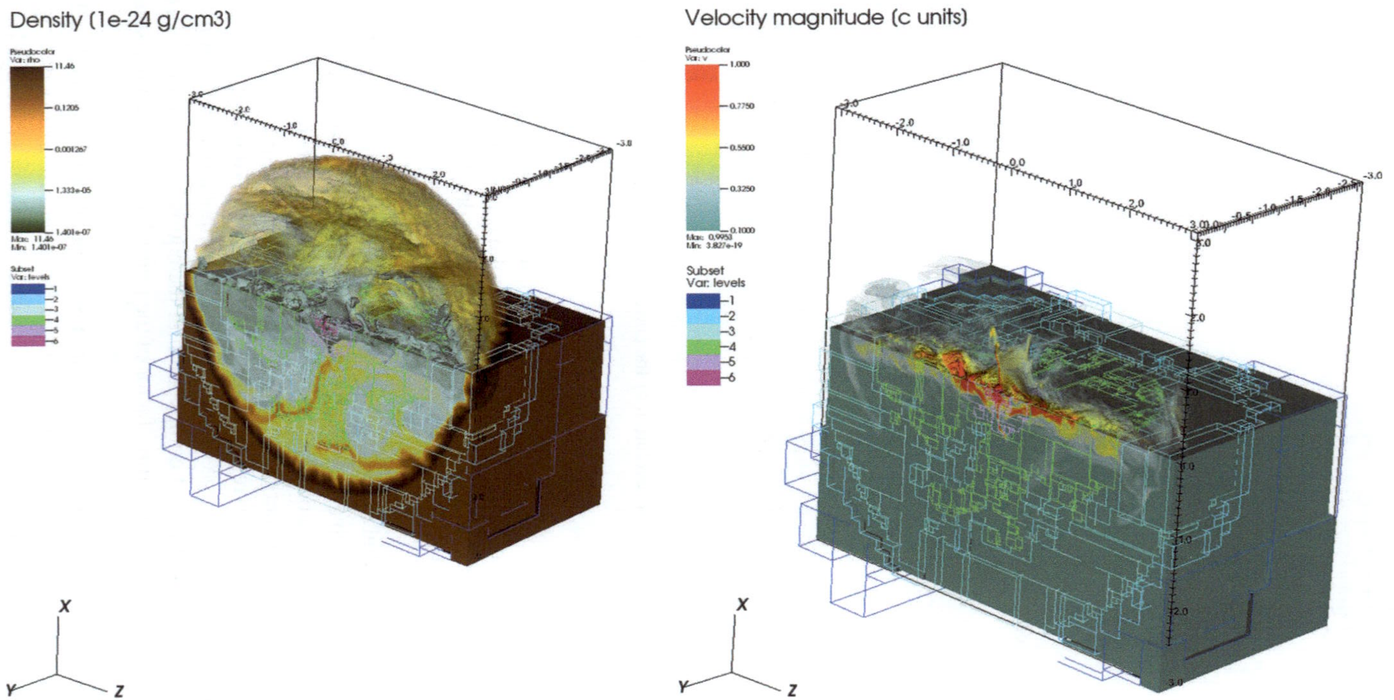

Fig. 9.12 Density (*left*) and velocity magnitude (*right*) maps showing the (*inner*) structure of the pulsar wind nebula at $t = 500$ years. Refinement levels are oveplotted. Figure courtesy of B. Olmi

grid of 128^3 zones and using 7 levels of refinements with consecutive grid jump ratio of 2 except for the finest level which has a jump of 4. This computation extends the results shown in Sect. 9.4.4 to the three-dimensional case and it is based on the same configuration parameters. For the present case, $\sigma = 1.0$, $\gamma = 10$ have been employed. The figure shows density and velocity structure of the nebula at $t = 500$ years as well as the nested level structure which is used to refine the (extremely small) central region around the relativistic cold wind. The gain factor is, in this case, enormous and a static grid approach would make the problem unbearable in terms of computational resources.

References

Anile, A.M.: Relativistic Fluids and Magneto-Fluids: With Applications in Astrophysics and Plasma Physics. Cambridge University Press, Cambridge (1989)

Anton, L., et al.: Relativistic magnetohydrodynamics: renormalized eigenvectors and full wave decomposition Riemann solver. Astrophys. J. Suppl. Ser. **188**, 1–31 (2010)

Balsara, D.: Total variation diminishing scheme for relativistic magnetohydrodynamics. Astrophys. J. Suppl. Ser. **132**, 83–101 (2001)

Batten, P., Clarke, N., Lambert, C., Causon, D.M.: On the choice of wavespeeds for the HLLC Riemann solver. SIAM J. Sci. Comput. **18**, 1553–157 (1997)

Bellan, P.M.: Fundamentals of Plasma Physics. Cambridge University Press, Cambridge (2006)

Berger, M.J., Colella, P.: Local adaptive mesh refinement for shock hydrodynamics. J. Comput. Phys. **82**, 67–84 (1989)

Boyd, T.J.M., Sanderson, J.J.: The Physics of Plasmas. Cambridge University Press, New York (2003)

Bryan, G.L., et al.: ENZO: an adaptive mesh refinement code for astrophysics. Astrophys. J. Suppl. Ser. **211**, 52 pp. (2014)

Cargo, P., Gallice, G.: Roe matrices for ideal MHD and systematic construction of roe matrices for systems of conservation laws. J. Comput. Phys. **136**, 446–466 (1997)

Chiuderi, C., Velli, M.: Basics of Plasma Astrophysics. Springer, Berlin (2015)

Davis, S.F.: Simplified second-order godunov-type methods. SIAM J. Sci. Stat. Comput. **9**(3), 445–473 (1988)

Del Zanna, L., Zanotti, O., Bucciantini, N., Londrillo, P.: ECHO: a Eulerian conservative high-order scheme for general relativistic magnetohydrodynamics and magnetodynamics. Astron. Astrophys. **473**, 11–30 (2007)

Dumbser, M., Zanotti, O.: Very high order $P_N P_M$ schemes on unstructured meshes for the resistive relativistic MHD equations. J. Comput. Phys. **228**, 6991–7006 (2009)

Einfeldt, B., Munz, C.D., Roe, P.L., Sjögreen, B.: On Godunov-type methods near low densities. J. Comput. Phys. **92**, 273–295 (1991)

Fryxell, B., et al.: FLASH: an adaptive mesh hydrodynamics code for modeling astrophysical thermonuclear flashes. Astrophys. J. Suppl. Ser. **131**, 273–334 (2000)

Giacomazzo, B., Rezzolla, L.: The exact solution of the Riemann problem in relativistic magnetohydrodynamics. J. Fluid Mech. **562**, 223–259 (2006)

Godunov, S.K.: A finite difference method for the computation of Discontinuous solutions of the equations of fluid dynamics. Mat. Sb. **47**, 357–393 (1959)

Gurski, K.F.: An HLLC-type approximate Riemann solver for ideal magnetohydrodynamics. SIAM J. Sci. Comput. **25**, 2165 (2004)

Harten, A., Lax, P.D., van Leer, B.: On upstream differencing and godunov-type schemes for hyperbolic conservation laws. SIAM Rev. **25**(1), 35–61 (1983)

Honkkila, V., Janhunen P.: HLLC solver for ideal relativistic MHD. J. Comput. Phys. **223**, 643–656 (2007)

Keppens, R., Meliani, Z., van Marle, A.J., Delmont, P., Vlasis, A., van der Holst, B.: Parallel, grid-adaptive approaches for relativistic hydro and magnetohydrodynamics. J. Comput. Phys. **231**, 718–744 (2012)

Khokhlov, A.M.: Fully threaded tree algorithms for adaptive refinement fluid dynamics simulations. J. Comput. Phys. **143**, 519–543 (1998)

Kim, J., Balsara, D.S.: A stable HLLC Riemann solver for relativistic magnetohydrodynamics. J. Comput. Phys. **270**, 634–639 (2014)

Koldoba, A.V., Kuznetsov, O.A., Ustyugova, G.V.: An approximate Riemann solver for relativistic magnetohydrodynamics. Mon. Not. R. Astron. Soc. **333**, 932–942 (2002)

LeVeque, R.J.: Finite volume methods for hyperbolic systems. Cambridge University Press, Cambridge (2002)

Li, S.: An HLLC Riemann solver for magneto-hydrodynamics. J. Comput. Phys. **203**, 344–357 (2005)

McCorquodale, P., Colella, P.: A high-order finite-volume method for conservation laws on locally refined grids. Commun. Appl. Math. Comput. Sci. **6**, 1–25 (2011)

Mignone, A.: A simple and accurate Riemann solver for isothermal MHD. J. Comput. Phys. **225**, 1427–1441 (2007)

Mignone, A.: High-order conservative reconstruction schemes for finite volume methods in cylindrical and spherical coordinates. J. Comput. Phys. **270**, 784–814 (2014)

Mignone, A., Bodo, G.: An HLLC Riemann solver for relativistic flows – II. magnetohydrodynamics. Mon. Not. R. Astron. Soc. **368**, 1040–1054 (2006)

Mignone, A., Massaglia, S., Bodo, G.: Astrophysical jet simulations: comparing different numerical methods. Astrophys. Space Sci. **293**, 199–207 (2004)

Mignone, A., Bodo, G., Massaglia, S., Matsakos, T., Tesileanu, O., Zanni, C., Ferrari, A.: PLUTO: a numerical code for computational astrophysics. Astrophys. J. Suppl. Ser. **170**, 228–242 (2007)

Mignone, A., Ugliano, M., Bodo, G.: A five-wave Harten-Lax-van Leer Riemann solver for relativistic magnetohydrodynamics. Mon. Not. R. Astron. Soc. **393**, 1141–1156 (2009)

Mignone, A., Bodo, G., Ugliano, M.: Approximate Harten-Lax-van Leer Riemann solvers for relativistic magnetohydrodynamics. In: Vázquez-Cendón, E., et al. (eds.) Numerical Methods for Hyperbolic Equations, pp. 219–226. CRC Press, Leiden (2012)

Mignone, A., Zanni, C., Tzeferacos, P., van Straalen, B., Colella, P., Bodo, G.: The PLUTO code for adaptive mesh computations in astrophysical fluid dynamics. Astrophys. J. Suppl. Ser. **198**, 31 pp. (2012)

Miyoshi, T., Kusano, K.: A multi-state HLL approximate Riemann solver for ideal magnetohydrodynamics. J. Comput. Phys. **208**, 315–344 (2005)

Olmi, B., Del Zanna, L., Amato, E., Bandiera, R., Bucciantini, N.: On the magnetohydrodynamic modelling of the Crab nebula radio emission. Mon. Not. R. Astron. Soc. **438**, 1518–1525 (2014)

Olmi, B., Del Zanna, L., Amato, E., Bucciantini, N.: Constraints on particle acceleration sites in the Crab nebula from relativistic magnetohydrodynamic simulations. Mon. Not. R. Astron. Soc. **449**, 3149–3159 (2015)

Olmi, B., Del Zanna, L., Amato, E., Bucciantini, N., Mignone, A.: Multi-D magnetohydrodynamic modelling of pulsar wind nebulae: recent progress and open questions. J. Plasma Phys. **82**, 30 pp. (2016)

Porth, O., Komissarov, S.S., Keppens, R.: Three-dimensional magnetohydrodynamic simulations of the Crab nebula. Mon. Not. R. Astron. Soc. **438**, 278–306 (2014)

Roe, P.L.: Approximate Riemann solvers, parameter vectors, and difference schemes. J. Comput. Phys. **135**, 250–258 (1981)

Shumlak, U., Loverich, J.: Approximate Riemann solver for the two-fluid plasma model. J. Comput. Phys. **187**, 620–638 (2003)

Teyssier, R.: Cosmology hydrodynamics with adaptive mesh refinement. A new high resolution code called RAMSES. Astron. Astrophys. **385**, 337–364 (2002)

Toro, E.: Riemann Solvers and Numerical Methods for Fluid Dynamics: A Practical Introduction. Springer, Berlin (1997)

Toro, E.F., Spruce, M., Speares, W.: Restoration of the contact surface in the HLL-Riemann solver. Shock Waves **4**, 25–34 (1994)

Vaidya, B., Mignone, A., Bodo, G., Massaglia, S.: Astrophysical fluid simulations of thermally ideal gases with non-constant adiabatic index: numerical implementation. Astron. Astrophys. **580**, A110 (2015)

van der Holst, B., et al.: A multidimensional grid-adaptive relativistic magnetofluid code. Comput. Phys. Commun. **179**, 617–627 (2008).

Zanotti, O., Fambri, F., Dumbser, M.: Solving the relativistic magnetohydrodynamics equations with ADER discontinuous Galerkin methods, a posteriori subcell limiting and adaptive mesh refinement. Mon. Not. R. Astron. Soc. **452**, 3010–3029 (2015)

Chapter 10
Multidimensional Relativistic MHD Simulations of Pulsar Wind Nebulae: Dynamics and Emission

Luca Del Zanna and Barbara Olmi

Abstract Pulsar Wind Nebulae, and the Crab nebula in particular, are the best cosmic laboratories to investigate the dynamics of magnetized relativistic outflows and particle acceleration up to PeV energies. Multidimensional MHD modeling by means of numerical simulations has been very successful at reproducing, to the very finest details, the innermost structure of these synchrotron emitting nebulae, as observed in the X-rays. Therefore, the comparison between the simulated source and observations can be used as a powerful diagnostic tool to probe the physical conditions in pulsar winds, like their composition, magnetization, and degree of anisotropy. However, in spite of the wealth of observations and of the accuracy of current MHD models, the precise mechanisms for magnetic field dissipation and for the acceleration of the non-thermal emitting particles are mysteries still puzzling theorists to date. Here we review the methodologies of the computational approach to the modeling of Pulsar Wind Nebulae, discussing the most relevant results and the recent progresses achieved in this fascinating field of high-energy astrophysics.

10.1 Introduction

A pulsar wind nebula (PWN) is a particular class of supernova remnant (SNR) in which the radiation is dominated at all wavelengths by non-thermal mechanisms, namely synchrotron and Inverse Compton (IC) emission by electrons and positrons accelerated to ultra-relativistic velocities. Compared to standard SNRs, their emission is stronger in the central regions rather than at the limbs, this is the reason why PWNe are also known as *plerions*. To date, more than one hundred PWNe have been discovered in our Galaxy and they are among the most luminous sources of the sky in the high-energy bands of emission, X and γ-rays. Due to its vicinity ($\sim$2 kpc) and relatively young age ($\sim$1000 years), the Crab nebula is considered as the PWN

L. Del Zanna (✉) • B. Olmi
Dipartimento di Fisica e Astronomia, Università degli Studi di Firenze, Via G. Sansone 1, 50019 Sesto Fiorentino (FI), Italy
e-mail: luca.delzanna@unifi.it; barbara.olmi@unifi.it

D.F. Torres (ed.), *Modelling Pulsar Wind Nebulae*, Astrophysics and Space Science Library 446, DOI 10.1007/978-3-319-63031-1_10

prototype, it is certainly the best observed object of its class, and probably the most studied astrophysical source beyond the solar system. Many exhaustive reviews on PWNe have been published in the last decade (Gaensler and Slane 2006; Hester 2008; Bühler and Blandford 2014; Kargaltsev et al. 2015), thus we refer the reader to these papers for observational details and phenomenology.

From a theoretical point of view, PWNe represent fantastic astrophysical laboratories, namely for the dynamics of relativistic plasmas, for high-energy conversion mechanisms, and for particle acceleration up to extreme (PeV) energies. They are the best investigated example of *cosmic accelerators*, and almost certainly the principal antimatter factories in the Galaxy. Moreover, PWNe serve as close and well observed benchmarks to test models for similar physics encountered in other classes of sources, like the engines of gamma-ray bursts (GRBs) or active galactic nuclei (AGNs), thus their modeling is of paramount importance for the whole field of high-energy astrophysics. For updated reviews on theoretical modeling of PWNe and open problems subject of current research see Arons (2012), Amato (2014), Bucciantini (2014) and Olmi et al. (2016).

As already suggested in the early days of the discovery of pulsars (Pacini 1967, 1968; Gold 1968), the Crab nebula and the other PWNe, especially young ones, are very efficiently powered (up to 30%) by the spin-down energy losses from a neutron star (NS) located at their center. PWNe are basically hot bubbles of magnetized plasma, arising from the confinement of the relativistic wind from a rapidly rotating and strongly magnetized pulsar, in the form of leptons and electromagnetic fields, by the external slowly expanding ejecta of the SNR. At the interface between the pulsar wind (PW) and the plerionic nebula, a reverse MHD shock forms, generally referred to as the termination shock (TS). This is the location where the wind bulk flow converts into disordered motion (heat), and where pairs (possibly even ions) are accelerated, becoming responsible for the synchrotron and IC non-thermal emission from the nebula. Outside the PWN, the SNR expands in the interstellar medium (ISM) at supersonic but sub-relativistic speed, as a blast wave.

Time-dependent one-zone models and steady-state relativistic hydro/MHD spherically symmetric models (Pacini and Salvati 1973; Rees and Gunn 1974; Kennel and Coroniti 1984a,b) reproduced quite well the observed evolution, spectrum, and global structure of PWNe, thus the theoretical picture arising from these works has been considered as rather successful up to the turn of the century. In Fig. 10.1 we sketch the classical radially symmetric model of the Crab nebula proposed the cited models, useful to recognize the main components of the PWN/SNR system.

A revival of interest in PWNe came after the first detailed images of the Crab nebula from the X-ray Chandra satellite (Weisskopf et al. 2000). Its inner structure revealed a wealth of fine details such as bright and variable features, close to the location of the pulsar, and an unexpected axisymmetric *jet-torus* structure. This was later found to be common to all other young PWNe where the inner morphology is discernible, most notably around the Vela pulsar (Helfand et al. 2001; Pavlov et al. 2001; Ng and Romani 2004). The torus of enhanced emission appears to be asymmetric in brightness (a clear signature of the Doppler boosting effect), and the

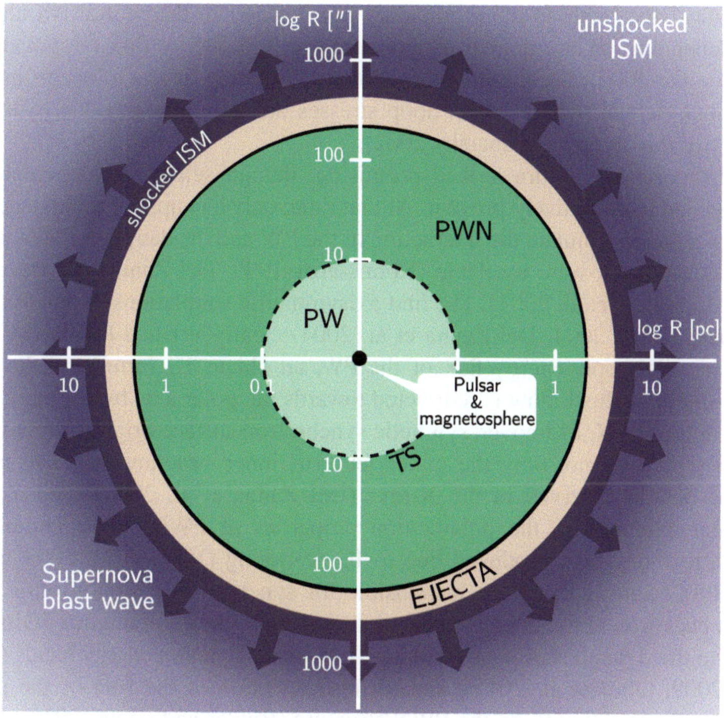

Fig. 10.1 The radial structure of a PWN/SNR system for the case of the Crab nebula, with distances in logarithmic scale expressed in pc and arcsec along the x and y axes, respectively (adapted from Kennel and Coroniti 1984a). The central region contains the ultra-relativistic, cold pulsar wind (PW), surrounded by the hot bubble of magnetized plasma, that is the PWN itself shining with synchrotron light. The nebula is bounded by the radii of the wind termination shock ($R_{\rm TS} \simeq 0.1$ pc) and of the contact discontinuity with the external dense material of the SNR ejecta ($R_{\rm PWN} \simeq 2$ pc). Beyond the freely expanding ejecta, the ISM material is shocked by the SNR blast wave propagating outwards

same is true for the two polar jets, mainly aligned with the symmetry axis. The torus is threaded by rings and the innermost one delimits a dark zone of absence of emission. Bright and variable *knots* are observed along the inner ring and close to the pulsar. Moving features with mildly relativistic speeds are observed along the jets (that *kink* at their extremes) and in the equatorial region before the torus, as outgoing arc-like features named *wisps*, also seen in the optical band (Scargle 1969; Hester et al. 2002).

This situation was felt by theoretical astrophysicists as a new, exciting challenge. In particular, the presence of jets originating very near the pulsar, apparently within the PW itself (identified as the dark region inside the inner ring), appeared puzzling, since plasma collimation by magnetic fields in a relativistic outflow is known to be inefficient. However, provided the torus of enhanced emission could arise from a higher energy input of the wind in the equatorial zone, the TS would then assume an oblate shape, with an inner boundary much closer to the central NS along the axis, depending on the degree of anisotropy in the energy flux (Bogovalov and Khangoulian 2002). Such a situation allows collimation of jets to occur inside the post-shock nebula, where MHD hoop stresses are now efficient due to the mild velocity of the outflow (Lyubarsky 2002).

The theoretical picture looked promising, though self-consistent calculations were needed to definitively prove it. At that time, only a couple of relativistic MHD codes, needed to simulate the interaction of the PW and the daughter PWN with the external remnants, were available (Komissarov 1999; Del Zanna and Bucciantini 2002; Del Zanna et al. 2003). The first axisymmetric simulations (Komissarov and Lyubarsky 2003, 2004; Del Zanna et al. 2004) clearly verified Lyubarsky's idea: given an anisotropic energy flux of the PW, an oblate TS forms and the mildly relativistic post-shock flow is redirected towards the polar axis by magnetic forces acting in the PWN. Moreover, synthetic synchrotron surface brightness maps were able to reproduce not only the major jet-torus inner structure, but also most of the fine details observed in the X-rays (Del Zanna et al. 2006). The modeling was later extended to the polarization properties of PWNe (Bucciantini et al. 2005a), to the γ-rays produced due to IC scattering (Volpi et al. 2008) (by the same ultra-relativistic leptons responsible for synchrotron emission onto various photon background fields), to the detailed modeling of radio emission (Olmi et al. 2014), to the dynamics of wisps variability in various photon energy bands (Camus et al. 2009; Olmi et al. 2015), and to the inspection of the role of the pulsar's magnetosphere obliquity on the PWN structure (Bühler and Giomi 2016). In spite of its simplicity and of the highly successful results obtained, axisymmetry was then abandoned in favor of full 3-D simulations (Porth et al. 2013, 2014a; Olmi et al. 2016), allowing for a more realistic distribution and dissipation of the nebular magnetic field.

It is not our intention to cover in the present review all topics related to PWN numerical modeling. For instance, PWNe will be used as benchmarks to test future X-ray polarimetry instruments, thus synchrotron diagnostic tools based on MHD modeling will be certainly needed (Del Zanna et al. 2006; Bucciantini et al. 2005a). Inside the hot and magnetized PWN bubble, the development of fluid and MHD instabilities are expected (Kelvin-Helmholtz around the TS, Rayleigh-Taylor at the contact discontinuity with ejecta, kink-like in jets, tearing in thin current sheets), and numerical simulations are a very useful tool to investigate their evolution in the relativistic regime (Bucciantini and Del Zanna 2006; Bucciantini et al. 2004a; Porth et al. 2014b; Mizuno et al. 2011; Mignone et al. 2013; Del Zanna et al. 2016). A rich and complex phenomenology comes from the modeling of evolved PWNe interacting with different environments, like the ISM in the case of bow-shock nebulae from pulsars in supersonic motion Bucciantini et al. (2005b) or stellar winds in the case of peculiar binary systems (Bogovalov et al. 2012; Bosch-Ramon et al. 2012; Dubus et al. 2015). A similar approach together with the tools employed in the investigation of PWNe have been also applied to the modeling of long and short GRBs: in this scenario the central engine is a millisecond proto-magnetar inflating a relativistic magnetized bubble confined by the stellar (or NS-NS merger) external envelope, through which jets collimated by hoop stresses will eventually penetrate, thus sharing a very similar dynamics with PWNe (Komissarov and Barkov 2007; Bucciantini et al. 2009, 2012).

In the following we shall mainly concentrate on the modeling of *standard* (young) PWNe, focussing on multidimensional (both 2-D axisymmetric and full 3-D) relativistic MHD simulations of the Crab nebula and similar objects. First the physical framework and the methods employed in the above cited works will be reviewed, including the recipes for computing the non-thermal emission on top of the numerical results. The main findings and achievements provided by relativistic MHD modeling will be described in separate sections, where (partial) answers to the open issues of PWN physics will be proposed. Finally, conclusions will be drawn and the possible future of MHD modeling will be discussed.

10.2 The Setup for Numerical Modeling

As outlined in the introduction and as will be described in details in the remainder of this chapter, the (relativistic) MHD assumption has proved to be very successful for PWNe modeling, starting from the seminal works (Kennel and Coroniti 1984a,b). We recall that the MHD assumption is valid when: (1) Larmor radii of particles are much smaller than the typical size of the nebula, in this case the TS radius; (2) a one-fluid approximation can be assumed, that is a pure pair plasma is considered, neglecting the possible presence of ions; (3) radiative losses are small, in the sense that only the distribution functions of non-thermal accelerated particles are affected, while preserving a Maxwellian shape for particles composing the bulk plasma (for which the synchrotron lifetime is longer than the age of the nebula). These three conditions appear to be all met in the Crab nebula and similar young sources.

The initialization of any MHD simulation of the PWN/SNR system is basically the same regardless of the symmetries imposed. A relativistic wind with radial flow must be injected from the inner boundary of the numerical box, whereas radially expanding SNR ejecta and a static and unperturbed ISM must be initialized elsewhere. Notice that the PWN itself is not defined at the initial time, while it is expected to be created self-consistently during the computation of the system evolution. In the following we shall discuss in some details the equations, methods, and basic initialization needed to perform relativistic MHD simulations of PWNe.

10.2.1 Relativistic MHD and Numerical Methods

The full and accurate description of the dynamics of a pulsar wind, from its origin within the NS magnetosphere to its interaction with the supernova remnant (and of the latter with the ISM), is certainly challenging not only for the very different spatial scales (a large factor separates the inner light cylinder radius from the TS radius, namely $R_{TS}/R_{LC} \sim 10^9$), but also due to the diverse physical regimes involved. This is important from the point of view of numerical modeling, as it is not easy to match different regions where the evolution is described by different set of equations.

The presence of the huge magnetic field anchored to the NS crust, $B \sim 10^{12}$ G for standard pulsars and even $B \sim 10^{15}$ G for magnetars, and of the electric fields induced by rotation, allows one to employ the *force-free electrodynamics* (FFE) approximation within the magnetosphere: electromagnetic forces dominate and balance themselves, while the plasma inertia and pressure forces may be neglected, as shown in the first 3-D simulations (Spitkovsky 2006). Moreover, a precise modeling of the NS magnetosphere would require to include general relativistic effects, either in the FFE regime (Pétri 2016), in the MHD fluid approximation (Pili et al. 2014; Bucciantini et al. 2015), or employing full particle-in-cell (PIC) simulations (Philippov et al. 2015). In the wind zone, the escaping particles (pairs and maybe protons) are accelerated to relativistic velocities, though their thermal speed remains negligible, so that a cold relativistic fluid-like description is more appropriate (Bogovalov 1999). After the TS the plasma is heated to relativistically hot temperatures and the magnetic field is amplified up to equipartition, allowing for a full relativistic MHD description. At the external boundary of the PWN, after the contact discontinuity radius R_{PWN}, the slowly expanding supernova ejecta follow standard Newtonian hydrodynamics, up to their outer regions where they merge into the static ISM (through a bow-shock transition in the case of pulsars in supersonic motion Bucciantini et al. 2005b).

The complexity of the physics described above clearly explains why numerical simulations are limited to either just the magnetosphere and the PW launching region, or just to the PWN and to its interaction with the environment. In the latter case, the wind before the TS is prescribed as an inner boundary condition, and the evolution of the PWN system is followed by solving the relativistic MHD equations (in a flat spacetime). Even if the external medium is non-relativistic, the standard equation of state for an ideal gas with constant adiabatic index is typically adopted.

The equations of (special) relativistic MHD, written in conservative form as appropriate for the use in shock-capturing numerical codes, express the conservation of mass, momentum, and energy as follows:

$$\frac{\partial}{\partial t}(\rho\gamma) + \nabla \cdot (\rho\gamma\,\mathbf{v}) = 0, \tag{10.1}$$

$$\frac{\partial}{\partial t}\left(\frac{w}{c^2}\gamma^2\mathbf{v} + \frac{\mathbf{E}\times\mathbf{B}}{4\pi c}\right) + \nabla \cdot \left[\frac{w}{c^2}\gamma^2\mathbf{v}\mathbf{v} - \frac{\mathbf{E}\mathbf{E}+\mathbf{B}\mathbf{B}}{4\pi} + \left(P + \frac{E^2+B^2}{8\pi}\right)\mathbf{I}\right] = 0, \tag{10.2}$$

$$\frac{\partial}{\partial t}\left(w\gamma^2 - P + \frac{E^2+B^2}{8\pi}\right) + \nabla \cdot \left[w\gamma^2\mathbf{v} + \frac{c}{4\pi}(\mathbf{E}\times\mathbf{B})\right] = 0. \tag{10.3}$$

Here ρ is the rest mass density of the fluid, P the pressure in the local rest frame, $\mathbf{v}$ is the bulk flow velocity (and $\gamma = 1/\sqrt{1 - v^2/c^2}$ is its Lorentz factor), $w = \rho c^2 + 4P$ is the relativistic enthalpy (for a relativistic ideal gas with adiabatic index $4/3$), $\mathbf{E}$, $\mathbf{B}$ are the electric and magnetic field as measured in the laboratory frame, $\mathbf{I}$ is the identity tensor. We recall that in the absence of dissipation, the ideal MHD approximation requires the vanishing of the comoving electric field, that is $\mathbf{E} = -(1/c)\mathbf{v} \times \mathbf{B}$ is used in the above equations. Thus, the electric field is a derived quantity and Maxwell equations just provide the evolution of the magnetic field. In particular, Faraday's law turns into the induction equation

$$\frac{\partial \mathbf{B}}{\partial t} = \nabla \times (\mathbf{v} \times \mathbf{B}), \tag{10.4}$$

to be supplemented with the no-monopole constraint $\nabla \cdot \mathbf{B} = 0$.

Modern shock-capturing codes employed for PWN simulations (Komissarov 1999; Del Zanna and Bucciantini 2002; Del Zanna et al. 2003, 2007; Mignone et al. 2007, 2012; Porth et al. 2014c) solve the above ideal relativistic MHD equations by using reconstruction-evolution methods based on finite-differences or finite-volumes schemes. Due to the presence of strong shocks and discontinuities, typically the evolution is only second-order accurate in space and time, based on the use of upwind limited reconstruction and Runge-Kutta methods, respectively. The Riemann problem at cell interfaces is usually approximated via a simple and robust two-wave method (see the cited papers for further numerical details), especially in the MHD case. When radially or axially symmetric simulations are performed, spherical coordinates are employed and a (static) grid stretching along the radial direction is enforced to better resolve the delicate wind region and its transition to the inner PWN. In the 3-D case, it is instead convenient to adopt Cartesian coordinates and *adaptive mesh refinement* (AMR) methods (with several levels needed to enforce the ultra-relativistic PW).

10.2.2 Model Setup and 1-D Dynamics

Any numerical simulation of the PWN system evolution is initialized by providing the wind, which must be also injected during the evolution at a given radius (which acts as the inner boundary condition). In addition, initial conditions beyond that radius must comprise the SNR density and velocity distribution, and eventually those pertaining the ISM in which the ejecta are expanding.

The steady-state structure of a PWN evolving inside a slowly expanding SNR can be modeled analytically by using relativistic MHD. The shock jump conditions and the radial profile of all quantities are seen to depend crucially on the wind magnetization parameter, calculated just ahead of the TS and defined as

$$\sigma_0 = \frac{B^2}{4\pi \rho c^2 \gamma^2}. \tag{10.5}$$

This is the ratio between Poynting to kinetic energy fluxes in the (cold) relativistic PW, with a Lorentz factor $\gamma \gg 1$ ($v \to c$), density ρ, and magnetic field B, supposed to be transverse to the flow (only the toroidal component survives at large distances from the pulsar). When 1-D radial models are adopted, the tension of the toroidal field B amplified post-shock builds up for steady energy injection by the PW, and the confinement by the external SNR, supposed to move at velocity $V_{\rm PWN}$, requires that the magnetization is small ($\sigma_0 \ll 1$). In that limit, assuming $B \sim r^{-1}$ and neglecting energy losses in the PWN, the characteristic radii bounding the PWN and its expansion velocity satisfy the two relations (Rees and Gunn 1974)

$$\frac{R_{\rm TS}}{R_{\rm PWN}} \simeq \sqrt{\frac{V_{\rm PWN}}{c}} \simeq \sqrt{\sigma_0}. \tag{10.6}$$

In the case of the Crab nebula an expansion velocity of $V_{\rm PWN} \simeq 1500$ km/s requires a low magnetization value of $\sigma_0 = 5 \times 10^{-3}$, yielding $R_{\rm TS}/R_{\rm PWN} \simeq 0.07$, slightly larger compared to what inferred from observations. The refined models in which the magnetic field is retrieved self-consistently from the MHD equations (Kennel and Coroniti 1984a) indicate as best-fit values $V_{\rm PWN} \simeq 2000$ km/s, $\sigma_0 = 3 \times 10^{-3}$, $R_{\rm TS}/R_{\rm PWN} \simeq 0.05$ (in accordance with the observed values, see Fig. 10.1), assuming a constant pulsar spindown luminosity $L_0 = 5 \times 10^{38}$ erg/s.

In the early stage of the evolution, the environment where the PWN bubble forms and propagates mainly consists of stellar ejecta (considered as cold and unmagnetized for simplicity) in free expansion into the ISM. Only later a reverse shock will form and the evolution enters the Sedov-Taylor phase. Spherically symmetric hydrodynamics and relativistic MHD simulations of the PWN/SNR system in the two phases can be found in van der Swaluw et al. (2001) and Bucciantini et al. (2003, 2004b). Most of the multidimensional MHD simulations of PWNe are usually limited to the initial free expansion phase (as relevant for the Crab nebula), and assume the same self-similar model for the cold ejecta as adopted in the cited works. This solution entails a Hubble-type profile for the velocity, a spatially constant density, and for $r \le r_{\rm ej}(t_0) = v_{\rm ej}\, t_0$ ($r_{\rm ej}$ is the external radius of the SNR at time $t = t_0$ after the explosion) we have

$$v(r, t_0) = v_{\rm ej} \frac{r}{r_{\rm ej}(t_0)}, \quad \rho_{\rm ej}(t_0) = \frac{3}{4\pi} \frac{M_{\rm ej}}{r_{\rm ej}^3(t_0)}, \quad v_{\rm ej} = \sqrt{\frac{10}{3} \frac{E_{\rm ej}}{M_{\rm ej}}}, \tag{10.7}$$

where the two latter quantities have been determined by computing the total mass and kinetic energy of the ejecta, respectively. Typical values commonly assumed for 1-D as well as multidimensional numerical models of the Crab nebula are $M_{\rm ej} = 3M_\odot = 6 \times 10^{33}$ g and $E_{\rm ej} = 10^{51}$ erg (Del Zanna et al. 2004; Porth et al. 2014a). Once t_0, or $r_{\rm ej}(t_0)$, is chosen, the above settings define the ambient initial condition in which the nascent PWN will propagate at later times of evolution $t > t_0$. For $r > r_{\rm ej}(t_0)$ standard static ISM conditions can be applied or, if the outer blast wave is not of primary interest, the ejecta solution can be extrapolated up to the external computational boundaries.

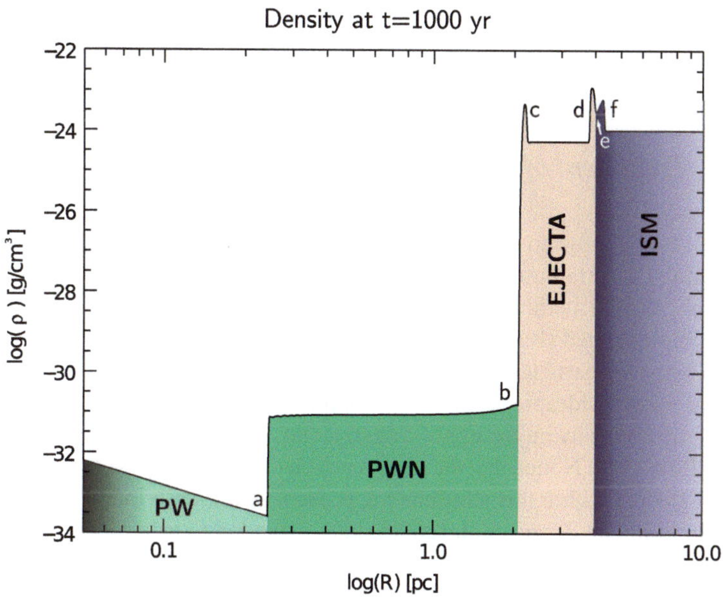

Fig. 10.2 The mass density profile from radially symmetric relativistic MHD simulations of the Crab nebula at the present age of $t = 1000$ yr. The adopted parameters are $L_0 = 5 \times 10^{38}$ erg/s, $\sigma_0 = 3 \times 10^{-3}$, $M_{ej} = 3M_{\odot} = 6 \times 10^{33}$ g and $E_{ej} = 10^{51}$ erg. The PW is shocked at a where the hot PWN is created, b is the contact discontinuity separating the leptonic wind plasma from the swept-up shell of ejecta material. The constant density profile of the SNR ejecta is apparent from c and d, where the latter is the position of the reverse shock arising from interaction with ISM. Beyond the contact discontinuity in e we have the shocked ISM, propagating through the blast wave in f into the unperturbed medium with standard conditions $\rho = 10^{-24}\,\mathrm{g\,cm^{-3}}$ and $T = 10^4$ K

The radial structure of the PWN/SNR system as arising from 1-D relativistic MHD simulation is shown in Fig. 10.2, where we adopt the same notation and colors as in the sketch of Fig. 10.1 and the same labeling for discontinuities as in Bucciantini et al. (2003). The mass density is plotted in logarithmic scale, for the already cited settings relevant for the Crab nebula as employed in semi-analytical models (Kennel and Coroniti 1984a), and for the ejecta profile in Eq. (10.7). The initial decrease as $\sim r^{-2}$ is the PW zone, the plateau between a (the TS) and b (the contact discontinuity) is the hot PWN. Notice the huge density jump between the light leptonic material and the swept-up shell of dense ejecta, confined by the shock in c, which is challenging to handle even for shock-capturing schemes. Given the self-similar solution for the ejecta employed here (and for constant L_0), the PWN is expected to expand as $R_{\mathrm{PWN}}(t) \sim t^{6/5}$ (Reynolds and Chevalier 1984). The external SNR blast wave propagates into the unperturbed ISM through the shock in f, heating the material and driving a shock backwards from e to d into the constant-density ejecta. When this reverse shock reaches the PWN, the free expansion stage ends, and after a transition phase in which ejecta are fully heated the SNR expansion enters the Sedov-Taylor stage, appropriate for the evolution of old PWN/SNR systems (e.g. the Vela nebula), not treated here.

10.2.3 Treatment of Non-thermal Emission

Here we discuss how to extract synthetic synchrotron surface brightness maps and integrated spectra based on the results of multidimensional MHD simulations. While sophisticated methods relying on the kinetic transport of the emitting particles could be implemented (Bai et al. 2015; Porth et al. 2016), simpler recipes that still take into account particles advection, adiabatic losses and synchrotron radiative losses are more widespread in the community. These simplified (though less accurate) methods rely on the use of *tracers* to be added as new evolution equations to the MHD system. Notice that the back reaction of energy losses of particles on the fluid is neglected within this simplified approach. A full description of the method can be found in Del Zanna et al. (2006), later extended to gamma-ray IC emission (Volpi et al. 2008), and further refined (Camus et al. 2009; Olmi et al. 2014).

The broadband, non-thermal spectrum of PWNe, and of the Crab nebula in particular, is known to be best reproduced if the contribution by two families of emitting electrons are considered: one responsible for the low energy emission (radio particles) and one for the high energy emission (optical/X-ray particles) (Atoyan and Aharonian 1996). While radio electrons have a very long radiative life and can be, indifferently, either evolved or uniformly distributed throughout the nebula (Olmi et al. 2014), the X-ray family must be injected at the TS and evolved accurately, since synchrotron burn-off is expected to result in the diverse appearance of PWNe at different photon energies. In addition, different species of electrons may be accelerated at different locations along the TS, so that it could be necessary to evolve more than one family of emitting particles (Olmi et al. 2015).

The form of the distribution function of freshly injected particles is usually taken as a power-law of index p (an exponential cut-off may be added), that is

$$f_0(\epsilon_0) \propto \epsilon_0^{-p}, \quad \epsilon_0^{\min} \le \epsilon_0 \le \epsilon_0^{\max}, \tag{10.8}$$

where ϵ_0 is the particle energy, here in units of mc^2 (that is its Lorentz factor). The local distribution function of wind particles, at any location and time of the PWN evolution, is determined by the conservation of particle number along the streamlines, taking into account adiabatic and synchrotron losses as anticipated. If we indicate with n the number density, for each species of accelerated particles, and with n_0 its value immediately downstream of the shock, the result is

$$f(\epsilon) = \left(\frac{n}{n_0}\right)^{4/3} \left(\frac{\epsilon_0}{\epsilon}\right)^2 f_0(\epsilon_0), \quad \epsilon_0 = \left(\frac{n}{n_0}\right)^{-1/3} \frac{\epsilon}{1 - \epsilon/\epsilon_\infty}, \tag{10.9}$$

where $n_0 = \int_{\epsilon_0^{\min}}^{\epsilon_0^{\max}} f_0(\epsilon_0) d\epsilon_0$.

In the above expressions, the factors with n/n_0 are those taking into account the *adiabatic* changes in volume of the fluid element. These quantities are evolved together with the MHD equations, in conservative form, as

$$\frac{\partial}{\partial t}(n\gamma) + \nabla \cdot (n\gamma\, \mathbf{v}) = 0, \quad \frac{\partial}{\partial t}(n_0\, n\gamma) + \nabla \cdot (n_0\, n\gamma\, \mathbf{v}) = 0, \tag{10.10}$$

so that n obeys to the same continuity equation as ρ, while n_0 is simply advected along streamlines after its definition post-shock of the TS. A simplified version of this procedure (Del Zanna et al. 2006), only feasible for $p \simeq 2$, is to neglect the volume changes in the above expression for ϵ_0 and to replace the $(n/n_0)^{4/3}$ term with a term proportional to the fluid pressure P, though in this way we lose the information about where a particle has been injected. Notice that if different families of particles are injected at the TS, two more evolution equations must be solved for each additional species.

A last evolution equation must be introduced for the quantity ϵ_∞, which represents the highest attainable (normalized) energy that any particle may have at any instant and location within the PWN, to be initialized to a large number at injection (say $> 10^{10}$). The ratio $\epsilon/\epsilon_\infty < 1$ provides the radiative losses by modifying the distribution function as in Eq. (10.9). The equation for ϵ_∞ is

$$\frac{\partial}{\partial t}(\rho^{2/3}\epsilon_\infty\gamma) + \nabla \cdot (\rho^{2/3}\epsilon_\infty\gamma\, \mathbf{v}) = -\frac{4e^4 B'^2}{9m_e^3 c^5}\rho^{2/3}\epsilon_\infty^2, \tag{10.11}$$

where the source term is due to synchrotron losses, B' is the magnetic field strength in the fluid rest frame, and isotropy in pitch angle distribution has been assumed.

The local emissivity function, at any time and position in the nebula, is given by the integral (primed quantities refer to the fluid rest frame)

$$j'_\nu(\nu', \mathbf{n}') = \int \frac{2e^4}{3m_e^2 c^3}|\mathbf{B}' \times \mathbf{n}'|^2 \epsilon^2 \mathcal{S}(\nu', \nu'_c) f(\epsilon) \mathrm{d}\epsilon, \tag{10.12}$$

where $\mathbf{n}$ is the direction pointing towards the observer (normal to the plane of the sky) and where the spectral density function $\mathcal{S}$ for synchrotron radiation can be either computed exactly (Rybicki and Lightman 1979), or it can be approximated to a delta function of the critical emission frequency

$$\nu'_c(\epsilon) = \frac{3e}{4\pi m_e c}|\mathbf{B}' \times \mathbf{n}'|\epsilon^2, \tag{10.13}$$

so that the integral above can be calculated analytically. In order to obtain the emissivity in the observer's fixed frame of reference, relativistic corrections must be taken into account, by computing $j_\nu = D^2 j'_\nu$ and $\nu = D\nu'$ in Eq. (10.12), where $D = [\gamma\,(1 - \mathbf{v} \cdot \mathbf{n}/c)]^{-1}$ is the Doppler boosting factor.

In the particular case where the power-law form in Eq. (10.8) is used for particles injected at the TS, and if monochromatic emission is assumed for simplicity, the spectral emissivity function in the laboratory frame becomes

$$j_\nu(\nu, \mathbf{n}) \propto n_0(n/n_0)^{\frac{2+p}{3}} D^{\frac{3+p}{2}} \, |\mathbf{B}' \times \mathbf{n}'|^{\frac{1+p}{2}} (1 - \sqrt{\nu/\nu_\infty})^{p-2} \, \nu^{-\frac{p-1}{2}}. \tag{10.14}$$

Here $\nu_\infty = D\nu_c'(\epsilon_\infty)$ is the cut-off frequency for synchrotron burn-off, and the emissivity will be different from zero only where $\nu < \nu_\infty$. Additional details and the recipes to compute surface brightness maps, spectral index maps, polarization maps, and integrated spectra from the emissivity j_ν can be found in Del Zanna et al. (2006).

10.3 The Flow Dynamics and the Jet-Torus Inner Structure

Let us now discuss the main results found by the numerical community in more than a decade of relativistic MHD simulations of PWNe. The first issue is certainly the inner jet-torus structure revealed by X-ray observations.

The detailed angular dependence of the PW is crucial to determine the shape that the PWN will assume during the evolution in multidimensional simulations. The energy flux is usually initialized as axially symmetric (with respect to the pulsar rotation axis), even in 3-D simulations, and it must be taken anisotropic in order to reproduce the observed jet-torus inner structure (Lyubarsky 2002). This behavior is roughly that predicted by the split-monopole model of aligned rotators (Michel 1973), where a radiation-like toroidal component $B \equiv B_\phi \sim r^{-1} \sin\theta$ dominates at large distances from the light cylinder, for updated reviews on pulsar magnetospheres and their winds see Pétri (2016) and Cerutti and Beloborodov (2016). The angular dependence of a radial, ultra-relativistic, cold wind ($v \equiv v_r \simeq c$, $\gamma = \sqrt{1 - v^2/c^2} \gg 1$, $P \ll \rho c^2$) of constant spindown luminosity L_0 can be thus introduced as

$$L_0 \mathcal{F}(\theta) = 4\pi r^2 c \left(\rho c^2 \gamma^2 + \frac{B^2}{4\pi} \right) = 4\pi r^2 \rho c^3 \gamma^2 [1 + \sigma(\theta)], \tag{10.15}$$

where $\frac{1}{4\pi} \int \mathcal{F}(\theta) d\Omega = 1$ and $\sigma(\theta)$ is the magnetization function. Once $\mathcal{F}(\theta)$ and $\sigma(\theta)$ are chosen, the angular dependence of $\rho\gamma^2$ and B^2 (both decaying as r^{-2}) are also known and the wind is fully characterized (equivalently, one can provide the angular shape of B directly). The wind Lorentz factor can be either taken as constant or dependent on θ, assuming an isotropic mass flux in this latter case (Del Zanna et al. 2004), in any case limited to the range $\gamma = 10$–100 to avoid computational problems.

Various choices are possible for the two angular functions. The radially symmetric case is retrieved with $\mathcal{F} = 1$ and $\sigma = \sigma_0$, where the latter is the usual pulsar

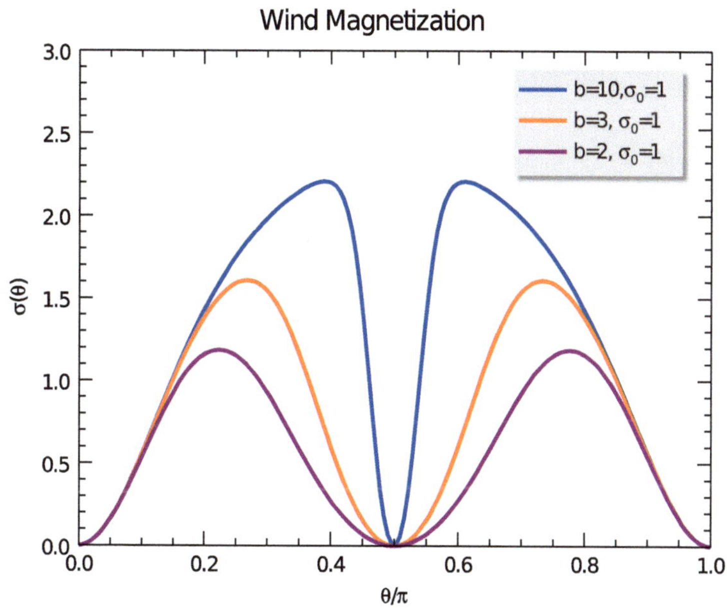

Fig. 10.3 The angular dependence of the wind magnetization function $\sigma(\theta)$, here for the reference value of $\sigma_0 = 1$ employed in Olmi et al. (2016), from which the figure is taken, and three values of the parameter b, shaping the toroidal magnetic field (Del Zanna et al. 2004; Olmi et al. 2014). The *upper (blue) curve* refers to the narrow striped-wind region case, used by the authors to reproduced the Crab nebula observations, the *bottom (purple)* case roughly reproduces the choice in the B3D run by Porth et al. (2014a)

wind magnetization parameter of Eq. (10.5) defined in Kennel and Coroniti (1984a). Already in the very first axisymmetric simulation (Komissarov and Lyubarsky 2003) the toroidal magnetic field was shaped to allow for a striped wind region of lower magnetization around the equatorial plane, see also Del Zanna et al. (2004) and Olmi et al. (2014) for the alternative definitions used by our group. Typical settings, also employed in recent 3-D simulations (Olmi et al. 2016), define a wind energy flux which is ten times higher in the equatorial direction with respect to the polar ones, and a magnetization function shaped as in Fig. 10.3, depending on σ_0 ($B \propto \sqrt{\sigma_0}$) and another parameter b providing the size of the equatorial striped region, in turn depending on the pulsar's magnetosphere inclination angle ($b \to \infty$ for the split-monopole solution without the equatorial region of lower magnetization). More complex assumptions, also valid for strongly inclined rotators, can be found in Komissarov (2013), Porth et al. (2014a), Philippov et al. (2015) and Tchekhovskoy et al. (2016).

The typical flow structure arising in axisymmetric simulations in the cases of low and intermediate magnetization is shown in Fig. 10.4, taken from Del Zanna et al. (2004) to which the reader is referenced for further details. In the left panel we show the flow and shock pattern arising around the TS in the upper meridional plane

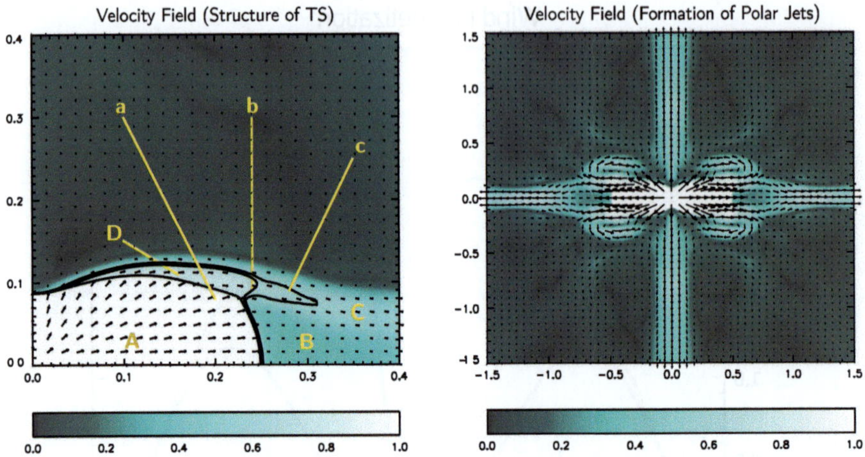

Fig. 10.4 Maps of flow velocity (in units of c) from the axisymmetric simulations in Del Zanna et al. (2004). In the *left panel* the flow structure around the TS, here for an early time $t = 100$ y, is shown for a low-magnetization run without the striped-wind equatorial region ($\sigma_0 = 0.003$, $b \to \infty$; labels are described in the text). In the *right panel* we show the flow structure arising at later times for higher magnetizations and for a case with a (*narrow*) striped wind region (Run A: $\sigma_0 = 0.03$, $b = 10$, here $t = 300$ y). Notice the simultaneous presence of supersonic equatorial flows and polar jets

(here for a run without the striped wind region and imposing equatorial symmetry). Labels refer to: (A) ultra-relativistic wind region; (B) subsonic equatorial outflow; (C) equatorial supersonic funnel; (D) super-fastmagnetosonic shocked outflow; (a) termination shock (TS) front; (b) *rim* shock; (c) fastmagnetosonic surface. The magnetization parameter is that assumed in spherical and stationary models $\sigma_0 = 0.003$, though, due to the anisotropic structure of the injected PW, only in the equatorial region we observe the expected transition to a shocked flow speed of $v \simeq c/3$ (region B). A complex shock pattern arises away from the equatorial belt, and a fast-speed channel with $v \simeq 0.7$ forms along the TS boundary from the poles (region D) and survives at large distances bending towards the equator (region C).

When even a narrow striped-wind region is allowed and the magnetization is high enough, a strong equatorial flow forms and part of it is readily diverted towards the polar axis by *hoop stresses* due the enhanced toroidal magnetic field in the post-shock nebula, as predicted analytically (Bogovalov and Khangoulian 2002; Lyubarsky 2002) and observed in the first relativistic MHD simulations (Komissarov and Lyubarsky 2003, 2004). The new flow pattern is shown in the right panel, for the so-called Run A parameters ($\sigma_0 = 0.03$, $b = 10$). We clearly see the simultaneous presence of supersonic outflows both in the equatorial plane and along the polar axis, in the form of narrow jets (or plumes). The complex pattern around the TS is expected to produce Doppler-boosted features when non-thermal emission is computed on top of the MHD quantities, and the presence of the fast

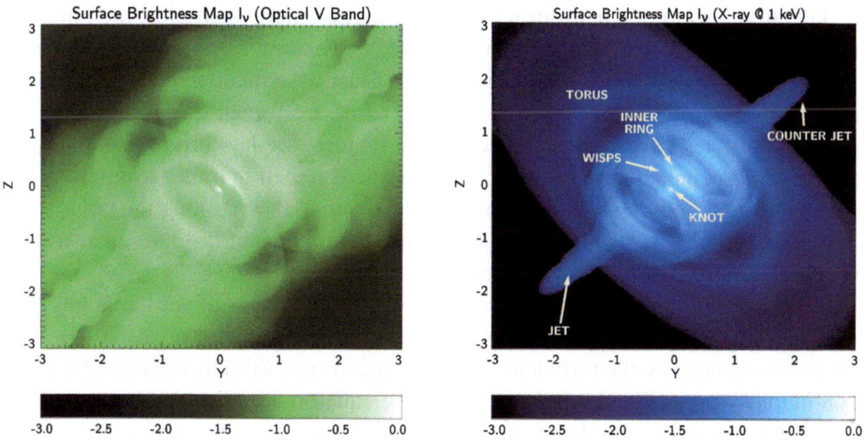

Fig. 10.5 Simulated non-thermal emission from an axisymmetric relativistic MHD model of the Crab nebula. In the *left (right) panel* the optical (X-ray at 1 keV) surface brightness map is shown, taken from Del Zanna et al. (2006). A logarithmic scale is employed (the intensity is normalized to its maximum) and distances from the central pulsar on the axes are measured in ly. In the X-ray image many features which are also present in real data can be identified

outflow will advect the emitting particles at large distances both in the equatorial plane and along the polar axis.

In order to verify whether the flow pattern arising from simulations with a moderate wind magnetization σ_0 is really the correct counterpart of the jet-torus structure observed in the X-rays (Weisskopf et al. 2000), a detailed calculation of the non-thermal emission based on the evolved MHD quantities and particle tracers must be then performed. This was achieved in Del Zanna et al. (2006), where various aspects of the non-thermal emission, computed following the recipes summarized in the previous section, were discussed. Here we report, in Fig. 10.5, the optical and X-ray simulated surface brightness maps for a set of parameters that match the Crab nebula emission properties at best within 2-D modeling (Run A). A square box of 6 ly $\times$ 6 ly is displayed, with intensity in logarithmic scale, normalized against the maximum in each band.

The optical image (left panel) appears to be elongated and rather uniform around the polar axis up to distances of 2–3 ly. No hints of an equatorial torus of enhanced emission are found, while Doppler-boosted arcs, the variable wisps, and a very bright knot are present close to the pulsar position, as observed in HST data (Hester et al. 2002). These are due to a higher value of the magnetic field close to the TS and to the presence of mildly relativistic velocities pointing towards the observer direction, as discussed in Komissarov and Lyubarsky (2004). The knot appears to be located at the base of the polar jet, a candidate site for the γ-ray flares observed in the Crab nebula (Komissarov and Lyutikov 2011; Yuan and Blandford 2015; Lyutikov et al. 2016).

Due to the shorter synchrotron lifetime of higher-energy particles, the inner jet-torus structure is instead clearly visible in the X-ray surface brightness map, which is indeed very similar to what observed, at least at a qualitative level. A polar (hollow) jet and an equatorial (rather faint) torus are found, due to the injection of fresh particles advected by the flow. Both features are less bright than expected for the Crab nebula, due to a lower magnetic field (recall that the assumption of a purely toroidal field in the axisymmetric case requires $\mathbf{B} = 0$ along the polar axis). However, the field is strong near the TS, and several arcs of brighter emission are seen, even more clearly than in the optical band.

Time variability of the high-energy emission and of wisps in particular has been also investigated (Volpi et al. 2008; Camus et al. 2009; Olmi et al. 2015). This motion is induced by the turbulent flow structure around the TS, where vortexes are continuously created and advected away. The equatorial outflow with $v \sim 0.5c$ is highly inhomogeneous and regions with different plasma magnetization are crossed, leading to the variable synchrotron emission observed in the form of expanding wisps. The period corresponding to the largest scale of the turbulent flow is 1–2 y, matching what is observed for the moving features in the Crab nebula (Hester et al. 2002). In the next section we will see how the motion of wisps can be used as a probe for the mechanisms and sites for particle acceleration in PWNe. An example of wisps as observed in synthetic X-ray maps and of the turbulent plasma, responsible for their appearance as quasi-periodical bright features inside regions of enhanced magnetic field, will be shown in Fig. 10.7.

The main parameters characterizing the PW, such as the degree of anisotropy, the magnetization, and the inclination of the pulsar magnetosphere (or equivalently the width of the striped wind region), are of course expected to be different for different sources, so that a detailed modeling based on 2-D MHD simulations can be carried out by varying the mentioned parameters. For instance, a wider striped wind region seems to be more appropriate for the Vela PWN, since the jet-torus structure is less apparent and two boosted arcs appear in the synthetic surface brightness maps (Del Zanna et al. 2006). By assuming wind prescriptions based on 3-D force-free models of the pulsar magnetosphere (Tchekhovskoy et al. 2016), the morphology of the Vela nebula was also reproduced in a recent work (Bühler and Giomi 2016), still based on axisymmetric simulations, though limited to a much shorter evolution time. This was achieved by choosing a very strong magnetization $\sigma_0 = 3$ and an inclination angle of 45°. In the same work, the modeling of the Crab nebula requires a much lower magnetization, regardless of the inclination angle assumed, namely the same $\sigma_0 = 0.03$ adopted in Del Zanna et al. (2006).

10.4 Probing Particle Acceleration Mechanisms

PWNe are known to be among the most efficient cosmic accelerators, given that the measured synchrotron γ-ray spectrum of the Crab nebula and similar sources can only be explained by emitting particles (pairs) which are accelerated up to PeV energies (Arons 2012). One-zone or radial 1-D models suggest that at least two families are responsible for the non-thermal emission: the so-called *radio particles* of lower energies, with a distribution $f_0(\varepsilon_0) \propto \varepsilon_0^{-p}$ of rather flat a power-law index $p \sim 1.5$, and a family of higher energy leptons with $p \sim 2.2$, the *X-ray particles* (Kennel and Coroniti 1984b; Atoyan and Aharonian 1996). The two families appear to be smoothly connected around a particle energy of ~100 GeV, which roughly corresponds to photon frequencies in the optical band for the inferred values of the nebular magnetic field of 200 μG (de Jager and Harding 1992), and in some models even a single family of accelerated particles in the form of a broken power-law is able to reproduce the entire spectrum of different sources, see e.g. Bucciantini et al. (2011).

A self-consistent and combined treatment of the synchrotron and IC radiation components based on relativistic MHD simulations is a very powerful diagnostic technique for PWN modeling, since strong constraints can be placed on the model parameters, like the wind anisotropy, its magnetization, or even the composition (possible presence of ions) (Amato 2014). Unfortunately, this task has been attempted so far on top of multidimensional simulations only in (Volpi et al. 2008), where the recipes for non-thermal emission were extended to the IC mechanism, including the effects of a Klein-Nishina differential cross-section, and different targets were considered (synchrotron, infrared, and cosmic microwave background photons).

The multi-band spectrum (the spectral energy distribution νL_ν) obtained in the cited paper is reported in Fig. 10.6 (right panel) for the Run A parameters, described in the previous section as those producing the surface brightness maps more similar

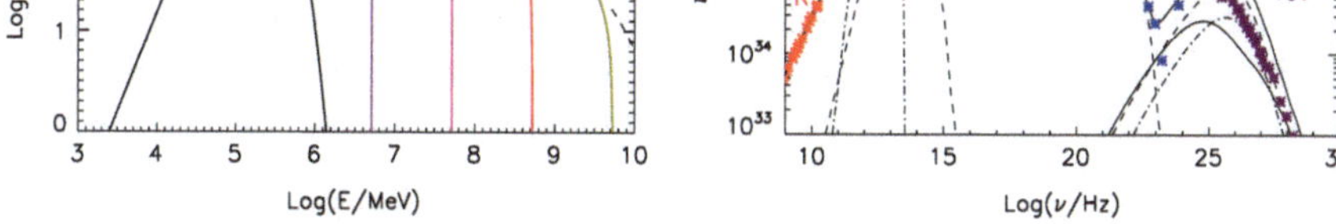

Fig. 10.6 Simulated non-thermal emission from an axisymmetric relativistic MHD model of the Crab nebula, taken from Volpi et al. (2008). In the *left panel* the two *initial* ($\epsilon_\infty \to \infty$) energy distribution functions are shown as functions of $E = \epsilon\, m_e c^2$ (*dashed black lines*), and the total evolved distribution function is shown for various values of the burn-off energy ϵ_∞ (see Eq. (10.11)). In the *right panel*, the resulting photon spectral energy distribution (SED) function is plotted and compared with observational data. The separated contribution to the synchrotron emission by the two families is again indicated by *dashed lines*, whereas the *dot-dashed line* refers to thermal dust emission. The contributions from different incident photon targets to the IC emission are reported with different colors and line-styles

to the images of the Crab nebula in the X-ray. The separate contributions to the synchrotron emission produced by the two families (power laws with exponential cutoff) of electrons are plotted, together with a thermal component needed to model infrared data (emission by local dust). In order to match the synchrotron X-ray and γ-ray data, a steep index of $p \sim 2.7$ (compared to stationary 1-D models) was found to be required for the higher energy particles accelerated at the TS. On the other hand, the electrons responsible for the radio emission could be taken as homogeneous in the whole nebula and constant in time. With these prescriptions, the spectral break observed in the Crab nebula's spectrum between the infrared and the optical bands naturally arises as due to the different spectral indexes assumed for the two distribution functions, separated between 100 and 300 GeV. A second break, at ultraviolet frequencies, and the further steepening of the spectrum, are instead due to burn-off effects.

The steep index required for the optical/X-ray particles is needed to compensate the reduced losses and consequent spectral softening produced by a lower magnetic field, compared to that predicted by one-zone or 1-D models. It was pointed out that, in axisymmetric simulations, the toroidal field is in fact compressed artificially around the TS and towards the polar regions, so that a small value for the average nebular field is found ($\sim$50–100 μG), while previous models aimed at reproducing the full spectrum indicate values which are 2–3 higher. At the same time, the number of emitting particles continuously injected at the TS must be enhanced up to values corresponding to an unnatural 100% of acceleration efficiency from the PW, and this reflects the fact that the computed IC emission is 2–3 times higher than observed. The γ-ray emission for TeV photon energies is found to be dominated by the self-synchrotron component, while minor contributions arise from the cosmic microwave background and the thermal dust emission, in agreement with radial or one-zone models (Atoyan and Aharonian 1996; Meyer et al. 2010).

In spite of the wealth of data concerning the Crab nebula and other PWNe, the origin of the two families is still debated. Radio particles are dominant by number, and if these are continuously originated inside the pulsar's magnetosphere via cascading processes and injected in the relativistic wind first and then in the nebula, an accurate modeling of PWNe emission is expected to provide precious information on their rate of production. As mentioned in the introduction, this has important consequences also on fundamental physical issues, as pulsars are considered to be the most efficient antimatter factories in our Galaxy, and thus they may provide an astrophysical explanation to the measured cosmic ray positron excess (PAMELA coll 2009; AMS-02 coll 2013; Serpico 2012).

Going into deeper details, it is well known that particles constituting the wind and the PWN are not those directly extracted from the neutron star surface, but rather they are originated in a cascade process inside the magnetosphere greatly increasing their number by an unknown *multiplicity* parameter κ. Estimates of this number crucially depend on whether the radio particles are part of the pulsar outflow or not. In the first case, stationary 1-D MHD models predict $\kappa \sim 10^4$ (Kennel and Coroniti 1984a; Atoyan and Aharonian 1996), while if radio particles have a different origin, time-dependent one-zone models predict $\kappa \sim 10^6$ (Bucciantini et al. 2011). In this latter case, the possible presence of ions in the wind would be irrelevant from an energetic point of view. It is thus clear that discriminating between the two scenarios is of fundamental importance for a correct modeling of the PW (Amato 2014).

In a recent study (Olmi et al. 2014), different possibilities for the origin of radio particles have been considered, testing whether it is possible or not to discriminate their origin based on the resulting emission morphology. The question is thus whether radio particles can be considered as part of the pulsar outflow, and accelerated at the TS together with high energy emitting particles, or if they have a different origin and are accelerated somewhere else in the nebula. In the latter case, corresponding to either a primordial burst (relic population, still present thanks to the very long lifetime against synchrotron losses) or to a possible continuous re-acceleration, for example in the external thermal filaments, radio particles can be considered as a steady-state and uniform population. Relativistic MHD axisymmetric simulations were used to test the various scenarios, and results are discussed in detail in the cited paper. Surface brightness maps in the radio band, calculated for the first time based on numerical simulations, appeared to be undistinguishable in the two cases, thus unfortunately no constraints could be placed on the origin of radio particles and on the value of the pair multiplicity parameter. The only scenario that could be rejected was that of a primordial burst of radio particles followed by pure advection, while a continuous replenishment of the nebula seems to be required.

Another important result found in Olmi et al. (2014) is that the appearance in simulated radio maps of bright and time-dependent features around the TS is basically independent on the adopted scenario. These features are named wisps when observed in the optical and X-ray bands, as previously discussed, and they were also found to be present in radio in the Crab nebula (Bietenholz et al. 2004). Long-term variability of radio wisps (a few years) was measured up to $\sim$3 ly from the central pulsar, where the inferred velocity is $\sim 0.002c$, whereas variability on a month scale and much quicker wisps with $\sim 0.4c$ were found in the equatorial torus. Simulations fully reproduce these observations, without invoking different origins for the two classes of radio wisps, thus confirming, on the one hand and once more, the validity of the fluid/MHD description, and on the other hand that these variable features simply trace the underlying plasma flow.

However, wisps in the inner region of the Crab nebula are actually observed in (varying) positions which are not precisely coincident at the different wavelengths of observation, from radio to X-rays (Bietenholz et al. 2004; Hester et al. 2002; Schweizer et al. 2013), suggesting in turn a difference in the acceleration sites of the particles responsible of such emission. Different acceleration mechanisms require very different physical conditions to be effective, and thus identifying the location at which particles have been accelerated can put strong constraints on the mechanism at work. The motion of wisps is then expected to be a powerful diagnostic tool to investigate whether radio and optical/X-ray particles are accelerated in the same site or not.

In another recent numerical work by the same authors (Olmi et al. 2015), different possibilities were considered for injection of particles: in case *I* all particles

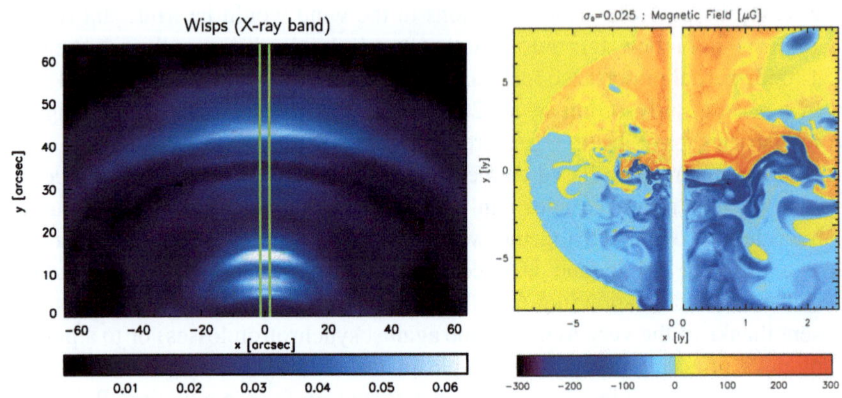

Fig. 10.7 *Left panel*: the simulated synchrotron X-ray surface brightness map at 1 keV, convolved with the *Chandra* point-spread function, in linear scale and expressed in units of mJy arcsec^{-1}, taken from Olmi et al. (2015) (only the upper hemisphere is shown, the *vertical lines* indicate the sector used for extraction of the radial profiles). *Right panel*: the turbulent magnetic field in the whole nebula and around the TS, responsible for the quasi-periodical wisp production, taken from Olmi et al. (2014)

are injected uniformly at the shock surface; in case *II* particles of one family are injected in a wide equatorial sector or in the complementary narrow polar one; in case *III* particles of one family are injected in a narrow equatorial region, or in the complementary wide polar one. The entire evolution of the Crab nebula was then simulated by introducing as many numerical tracers as necessary in order to follow the evolution of particles injected at the different locations. Following the analysis by Schweizer et al. (2013), intensity peaks were extracted from a 3″ wide slice around the polar axis in the upper hemisphere of each surface brightness map, where wisps are more prominent (see Fig. 10.7).

In Fig. 10.8 wisps obtained in the case of uniform injection (case *I*) or injection of X-ray particles in a narrow equatorial region and radio particles in the complementary sector (case *III*) are compared. As expected, when particles of different families are injected at the same location, i.e. across the entire shock surface, wisps appear to be coincident at the different wavelengths. But as soon as they are injected at distinct locations, wisps at radio, optical and X-ray frequencies are no more coincident. The main result of the analysis is that, in order to reproduce the absence of X-ray wisps in the region within ∼6″ from the pulsar, as observed in Schweizer et al. (2013), X-ray particles must be injected in a narrow equatorial sector (the case displayed in the right panel), approximately coincident with the striped region of the wind where dissipation of magnetic field is expected to be most efficient. This may indicate that X-ray particles are produced via Fermi I acceleration in the striped zone, where magnetization can be lowered enough by dissipation so as to allow this mechanism to be effective (Sironi and Spitkovsky 2009). The same conclusion was reached by the analysis of wisps motion in 3-D simulations (Porth et al. 2014a).

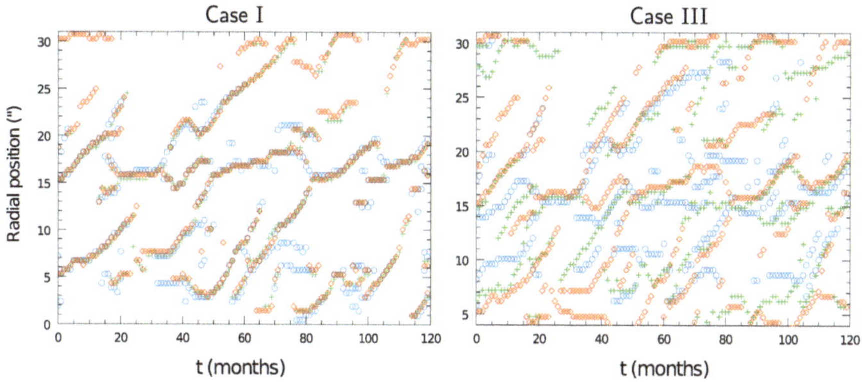

Fig. 10.8 Radial positions of the local intensity maxima (in arcsec) as a function of time (in months) with *orange diamonds* identifying radio wisps (at 5 GHz), *green crosses* optical ones (at 3.75×10^{14} Hz) and *light-blue circles* for X-rays (at 1 keV). On the *left case I* is shown. On the *right case III* is shown, with X-ray particles injected in a narrow equatorial zone and radio ones injected in the complementary sector. Further details in the text and in Olmi et al. (2015)

Moving to lower frequencies, strong constraints on radio emission are again difficult to derive: the only case that can be easily excluded is the one in which radio particles are injected in a narrow polar cone, since no wisps appear to be produced, while even a steady and uniform distribution of radio emitting particles throughout the nebula does not change the overall picture. A possible interpretation of this finding is that radio particles could result from driven magnetic reconnection primarily occurring at moderately high latitudes, where conditions for driven magnetic reconnection to operate as an acceleration mechanism might be locally satisfied (Sironi and Spitkovsky 2011), or even that radio particles are uniformly distributed in the nebula, possibly due to some continuous and ubiquitous re-acceleration mechanism, for instance induced at the Rayleigh-Taylor fingers protruding throughout the nebula (Komissarov 2013).

10.5 From Axisymmetry to Full 3-D: Solution to the σ-Problem?

The so-called σ-problem, or σ-paradox, has puzzled theorists since the very beginning of MHD modeling of PWNe. In addition to the latitudinal dependence of the wind luminosity, needed to provide the jet-torus structure of the PWN arising from the confinement of the wind itself, also the magnetization σ of the PW is expected to vary (by several orders of magnitude) with the radial distance: at the pulsar magnetosphere the relativistic wind is certainly Poynting-dominated ($\sigma \gg 1$ in Eq. (10.5)), whereas at the TS, something like a billion of light cylinder radii farther out, steady-state radially symmetric models predict $\sigma_0 = \text{few} \times 10^{-3}$ to match the PWN expansion (see Eq. (10.6) and comments below). An initial conversion of energy from the electromagnetic field to the bulk inertial flow is expected to occur during the acceleration process within the fast-magnetosonic point (Kirk et al. 2009), whereas beyond that radius, up to the TS, both $\sigma \equiv \sigma_0 \gg 1$ and $\gamma \gg 1$ should remain basically constant, at least for relativistic MHD radial models.

For non-aligned rotators, that is when the magnetosphere of the pulsar is inclined with respect to the rotation axis, the equatorial current sheet where the toroidal radiation-like field reverses its sign wobbles and produces a striped-wind region of reconnected field and lower magnetization, as already discussed. Its (half) angular extent is given by the pulsar inclination angle, and the periodicity of this large-amplitude wave is that of the pulsar rotation (Coroniti 1990). Detailed models of energy conversion within the striped wind region have been proposed, though it is not clear wether the rate of injection of pairs in the wind and the efficiency of the dissipation mechanism (e.g. the tearing instability) are enough to reach a condition of $\sigma_0 \ll 1$ at the TS (Lyubarsky and Eichler 2001; Kirk and Skjæraasen 2003). Additional magnetic dissipation can also arise at the shock location, due to kinetic effects arising when the Larmor radii of particles become comparable with the TS size itself (Lyubarsky 2003; Sironi and Spitkovsky 2011). In any case, even if magnetic to kinetic energy conversion can be efficient around the equator, certainly the polar regions are expected to remain Poynting-dominated, thus leaving the σ-paradox mostly unsolved (Komissarov 2013).

A hint towards the solution comes from the relativistic MHD simulations object of the present review. We have already discussed that when 1-D radial models are abandoned in favor of 2-D axisymmetric ones, the lower tension of the post-shock toroidal field allows to reproduce the observed jet-torus inner structure of PWNe with a moderate magnetization $\sigma_0 \sim$ few $\times 10^{-2}$. However, the magnetic field assumed in these models is purely toroidal, and subject to (slow) diffusion only at the equatorial current sheet. Moreover, in axisymmetric models the nebular magnetization is expected to remain low at all radii, otherwise the pinching effect would create excessive axial elongation (Begelman and Li 1992). Indeed, 2-D simulations show a pile up of magnetic field around the TS and along the poles. But how realistic is the assumption of a purely toroidal field?

The stability of such MHD structure (the *Z-pinch* equilibrium in controlled fusion literature) was investigated by Begelman (1998) in the context of PWNe, who found that it would be very likely destabilized by 3-D kink-type instabilities very close to the TS radius. In particular, magnetized jets endowed with toroidal fields are subject to these current-driven instabilities, as confirmed by relativistic MHD simulations (Mizuno et al. 2011; Mignone et al. 2013), and as even seen in X-ray images of the Crab and Vela PWNe, where time-variable kinks in the polar jets are actually observed (Weisskopf et al. 2000; Hester et al. 2002; Pavlov et al. 2003).

The first full 3-D relativistic magnetohydrodynamic simulations of PWNe (Porth et al. 2013, 2014a) settled the issue. By using the same setup of axisymmetric runs, the authors were able to produce models with $\sigma_0 = 1$–3, that is up to a thousand times the magnetization employed in radial MHD models. The inner jet-torus structure is preserved, but jets are now subject to kink instabilities and a high level of magnetic randomization and dissipation occurs in the main body the PWN, simply due to the extra channels of dynamics made possible by the 3-D setup. The size of the TS is correctly retrieved and it is basically the same as in lower magnetization models, thus confirming Begelman's idea that lower dimensionality models are not adequate to put constraints on σ_0, due the excessive rigidity of purely toroidal fields. PWNe can thus be modeled by assuming $\sigma_0 > 1$ at the TS, and the long-standing paradox appears to be finally solved.

Is this the end of the story? Unfortunately these first 3-D simulations have been run for just $\simeq$100 y, 1/10 of the Crab lifetime, when the self-similar expansion phase is still far from being reached. In particular, it is not clear wether the average nebular field strength will be close to that inferred from observations at the final stage of the evolution ($\simeq$100–200 μG), and if the synthetic non-thermal emission will match observational data in all spectral bands, including the IC component. Moreover, magnetic dissipation is entirely of numerical origin in the ideal MHD code employed, a problem that can be partly overcome in 2-D simulations by simply increasing the resolution, but certainly not in 3-D runs using AMR, which are extremely demanding from a computational point of view.

Nevertheless, it is clear that this kind of 3-D simulations are the primary tool to reach a comprehensive description of PWNe in all their aspects. Here we report preliminary results from 3-D AMR simulations of the Crab nebula by our group (Olmi et al. 2016), where we successfully managed to reach, if not the proper lifetime of the Crab, at least the self-similar expansion phase ($t = 250$ y, corresponding to an effective age of 500 y for the given initial PWN radius). The setup of the PW was the same for 2-D models, and we chose $\sigma_0 = 1$ and $b = 10$ (the blue curve in Fig. 10.3).

In Fig. 10.9 the structure of magnetic field lines at $t = 250$ y is shown, with the color bar indicating the ratio of the toroidal to the total field magnitude, measured by the quantity $\alpha_B = B_{\rm tor}/B$. As expected, the magnetic field structure is complex: the field is mainly toroidal in the inner part of the nebula, as only this component is injected in the wind and amplified post-shock, while the poloidal component is mostly important outside, in the polar and equatorial external regions, where the field structure is more efficiently modified by axial high-speed flow or by the turbulent motions, respectively. In the same figure, we also show iso-surfaces of constant velocity, dark blue indicating the wind region and light blue to highlight the jets, for which we have chosen a reference value of $v = c/2$.

In a first phase, say up to 100 y of evolution, the average nebular magnetic field strength is found to decrease down to values of the order of $\sim$100 μG, which means that the average field in the nebula is about a factor of 2 lower than expected. However, the trend seems to indicate that the strong magnetic dissipation of the first period of evolution slows down after that time: the field strength reaches a minimum value and afterwards stays almost constant, or even slightly increases. This might indicate that magnetic dissipation progressively becomes less important, leading the magnetic field to saturate to the equipartition value at some point of the evolution, so we are confident that the emission properties derived from this simulation can in principle match the observations.

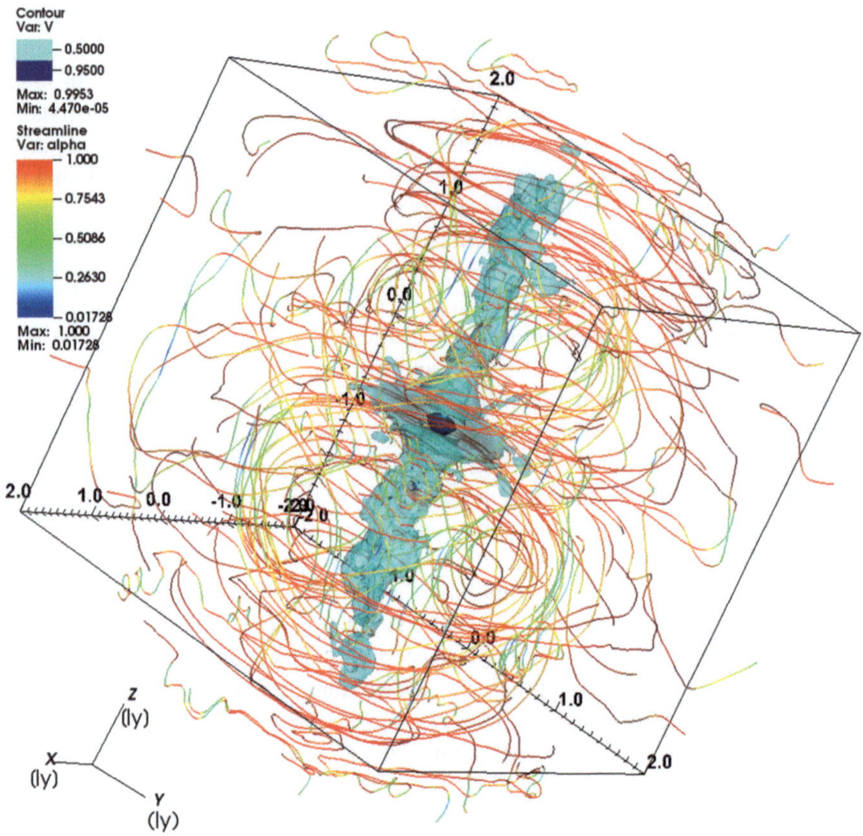

Fig. 10.9 Magnetic field lines drawn with seed points laying on a sphere of radius 2.5 ly, at the time $t = 250$ y, taken from Olmi et al. (2016). The contrast between toroidal and poloidal components is highlighted by showing the quantity $\alpha = B_{\rm tor}/B$ in different colors: *red* means that the field is completely toroidal and *blue* that it is predominantly poloidal. The velocity field is also represented as a 3-D contour plot, with color levels corresponding to $0.95c$ and $0.5c$. Dimensions of the nebula are indicated by the box axes, in units of ly

Surface brightness maps at optical frequencies and in the X-rays are shown in Fig. 10.10, corresponding to the end of the simulation at $t = 250$ y. As can be seen, the synchrotron burn-off effect is clearly apparent at increasing photon energies, and emission maps are encouraging. Polar jets are well formed and bright, contrary to what found in Porth et al. (2014a), but the torus is quite under-luminous. Most of the emission comes indeed from the base of the jets, where the particle density is high, while in the torus zone, in spite of a still substantial magnetic field, the number of particles is considerably lower and the non-thermal emission is reduced.

Investigations on wisps and a study to infer their motion and outflow properties were also achieved. As expected, wisps appear to be more prominent in the upper hemisphere of the nebula (with respect to the pulsar equatorial plane) than in the

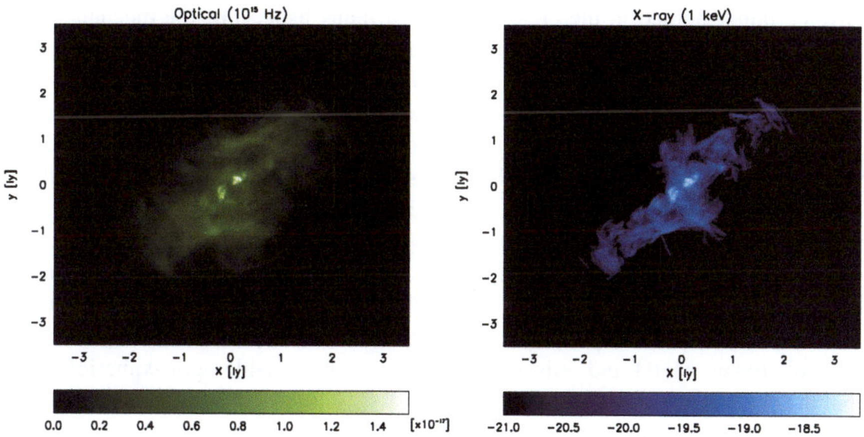

Fig. 10.10 Surface brightness maps from 3-D simulation of the Crab nebula at $t = 250$ y, taken from Olmi et al. (2016): the optical map at $\nu = 10^{15}$ Hz is displayed in the *left panel*, the X-ray one at 1 keV is displayed in the *right panel*. In both maps the intensity is expressed in mJy/arcsec^2 units and distance from the pulsar is in ly

lower one. Properties of wisps were obtained with the same technique as in the 2-D case, by subtracting the different intensities in emission maps taken at different times. We chose to study the variability in the radio band, for which the expected periodicity of wisps is of the order of years. A number of non-axisymmetric structures appear on the large scales, and the difference map shows a much closer resemblance to the time variability data reported in Bietenholz et al. (2004) with respect to the previous axisymmetric study. However, in the inner nebula, results appear to be rather similar, supporting the conclusion that the 2-D description with a dominant toroidal field is still adequate in that region.

To conclude, the first 3-D relativistic MHD simulations are promising and seem to provide a reasonable solution to the long-standing σ-problem. The inner structure is basically unaffected by the increased dimensionality, and the condition for $\sigma_0 \ll 1$ at the TS can be finally relaxed. The new channels for 3-D dynamics allow for field randomization and dissipation either in the polar jets (kink instabilities) and in the main body of the nebula. However, the final word can only come from an accurate modeling of the combined synchrotron and IC emission, with the same prescriptions for the evolution of emitting particles as (partially) successfully applied in the 2-D case. This research is currently ongoing.

10.6 Summary and Conclusions

Multidimensional MHD simulations have been widely recognized as a powerful tool for the investigation of the physics of the relativistic magnetized plasma in PWNe. More than a decade of axisymmetric models and the recent full 3-D approach have certainly helped in the understanding of many aspects of the mechanisms at work in pulsar winds, in the associated nebulae, and those responsible for particle acceleration, though a few open questions still remain.

The turn of the millennium presented a new challenge to theorists: an unexpected *jet-torus* structure was found in the X-ray emission of a number of nebulae, and this is most spectacularly observed in their class prototype, the Crab nebula (Weisskopf et al. 2000). Axisymmetric simulations were employed to compute the flow structure and other physical quantities downstream of the TS, allowing one to reproduce the emission of PWNe at different frequencies. The synthetic surface brightness maps of synchrotron radiation obtained on top of simulations were so strikingly similar to observations, down to the very finest details, that modelers could safely conclude that relativistic MHD and axisymmetry are very reasonable approximations, and that the challenge was won!

In particular, the jet-torus structure arises as the result of the PW structure and in particular of the anisotropy of its energy flux, with a higher power injected in the equatorial region as previously predicted analytically (Lyubarsky 2002). Jets are stronger when hoop stresses are able to redirect the equatorial flow towards the polar axis, thus depending on the wind magnetization parameter σ_0 (measured at the TS). In the innermost region, bright arcs (the *wisps*) and dots (and the *knot* in particular) appear as Doppler-boosted and time-variable features, determined by the highly turbulent flow structure around the TS and tracing the local magnetic field, which is amplified post-shock (Komissarov and Lyubarsky 2003, 2004; Del Zanna et al. 2004, 2006; Camus et al. 2009).

The open problems of PWN physics, still not completely and satisfactory solved by MHD numerical modeling, are the following (Olmi et al. 2016): the wind magnetization paradox, the origin of radio emitting particles and the related implications for the multiplicity of pairs, the precise location and mechanisms for particle acceleration at the TS, the origin of the γ-ray flares unexpectedly observed in the Crab nebula. The first three problems have been addressed here in the context of ideal MHD simulations, while the last one is likely to involve dissipative processes, namely relativistic tearing instabilities and fast reconnection, that have been studied so far just in limited numerical domains and never in global axisymmetric or 3-D simulations of PWNe (Cerutti et al. 2013, 2014; Sironi and Spitkovsky 2014; Del Zanna et al. 2016).

As far as the σ-problem is concerned, we have seen how low-dimensionality models do not allow for enough dissipation of the toroidal magnetic field inside the nebula, so that the tension remains strong and, in order to match the observed TS and PWN radii, the magnetization σ_0 at the TS can be $\sim$0.01–0.1 at most. On the other hand, 3-D models relax this constraint: magnetic structures can kink and dissipate efficiently throughout the nebula, as predicted by Begelman (1998), and equipartition or even conditions of $\sigma_0 > 1$ can be reached in the far region of the wind (Porth et al. 2013, 2014a; Olmi et al. 2016). While the structures in the inner region are basically unchanged, since the toroidal field is still the dominant component there, the situation is drastically different in the polar jets, now prone to non-axisymmetric kink instabilities (Mizuno et al. 2011; Mignone et al. 2013), and far from the torus. Poloidal components are created, and new channels for efficient magnetic dissipation are opened. However, what is still puzzling is how the strongly Poynting-dominated wind originating in the pulsar magnetosphere can convert most of its electromagnetic energy into the kinetic bulk flow component, which may still be dominant in the equatorial striped wind region. Moreover, the final word on the validity of the 3-D models, to date still be run up to realistic ages, can only come from the spectral properties, from the radio up to the TeV emission by IC scattering, in order to constrain both the kinetic and magnetic content of the nebula (Volpi et al. 2008).

The radio emitting particles are an important component of the non-thermal energy budget, and it is fundamental to understand whether they are part of the pulsar outflow or not, but they are unfortunately challenging to be modeled. Surface brightness maps and the appearance of wisps in numerical (axisymmetric) simulations seem to be insensitive on their origin: due to the long synchrotron lifetime, radio particles can be either accelerated at the TS and advected downstream, like higher energy particles (in this case they are part of the pulsar outflow), or they can be simply modeled as a steady and uniform distribution in the whole PWN. Therefore, no conclusion on the pulsar multiplicity can be drawn at this stage. The only hypothesis that can be rejected is that of a relic population due to an early outburst, because some sort of continuous re-acceleration would be required anyway (Olmi et al. 2014).

Finally, by investigating the motion of wisps, which appears to be slightly different at various photon frequencies, the site of particle acceleration can be in principle determined by comparing the numerical results against multi-wavelength observations of the Crab nebula (Bietenholz et al. 2004; Schweizer et al. 2013). Both axisymmetric and 3-D simulations seem to suggest that the best match is obtained when high-energy particles are accelerated in a narrow belt in the equatorial region of the TS (Porth et al. 2014a; Olmi et al. 2015), whereas once again no stringent conclusion can be drawn on radio emitting particles. The equatorial region is believed to be the lowest magnetization site, if effective dissipation takes place in the striped wind, and this conclusion is also supported by the MHD modeling of the knot of the Crab nebula (Lyutikov et al. 2016). The most natural mechanism responsible for the acceleration of those highest energy particles should be then Fermi I, which provides a particle spectral index close to that inferred from X-ray observations and it has been proven to be very effective at relativistic shocks of low enough magnetization (Spitkovsky 2008; Sironi and Spitkovsky 2011).

In spite of the success of axisymmetric simulations, the future is clearly in 3-D models. Such runs are extremely demanding in terms of computational time, since several refinement levels are required for the adaptive mesh to treat simultaneously the injection region, with the ultra-relativistic PW, as well as the external regions up to the supernova ejecta. While the first simulations reached just $\sim 1/10$ of the Crab nebula lifetime, a first attempt towards reaching the self-similar phase of expansion was recently made (Olmi et al. 2016). A slowing down of magnetic dissipation was observed as soon as self-similar expansion begins, reaching a value for the average magnetic field of about 100 μG, only a factor 2 lower than what observed. Emission at radio, optical and X-ray frequencies was computed and the synchrotron burn-off effect was observed through the decrease of the size of the emitting area with increasing observation frequency, as in 2-D. However, the brightness contrast of the torus is rather low when compared against observations, the brightest features lying at the bases of the jets, where kinetic pressure is higher. Doppler-boosted dots and outgoing wisps are also retrieved in the innermost region, once again confirming the validity of 2-D results.

What are the next frontiers for PWN simulations? Most probably the multidimensional MHD modeling should be extended to non-ideal plasma effects, like a finite resistivity allowing for reconnection in thin current sheets (Del Zanna et al. 2016; Landi et al. 2015), arising either at the TS, throughout the nebula, or inside the kinking jets (Mignone et al. 2013; Striani et al. 2016). Reconnection, or at least fast magnetic dissipation, is already present in simulations, though of numerical origin alone, arising whenever small scales are reached by the turbulent motions in the simulated plasma. The proper introduction of resistivity in the Ohm's law employed in relativistic MHD global simulations of PWNe would certainly represent an important upgrade.

On the other hand, one could think to improve the numerical treatment of the non-thermal particles responsible for emission, by adopting the hybrid MHD/kinetic approach first introduced for cosmic-ray acceleration at non-relativistic shocks (Bai et al. 2015) and later used to model the leptonic radiation from the magnetized plasma in AGN jets (Vaidya et al. 2016). A first attempt has been already achieved by studying the diffusion of particles in evolved PWNe (Porth et al. 2016), though in this work test particles are simply transported on the unperturbed MHD background provided by 3-D simulations (Porth et al. 2014a). A major breakthrough will come when the self-consistent hybrid MHD/kinetic approach for the Lagrangian test particles will be first employed to the transport and evolution of the families of accelerated leptons in PWNe, either at shock locations or in reconnecting current sheets.

To conclude, while numerical MHD modeling of PWNe has progressed impressively in more than a decade of simulations, the next years could witness newer and even more exciting achievements along the lines mentioned above. As already happened in the past, the future progresses in the numerical modeling of PWNe will surely also reflect in a better understanding of the physics of other astrophysical sources powered by magnetized relativistic outflows, most notably AGNs an GRBs.

Acknowledgements The authors sincerely thank the other members of the Arcetri high-energy astrophysics group, E. Amato, N. Bucciantini, R. Bandiera and D. Volpi, without whom most of the results reported in this review would have never been achieved. LDZ and BO also acknowledge support from INAF—Arcetri, from the INFN—TEONGRAV initiative, from MIUR (grant PRIN 2015: *Multi-scale Simulations of High-Energy Astrophysical Plasmas*), and from CINECA for computational resources (*CRAB3D: Three-dimensional relativistic MHD simulations of the dynamics and non-thermal emission of the Crab nebula*).

References

Amato, E.: Int. J. Modern Phys. Conf. Ser. **28**, 1460160 (2014) (Preprint, arXiv 1312.5945)
AMS-02 coll: Phys. Rev. Lett. **110**, 141102 (2013)
Arons, J.: Space Sci. Rev. **173**, 341–367 (2012) (Preprint, arXiv 1208.5787)
Atoyan, A.M., Aharonian, F.A.: Mon. Not. R. Astron. Soc. **278**, 525–541 (1996)
Bai, X.N., Caprioli, D., Sironi, L., Spitkovsky, A.: Astrophys. J. **809**, 55 (2015) (Preprint, arXiv:1412.1087)
Begelman, M.C.: Astrophys. J. **493**, 291 (1998) (Preprint, arXiv:astro-ph/9708142)
Begelman, M.C., Li, Z.Y.: Astrophys. J. **397**, 187–195 (1992)
Bietenholz, M.F., Hester, J.J., Frail, D.A., Bartel, N.: Astrophys. J. **615**, 794–804 (2004) (Preprint, arXiv:astro-ph/0408061)
Bogovalov, S.V.: Astron. Astrophys. **349**, 1017–1026 (1999) (Preprint, arXiv:astro-ph/9907051)
Bogovalov, S.V., Khangoulian, D.V.: Mon. Not. R. Astron. Soc. **336**, L53–L55 (2002) (Preprint, arXiv:astro-ph/0209269)
Bogovalov, S.V., Khangulyan, D., Koldoba, A.V., Ustyugova, G.V., Aharonian, F.A.: Mon. Not. R. Astron. Soc. **419**, 3426–3432 (2012) (Preprint, arXiv:1107.4831)
Bosch-Ramon, V., Barkov, M.V., Khangulyan, D., Perucho, M.: Astron. Astrophys. **544**, A59 (2012) (Preprint, arXiv:1203.5528)
Bucciantini, N.: Int. J. Modern Phys. Conf. Ser. **28**, 1460162 (2014) (Preprint, arXiv 1311.6946)
Bucciantini, N., Del Zanna, L.: Astron. Astrophys. **454**, 393–400 (2006) (Preprint, arXiv astro-ph/0603481)
Bucciantini, N., Blondin, J.M., Del Zanna, L., Amato, E.: Astron. Astrophys. **405**, 617–626 (2003) (Preprint, arXiv astro-ph/0303491)
Bucciantini, N., Amato, E., Bandiera, R., Blondin, J.M., Del Zanna, L.: Astron. Astrophys. **423**, 253–265 (2004a) (Preprint, arXiv astro-ph/0405276)
Bucciantini, N., Bandiera, R., Blondin, J.M., Amato, E., Del Zanna, L.: Astron. Astrophys. **422**, 609–619 (2004b) (Preprint, arXiv astro-ph/0404328)
Bucciantini, N., del Zanna, L., Amato, E., Volpi, D.: Astron. Astrophys. **443**, 519–524 (2005a) (Preprint, arXiv arXiv:astro-ph/0508212)
Bucciantini, N., Amato, E., Del Zanna, L.: Astrono. Astrophys. **434**, 189–199 (2005b) (Preprint, arXiv astro-ph/0412534)
Bucciantini, N., Quataert, E., Metzger, B.D., Thompson, T.A., Arons, J., Del Zanna, L.: Mon. Not. R. Astron. Soc. **396**, 2038–2050 (2009) (Preprint, arXiv 0901.3801)
Bucciantini, N., Arons, J., Amato, E.: Mon. Not. R. Astron. Soc. **410**, 381–398 (2011)
Bucciantini, N., Metzger, B.D., Thompson, T.A., Quataert, E.: Mon. Not. R. Astron. Soc. **419**, 1537–1545 (2012) (Preprint, arXiv 1106.4668)
Bucciantini, N., Pili, A.G., Del Zanna, L.: Mon. Not. R. Astron. Soc. **447**, 3278–3290 (2015) (Preprint, arXiv 1412.5347)
Bühler, R., Blandford, R.: Rep. Prog. Phys. **77**, 066901 (2014) (Preprint, arXiv 1309.7046)
Bühler, R., Giomi, M.: Mon. Not. R. Astron. Soc. **462**, 2762–2776 (2016) (Preprint, arXiv 1607.04277)
Camus, N.F., Komissarov, S.S., Bucciantini, N., Hughes, P.A.: Mon. Not. R. Astron. Soc. **400**, 1241–1246 (2009) (Preprint, arXiv 0907.3647)
Cerutti, B., Beloborodov, A.M.: Space Sci. Rev. (2016) (Preprint, arXiv 1611.04331)
Cerutti, B., Werner, G.R., Uzdensky, D.A., Begelman, M.C.: Astrophys. J. **770**, 147 (2013) (Preprint, arXiv 1302.6247)
Cerutti, B., Werner G.R., Uzdensky D.A., Begelman, M.C.: Phys. Plasmas **21**, 056501 (2014) (Preprint, arXiv 1401.3016)
Coroniti, F.V.: Astrophys. J. **349**, 538–545 (1990)
Del Zanna, L., Bucciantini, N.: Astron. Astrophys. **390**, 1177–1186 (2002) (Preprint, arXiv astro-ph/0205290)

Del Zanna, L., Bucciantini, N., Londrillo, P.: Astron. Astrophys. **400**, 397–413 (2003) (Preprint, arXiv astro-ph/0210618)
Del Zanna, L., Amato, E., Bucciantini, N.: Astron. Astrophys. **421**, 1063–1073 (2004) (Preprint, arXiv astro-ph/0404355)
Del Zanna, L., Volpi, D., Amato, E., Bucciantini, N.: Astron. Astrophys. **453**, 621–633 (2006) (Preprint, arXiv astro-ph/0603080)
Del Zanna, L., Zanotti, O., Bucciantini, N., Londrillo, P.: Astron. Astrophys. **473**, 11–30 (2007) (Preprint, arXiv 0704.3206)
Del Zanna, L., Landi, S., Papini, E., Pucci, F., Velli, M.: J. Phys. Conf. Ser. **719**, 012016 (2016) (Preprint, arXiv 1603.04995)
Dubus, G., Lamberts, A., Fromang, S.: Astron. Astrophys. **581**, A27 (2015) (Preprint, arXiv 1505.01026)
Gaensler, B.M., Slane, P.O.: Ann. Rev. Astron. Astrophys. **44**, 17–47 (2006) (Preprint, arXiv arXiv:astro-ph/0601081)
Gold, T.: Nature **218**, 731–732 (1968)
Helfand, D.J., Gotthelf, E.V., Halpern, J.P.: Astrophys. J. **556**, 380–391 (2001) (Preprint, arXiv arXiv:astro-ph/0007310)
Hester, J.J.: Ann. Rev. Astron. Astrophys. **46**, 127–155 (2008)
Hester, J.J., Mori, K., Burrows, D., Gallagher, J.S., Graham, J.R., Halverson, M., Kader, A., Michel, F.C., Scowen, P.: Astrophys. J. **577**, L49–L52 (2002)
de Jager, O.C., Harding, A.K.: Astrophys. J. **396**, 161–172 (1992)
Kargaltsev, O., Cerutti, B., Lyubarsky, Y., Striani, E.: Space Sci. Rev. **191**, 391–439 (2015) (Preprint, arXiv 1507.03662)
Kennel, C.F., Coroniti, F.V.: Astrophys.J. **283**, 694–709 (1984a)
Kennel, C.F., Coroniti, F.V.: Astrophys. J. **283**, 710–730 (1984b)
Kirk, J.G., Skjæraasen, O.: Astrophys. J. **591**, 366–379 (2003) (Preprint, arXiv astro-ph/0303194)
Kirk, J.G., Lyubarsky, Y., Petri, J.: In: Becker, W. (ed.) Astrophysics and Space Science Library, vol. 357, p. 421. Springer, New York (2009) (Preprint, arXiv arXiv:astro-ph/0703116)
Komissarov, S.S.: Mon. Not. R. Astron. Soc. **303**, 343–366 (1999)
Komissarov, S.S.: Mon. Not. R. Astron. Soc. **428**, 2459–2466 (2013) (Preprint, arXiv 1207.3192)
Komissarov, S.S., Barkov, M.V.: Mon. Not. R. Astron. Soc. **382**, 1029–1040 (2007) (Preprint, arXiv 0707.0264)
Komissarov, S.S., Lyubarsky, Y.E.: Mon. Not. R. Astron. Soc. **344**, L93–L96 (2003) (Preprint, arXiv astro-ph/0306162)
Komissarov, S.S., Lyubarsky, Y.E.: Mon. Not. R. Astron. Soc. **349**, 779–792 (2004)
Komissarov, S.S., Lyutikov, M.: Mon. Not. R. Astron. Soc. **414**, 2017–2028 (2011) (Preprint, arXiv 1011.1800)
Landi, S., Del Zanna, L., Papini, E., Pucci, F., Velli, M.: Astrophys. J. **806**, 131 (2015) (Preprint, arXiv 1504.07036)
Lyubarsky, Y.E.: Mon. Not. R. Astron. Soc. **329**, L34–L36 (2002)
Lyubarsky, Y.E.: Mon. Not. R. Astron. Soc. **345**, 153–160 (2003) (Preprint, arXiv arXiv:astro-ph/0306435)
Lyubarsky, Y., Eichler, D.: Astrophys. J. **562**, 494–498 (2001) (Preprint, arXiv arXiv:astro-ph/0204481)
Lyutikov, M., Komissarov, S.S., Porth, O.: Mon. Not. R. Astron. Soc. **456**, 286–299 (2016) (Preprint, arXiv 1506.07282)
Meyer, M., Horns, D., Zechlin, H.S.: Astron. Astrophys. **523** A2 (2010) (Preprint, arXiv 1008.4524)
Michel, F.C.: Astrophys. J. **180**, L133 (1973)
Mignone, A., Bodo, G., Massaglia, S., Matsakos, T., Tesileanu, O., Zanni, C., Ferrari, A.: Astrophys. J. Suppl. Ser. **170**, 228–242 (2007) (Preprint, arXiv astro-ph/0701854)
Mignone, A., Zanni, C., Tzeferacos, P., van Straalen, B., Colella, P., Bodo, G.: Astrophys. J. Suppl. Ser. **198**, 7 (2012) (Preprint, arXiv 1110.0740)

Mignone, A., Striani, E., Tavani, M., Ferrari, A.: Mon. Not. R. Astron. Soc. (2013) (Preprint, arXiv 1309.0375)
Mizuno, Y., Lyubarsky, Y., Nishikawa, K.I., Hardee, P.E.: Astrophys. J. **728**, 90 (2011) (Preprint, arXiv 1012.2770)
Ng, C.Y., Romani, R.W.: Astrophys. J. **601**, 479–484 (2004) (Preprint, arXiv astro-ph/0310155)
Olmi, B., Del Zanna, L., Amato, E., Bandiera, R., Bucciantini, N.: Mon. Not. R. Astron. Soc. **438**, 1518–1525 (2014) (Preprint, arXiv 1310.8496)
Olmi, B., Del Zanna, L., Amato, E., Bucciantini, N.: Mon. Not. R. Astron. Soc. **449**, 3149–3159 (2015) (Preprint, arXiv 1502.06394)
Olmi, B., Del Zanna, L., Amato, E., Bucciantini, N., Mignone, A.: J. Plasma Phys. **82**, 635820601 (2016) (Preprint, arXiv 1610.07956)
Pacini, F.: Nature **216**, 567–568 (1967)
Pacini, F.: Nature **219**, 145–146 (1968)
Pacini, F., Salvati, M.: Astrophys. J. **186**, 249–266 (1973)
PAMELA coll: Nature **458**, 607–609 (2009) (Preprint, arXiv 0810.4995)
Pavlov, G.G., Kargaltsev, O.Y., Sanwal, D., Garmire, G.P.: Astrophys. J. **554**, L189–L192 (2001) (Preprint, arXiv arXiv:astro-ph/0104264)
Pavlov, G.G., Teter, M.A., Kargaltsev, O., Sanwal, D.: Astrophys. J. **591**, 1157–1171 (2003) (Preprint, arXiv arXiv:astro-ph/0305510)
Pétri, J.: J. Plasma Phys. **82**, 635820502 (2016) (Preprint, arXiv 1608.04895)
Pétri, J.: Mon. Not. R. Astron. Soc. **455**, 3779–3805 (2016) (Preprint, arXiv 1511.01337)
Philippov, A.A., Cerutti, B., Tchekhovskoy, A., Spitkovsky, A.: Astrophys. J. **815**, L19 (2015) (Preprint, arXiv 1510.01734)
Pili, A.G., Bucciantini, N., Del Zanna, L.: Mon. Not. R. Astron. Soc. **439**, 3541–3563 (2014) (Preprint, arXiv 1401.4308)
Porth, O., Komissarov, S.S., Keppens, R.: Mon. Not. R. Astron. Soc. **431**, L48–L52 (2013) (Preprint, arXiv 1212.1382)
Porth, O., Komissarov, S.S., Keppens, R.: Mon. Not. R. Astron. Soc. **438**, 278–306 (2014a) (Preprint, arXiv 1310.2531)
Porth, O., Komissarov, S.S., Keppens, R.: Mon. Not. R. Astron. Soc. **443**, 547–558 (2014b) (Preprint, arXiv 1405.4029)
Porth, O., Xia, C., Hendrix, T., Moschou, S.P., Keppens, R.: Astrophys. J. Suppl. Ser. **214**, 4 (2014c) (Preprint, arXiv 1407.2052)
Porth, O., Vorster, M.J., Lyutikov, M., Engelbrecht, N.E.: Mon. Not. R. Astron. Soc. **460**, 4135–4149 (2016) (Preprint, arXiv 1604.03352)
Rees, M.J., Gunn, J.E.: Mon. Not. R. Astron. Soc. **167**, 1–12 (1974)
Reynolds, S.P., Chevalier, R.A.: Astrophys. J. **278**, 630–648 (1984)
Rybicki, G.B., Lightman, A.P.: Radiative Processes in Astrophysics (Wiley-Interscience, New York, 1979)
Scargle, J.D.: Astrophys. J. **156**, 401 (1969)
Schweizer, T., Bucciantini, N., Idec, W., Nilsson, K., Tennant, A., Weisskopf, M., Zanin, R.: ArXiv e-prints (2013) (Preprint, arXiv 1301.1321)
Serpico, P.D.: Astropart. Phys. **39**, 2–11 (2012) (Preprint, arXiv 1108.4827)
Sironi, L., Spitkovsky, A.: Astrophys. J. **698**, 1523–1549 (2009) (Preprint, arXiv 0901.2578)
Sironi, L., Spitkovsky, A.: Astrophys. J. **741**, 39 (2011) (Preprint, arXiv 1107.0977)
Sironi, L., Spitkovsky, A.: Astrophys. J. **783**, L21 (2014) (Preprint, arXiv 1401.5471)
Spitkovsky, A.: Astrophys. J. **648**, L51–L54 (2006) (Preprint, arXiv astro-ph/0603147)
Spitkovsky, A.: Astrophys.J. **682**, L5–L8 (2008) (Preprint, arXiv 0802.3216)
Striani, E., Mignone, A., Vaidya, B., Bodo, G., Ferrari, A.: Mon. Not. R. Astron. Soc. **462**, 2970–2979 (Preprint, arXiv 1607.07323) (2016)
van der Swaluw, E., Achterberg, A., Gallant, Y.A., Tóth, G.: Astron. Astrophys. **380**, 309–317 (2001)
Tchekhovskoy, A., Philippov, A., Spitkovsky, A.: Mon. Not. R. Astron. Soc. **457**, 3384–3395 (2016) (Preprint, arXiv 1503.01467)

Volpi, D., Del Zanna, L., Amato, E., Bucciantini, N.: Astron. Astrophys. **485**, 337–349 (2008) (Preprint, arXiv 0804.1323)
Vaidya, B., Mignone, A., Bodo, G., Massaglia, S.: J. Phys. Conf. Ser. **719**, 012023 (2016)
Weisskopf, M.C., Hester, J.J., Tennant, A.F., Elsner, R.F., Schulz, N.S., Marshall, H.L., Karovska, M., Nichols, J.S., Swartz, D.A., Kolodziejczak, J.J., O'Dell, S.L.: Astrophys. J. **536**, L81–L84 (2000) (Preprint, arXiv arXiv:astro-ph/0003216)
Yuan, Y., Blandford, R.D.: Mon. Not. R. Astron. Soc. **454**, 2754–2769 (2015) (Preprint, arXiv 1506.08121)

Chapter 11
Particle Acceleration in Pulsar Wind Nebulae: PIC Modelling

Lorenzo Sironi and Benoît Cerutti

Abstract We discuss the role of PIC simulations in unveiling the origin of the emitting particles in PWNe. After describing the basics of the PIC technique, we summarize its implications for the quiescent and the flaring emission of the Crab Nebula, as a prototype of PWNe. A consensus seems to be emerging that, in addition to the standard scenario of particle acceleration via the Fermi process at the termination shock of the pulsar wind, magnetic reconnection in the wind, at the termination shock and in the Nebula plays a major role in powering the multi-wavelength signatures of PWNe.

11.1 Introduction

In recent years, multi-dimensional magnetohydrodynamic (MHD) models of Pulsar Wind Nebulae (PWNe) have been able to reproduce the nebular morphology down to intricate details (see contributions by A. Mignone, L. del Zanna in this volume). In order to compare the results of MHD simulations to the multi-wavelength observations of PWNe (most notably, of the prototypical Crab Nebula), it is usually assumed that the termination shock of the pulsar wind—where the ram pressure of the ultra-relativistic wind emanating from the pulsar balances the thermal pressure of the surrounding nebula—is an efficient site of particle acceleration. This assumption cannot be tested directly within the MHD framework (which bears no information on the properties of the accelerated particles), but it requires fully-kinetic particle-in-cell (PIC) simulations. By capturing the interplay of charged particles and electromagnetic fields from first principles, PIC simulations allow to

Both authors contributed equally to this chapter.

L. Sironi (✉)
Columbia University, Pupin Hall, 550 West 120th Street, New York, NY 10027, USA
e-mail: lsironi@astro.columbia.edu

B. Cerutti
Univ. Grenoble Alpes, CNRS, IPAG, F-38000, Grenoble, France
e-mail: benoit.cerutti@univ-grenoble-alpes.fr

D.F. Torres (ed.), *Modelling Pulsar Wind Nebulae*, Astrophysics and Space Science Library 446, DOI 10.1007/978-3-319-63031-1_11

identify potential locations of particle acceleration in PWNe. On the other hand, due to computational constraints, PIC simulations are typically confined to a local description of the system, on scales much smaller than the nebular size. It is only by integrating the PIC results with a global MHD model of the nebula that we can properly reproduce the multi-wavelength signatures of PWNe.

In this chapter, we summarize the role of PIC simulations in understanding the origin of high-energy particles in PWNe. In Sect. 11.2 we describe the basics of the PIC method. In Sect. 11.3 we summarize the implications of PIC results on the quiescent and flaring emission from PWNe. We conclude in Sect. 11.4.

11.2 The Particle-in-Cell Technique

This section is intended to provide a brief overview of the most common methods and algorithms used in explicit PIC codes. A detailed presentation of this technique can be found in Hockney and Eastwood (1988) and Birdsall and Langdon (1991).

11.2.1 Collisionless Plasmas

A necessary condition for non-thermal particle acceleration is the absence of Coulomb collision in the plasma of interest. This is the case for most high-energy astrophysical systems, and in particular pulsar wind nebulae, where plasmas are very diluted. Roughly speaking, a plasma can be considered as "collisionless" if the frequency of Coulomb collision (ν) is much smaller than the plasma frequency, $\omega_{\rm pe} \gg \nu$. This condition implies that the number of particles per Debye sphere must be large, i.e., $N_{\rm D} \gg 1$. The dynamics of individual particle is driven by collection plasma phenomena rather than binary collisions at the sub-Debye length and plasma frequency scales, or simply referred below as the "kinetic" scale. As we will see in Sect. 11.3, these microscopic scales are involved in the particle acceleration processes and, thus, they must be well-resolved by simulations in contrast to the magnetohydrodynamic approach. To obtain meaningful astrophysical results, particle-based simulations must also capture large scale features, i.e., system size and long integration time.

The evolution of a collisionless plasma is governed by the Vlasov equation

$$\frac{\partial f}{\partial t} + \frac{\mathbf{p}}{\gamma m} \cdot \frac{\partial f}{\partial \mathbf{r}} + q\left(\mathbf{E} + \frac{\mathbf{v} \times \mathbf{B}}{c}\right) \cdot \frac{\partial f}{\partial \mathbf{p}} = 0, \tag{11.1}$$

where $f \equiv d\mathrm{N}/d\mathbf{r}d\mathbf{p}$ is the particle distribution function defined in 6D phase space $(\mathbf{r}, \mathbf{p})$ and 1D in time, with $\mathbf{r}$ is the position and $\mathbf{p} = \gamma m\mathbf{v}$ is the momentum, and q is the electric charge. Along with Maxwell's equations for the fields ($\mathbf{E}$ and $\mathbf{B}$), this is the full set of equations to model a collisionless plasma from first principles.

11.2.2 The Particle Approach

Analytical solutions to the Vlasov equation are known for a few idealized situations only. In most cases, it must be solved numerically. They are at least two ways to solve this equation. In the first approach, phase space is treated as a continuous fluid and Vlasov equation is solved directly using semi-Lagrangian or Eulerian methods (Cheng and Knorr 1976; Elkina and Büchner 2006). This method has the advantage to be insensitive to particle noise, and hence can capture well weak plasma phenomena and broad particle distribution functions. In theory this is the most appropriate approach to follow, but in practice the usage of Vlasov codes is currently limited due to prohibitive numerical costs for multidimensional problems (6D). The second approach is the PIC method which is the main focus of this chapter.

In PIC, Vlasov equation is solved indirectly by integrating discrete particle trajectories. This approach is equivalent to the direct method, and an easy way to see this is to rewrite Vlasov equation as a usual advection equation: $\partial f/\partial t + \nabla_{\mathbf{r},\mathbf{p}} \cdot (f\mathbf{U}) = \mathbf{0}$, where $\nabla_{\mathbf{r},\mathbf{p}} = (\partial/\partial\mathbf{r}, \partial/\partial\mathbf{p})$ and $\mathbf{U} = (\mathbf{p}/\gamma m, q\,(\mathbf{E} + \mathbf{v} \times \mathbf{B}/c))$. Thus, using the methods of characteristics, this first-order partial differential equation can be rewritten as a sets of ordinary differential equations (Newton's law) along characteristic curves which corresponds here to particle trajectories. For point-like particles, the particle distribution function is then approximated as

$$f\,(\mathbf{r},\mathbf{p},t) \approx \sum_{k=1}^{N} w_k \delta\,(\mathbf{r} - \mathbf{r}_k(t))\,\delta\,(\mathbf{p} - \mathbf{p}_k(t))\,, \tag{11.2}$$

where δ is the Dirac delta function and w_k is the particle weight. The number of particles must be very high for a good sampling of phase space and to be close to the exact solution of Vlasov equation. In practice, however, this number will be limited by computing resources and is always much smaller than the number of particles contained in real plasmas. To overcome this difficulty, a PIC particle represents a large number (given by the weight w_k) of physical particles that would follow the same trajectory in phase space (with the same q/m ratio). For this reason, the simulation particles are usually called "macroparticles".

Even though the plasma is collisionless, particles feels each other via long-range interactions. Summing over all particle-particle binary interactions, i.e. $N\,(N-1)\,/2 \approx N^2$, is numerically expansive and hard to implement. Instead, in PIC, particles do not feel each other directly but via the electromagnetic fields known on the grid which result from the evolution of the plasma. In this case, the number of operations scales as the number of particles N instead of N^2.

The PIC method has become increasingly popular in high-energy astrophysics to model non-thermal particle acceleration phenomena. PIC codes are much cheaper in comparison to Vlasov codes, and they are also conceptually simple, robust and easy to implement and parallelize efficiently to a large number of cores. This simplicity comes at the cost of significant particle noise which can lead to poor sampling

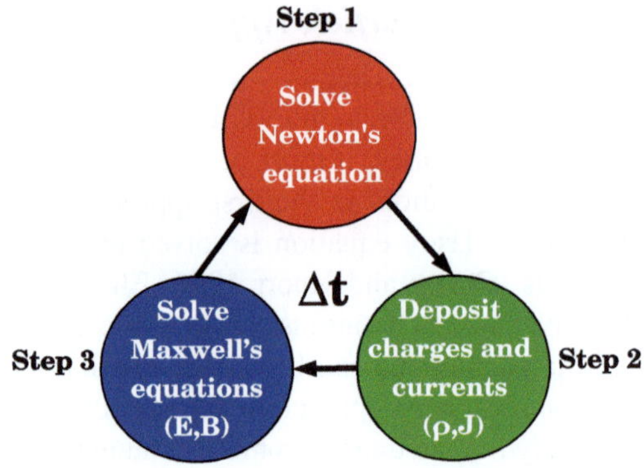

Fig. 11.1 Computation procedure per time step Δt in PIC

of the particle distribution (e.g., steep power-law tails), difficulty in capturing subtle or weak phenomena, artificial collisions, and load-balancing issues in parallel computing.

11.2.3 Main Computing Procedures in PIC

Figure 11.1 describes the three main operations performed per timestep Δt of an explicit PIC code: (1) Solve Newton's equation for each particle to evolve velocities and positions (2) Collect charge and current densities from all particles and deposit them on the grid, and (3) Solve Maxwell's equations to update the fields on the grid. Below is a brief technical description of each step:

Step 1: Particle Push

The set of equations to solve are

$$\frac{d\mathbf{u}}{dt} = \frac{q}{mc}\left(\mathbf{E} + \frac{\mathbf{u}\times\mathbf{B}}{\gamma}\right) \tag{11.3}$$

$$\frac{d\mathbf{r}}{dt} = \frac{c\mathbf{u}}{\gamma}, \tag{11.4}$$

where $\mathbf{u} = \gamma\mathbf{v}/c$ is the particle 4-velocity vector divided by the speed of light and $\gamma = 1/\sqrt{1-(v/c)^2}$ is the Lorentz factor. One of the most successful and most common method used in PIC to solve Newton's equation is the Boris push.[1] It has all the desirable numerical features one might think of: it is fast, stable, second order

[1]Other efficient methods exists as for instance the particle pusher developed by Vay (2008).

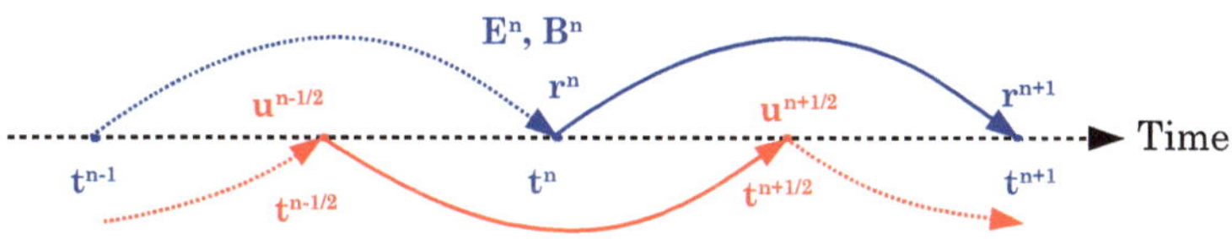

Fig. 11.2 The leapfrog scheme of the Boris method to solve Newton's equation

accurate, and conserves well the particle energy. The algorithm is based on the usual leapfrog integration method, i.e., 4-velocities $\mathbf{u}$ and positions $\mathbf{r}$ are staggered in time by half a timestep (Fig. 11.2). If particle positions and fields are known at time t^n ($\mathbf{r}^n$, $\mathbf{E}^n$, $\mathbf{B}^n$) and velocities at time $t^{n-1/2}$ ($\mathbf{u}^{n-1/2}$), the finite-difference time-centered expression of Eq. (11.3) is

$$\frac{\mathbf{u}^{n+1/2}-\mathbf{u}^{n-1/2}}{\Delta t}=\frac{q\mathbf{E}^n}{mc}+\frac{q}{mc}\left(\frac{\mathbf{u}^n\times\mathbf{B}^n}{\gamma^n}\right). \quad (11.5)$$

Now, the trick is to rewrite $\mathbf{u}^n$ appearing on the right-hand side of the equation as $\mathbf{u}^n = \left(\mathbf{u}^{n-1/2}+\mathbf{u}^{n+1/2}\right)/2$. Assuming that $\mathbf{E}^n$ and $\mathbf{B}^n$ are known, $\mathbf{u}^{n+1/2}$ can be extracted. It is convenient to define the following intermediate variables

$$\mathbf{u}^- = \mathbf{u}^{n-1/2}+\frac{q\Delta t\mathbf{E}^n}{2mc} \quad (11.6)$$

$$\mathbf{u}^+ = \mathbf{u}^{n+1/2}-\frac{q\Delta t\mathbf{E}^n}{2mc}. \quad (11.7)$$

Then, using Eqs. (11.6) and (11.7) and after a few algebraic manipulations one finds

$$\mathbf{u}^+ = \mathbf{u}^- + \mathbf{u}^-\times\mathbf{s} + (\mathbf{u}^-\times\mathbf{w})\times\mathbf{s}, \quad (11.8)$$

where

$$\mathbf{w}=\frac{q\Delta t\mathbf{B}^n}{2mc\gamma^n},\qquad \mathbf{s}=\frac{2\mathbf{w}}{1+\mathbf{w}^2},\qquad \gamma^n=\sqrt{1+(\mathbf{u}^-)^2}. \quad (11.9)$$

It is important to notice that the fields appearing in these equations are those felt at the particle position, not at the grid point where the fields are known. The fields must be interpolated to the particle positions. A linear interpolation scheme is usually sufficient. The final step is to update the particle positions

$$\mathbf{r}^{n+1} = \mathbf{r}^n + c\Delta t\frac{\mathbf{u}^{n+1/2}}{\gamma^{n+1/2}}, \quad (11.10)$$

where $\gamma^{n+1/2}=\sqrt{1+\left(\mathbf{u}^{n+1/2}\right)^2}$.

Step 2: Charge and Current Deposition

To solve Maxwell's equations, we need the source terms ρ and $\mathbf{J}$ that are given by the particles. In a continuous space, these macroscopic quantities can be recovered by summing over the contribution from all particles

$$\rho(\mathbf{r}) = \sum_{k=1}^{N} q_k w_k \delta(\mathbf{r} - \mathbf{r}_k), \qquad \mathbf{J}(\mathbf{r}) = \sum_{k=1}^{N} q_k w_k \mathbf{v}_k \delta(\mathbf{r} - \mathbf{r}_k), \tag{11.11}$$

where q_k, $\mathbf{v}_k$ are respectively the electric charge and the 3-velocity of the particle k. In PIC, charges and currents from the particles must be collected and dispatched among the nearest grid points. Charge and current densities at the grid point $\mathbf{r}_i$ can be written as

$$\rho(\mathbf{r}_i) = \sum_{k=1}^{N} q_k w_k S(\mathbf{r}_i - \mathbf{r}_k), \qquad \mathbf{J}(\mathbf{r}_i) = \sum_{k=1}^{N} q_k w_k \mathbf{v}_k S(\mathbf{r}_i - \mathbf{r}_k), \tag{11.12}$$

where S is a shape function which depends on the desired deposition scheme. Even though the particles are point-like, they have an effective size that is felt through the deposition of currents on the grid.

Figure 11.3 shows the example of a first order linear deposition method in a 2D Cartesian grid cell (or area-weighting method). The contributions from all the particles contained in the cell (x_i, y_j) to the current $\mathbf{J}$ are given by

$$\mathbf{J}_{i,j} = \sum_{k=1}^{N_{\text{cell}}} q_k w_k \mathbf{v}_k (1 - a_k)(1 - b_k) \tag{11.13}$$

$$\mathbf{J}_{i+1,j} = \sum_{k=1}^{N_{\text{cell}}} q_k w_k \mathbf{v}_k a_k (1 - b_k) \tag{11.14}$$

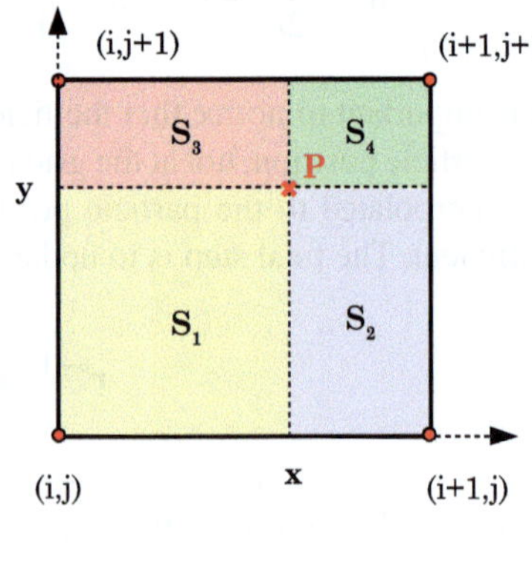

Fig. 11.3 The area-weighting technique to interpolate fields or deposit charges and currents on a 2D Cartesian cell (x_i, y_i) for a particle located in $P(x, y)$. The contribution to node (i, j) is given by S_4/S_{tot}, to $(i + 1, j)$ is S_3/S_{tot}, to $(i, j + 1)$ is S_2/S_{tot} and to $(i + 1, j + 1)$ is S_1/S_{tot}

$$\mathbf{J}_{i,j+1} = \sum_{k=1}^{N_{\text{cell}}} q_k w_k \mathbf{v}_k (1 - a_k) b_k \tag{11.15}$$

$$\mathbf{J}_{i+1,j+1} = \sum_{k=1}^{N_{\text{cell}}} q_k w_k \mathbf{v}_k a_k b_k, \tag{11.16}$$

where

$$a_k = \frac{x_k - x_i}{x_{i+1} - x_i}, \qquad b_k = \frac{y_k - y_j}{y_{j+1} - y_j} \tag{11.17}$$

are the usual bilinear interpolation coefficients. In this particular example, particles have a triangular shape.

Step 3: Fields Evolution

The last step is to update the fields on the grid. In principle, one needs solely to solve the time-dependent equations

$$\frac{\partial \mathbf{E}}{\partial t} = c\nabla \times \mathbf{B} - 4\pi \mathbf{J} \tag{11.18}$$

$$\frac{\partial \mathbf{B}}{\partial t} = -c\nabla \times \mathbf{E}, \tag{11.19}$$

because the current density is already given by the particles as we have seen in the previous paragraph. The other two should be automatically satisfied, but this is not necessarily true due to truncation errors in the discretization of space and time derivatives. The current deposition procedure does not always guarantee charge conservation[2] (i.e., $\nabla{\cdot}\mathbf{J} \neq -\partial\rho/\partial t$), but some solutions exist to enforce it to machine roundoff precision (Villasenor and Buneman 1992; Esirkepov 2001). Alternatively, Poisson equation should be solved to correct the electric field to make sure $\nabla \cdot \mathbf{E} = 4\pi\rho$. Parabolic and hyberbolic divergence cleaning methods also exist in the literature (Marder 1987; Munz et al. 2000).

The finite difference time domain (FDTD) method proposed by Yee (1966) for solving the time-dependent Maxwell equations enforces $\nabla \times \mathbf{B} = 0$ to machine roundoff precision. This is the most commonly used method in explicit PIC codes. Like the Boris push, the FDTD method combines stability, efficiency and second order accuracy (here in both space and time). To achieve this, fields must be staggered in time and in space. Figure 11.4 shows the order in time (top panel), as well as the spatial configuration of the fields within a Cartesian cell[3] in 2D (bottom-left panel) and in 3D (bottom-right panel). For illustrative purposes, within

[2]The total particle charge is conserved, but not necessarily the charge deposited on the grid.

[3]For a spherical geometry, see Holland (1983); Cerutti et al. (2015, 2016), and Belyaev (2015).

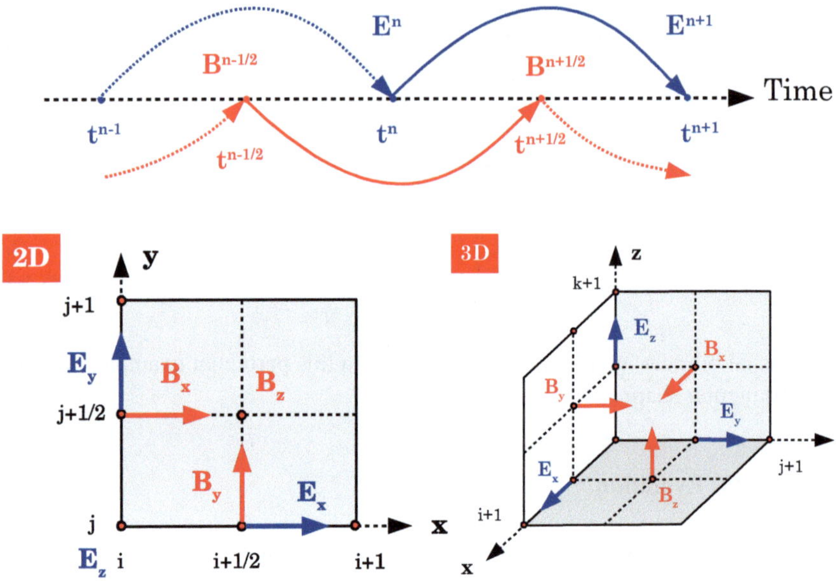

Fig. 11.4 *Top*: Leapfrog scheme for the fields in time. *Bottom*: Staggered mesh proposed by Yee (1966) in 2D (*left*) and in 3D (*right*)

this framework the z-component of Eq. (11.19) is

$$\frac{(B_z)^{n+1/2}_{i+1/2,j+1/2,k} - (B_z)^{n-1/2}_{i+1/2,j+1/2,k}}{\Delta t} = -c\frac{(E_y)^n_{i+1,j+1/2,k} - (E_y)^n_{i,j+1/2,k}}{\Delta x} + c\frac{(E_x)^n_{i+1/2,j+1,k} - (E_x)^n_{i+1/2,j,k}}{\Delta y}, \quad (11.20)$$

where Δx, Δy are the spatial step size along x and y. The FDTD method is stable under the usual Courant-Friedrichs-Lewy condition, i.e.,

$$\left(\frac{c\Delta t}{\Delta x}\right)^2 < 1 \text{ (1D)}, \quad (11.21)$$

$$(c\Delta t)^2\left(\frac{1}{\Delta x^2} + \frac{1}{\Delta y^2}\right) < 1 \text{ (2D)},$$

$$(c\Delta t)^2\left(\frac{1}{\Delta x^2} + \frac{1}{\Delta y^2} + \frac{1}{\Delta z^2}\right) < 1 \text{ (3D)}.$$

This is a purely numerical requirement, but physics imposes other constraints on the size of the steps, namely that the Debye length and the plasma frequency are well resolved by the code ($\Delta x/\Lambda_D \ll 1$ and $\omega_{pe}\Delta t \ll 1$), the latter condition being more stringent.

11.2.4 Boundary Conditions

Periodic boundary conditions are robust, easy to implement and physically useful in many case studies but there are not always appropriate. Below is a brief description of some other boundary conditions usually employed in PIC simulations.

11.2.4.1 Particles

- A perfectly reflective wall scatters the particle with no loss of momentum and energy. It can be useful in the context of a perfectly conducting wall for the fields (see below).
- In the context of an absorbing wall or of an open boundary, it is appropriate to absorb particles at the boundary. In such a case, particles are simply removed from the simulation.
- Conversely, new particles can be injected in simulations. This can be motivated by the physics involved in the problem, as for instance in pulsars where pair creation is important (Timokhin and Arons 2013; Chen and Beloborodov 2014; Philippov et al. 2015), or by the numerical setup if for example there is an inflow from one side of the box as in PIC simulations of relativistic shocks (Spitkovsky 2008b; Sironi and Spitkovsky 2011a). In this case, an expanding box with an injector receding at the speed of light can be desirable to reduce numerical cost (see 11.3 below).

11.2.4.2 Fields

- Perfectly conducting walls allow to reflect electromagnetic waves. They are easily implemented by applying the usual boundary conditions, namely that the tangential component of **E** and the perpendicular component of **B** vanishes at the interface. Semi-reflective medium can also be easily coded using surface current and charge densities.
- It is sometimes useful to absorb all electromagnetic waves leaving the box to simulate an open boundary, as for instance in pulsar winds (Cerutti et al. 2015; Belyaev 2015). In this case, the open boundary is coated with an absorbing medium of several cell thick where resistive terms are added to Maxwell's equations

$$\frac{\partial \mathbf{E}}{\partial t} + \lambda \mathbf{E} = c\nabla \times \mathbf{B} - 4\pi \mathbf{J} \tag{11.22}$$

$$\frac{\partial \mathbf{B}}{\partial t} + \lambda^{\star} \mathbf{B} = -c\nabla \times \mathbf{E}, \tag{11.23}$$

where λ and $\lambda^{\star}$ are artificial electric and magnetic "conductivities". The transition between the working domain and the absorbing layer should be gradual to avoid undesired reflections at the boundary. Conductivities usually are increasing function from the inner edge to the outer edge of the damping layer to make sure waves are completely absorbed. Equations (11.22–11.23) are valid for 1D layer. A perfectly matched layer is a generalization of these formulae to a multidimensional damping layer (Berenger 1994, 1996). In this framework, fields must be split into two subcomponents and hence the number of equations to solve is doubled (up to 12 in 3D, for an application to pulsars see e.g., Kalapotharakos and Contopoulos 2009 in the context of force-free MHD simulations).

11.2.5 Parallelization

PIC codes must be efficiently parallelized to model large system size and long integration time to have meaningful astrophysical applications. A common practice is to use the domain decomposition technique. It consists in dividing the computational box into smaller domains where one or more cores are assigned. Each CPU goes through the main steps described in Sect. 11.2.3 every timestep and exchanges information with the neighbouring processes to send particle and field data at the interface between subdomains. Communications between an arbitrary number of processes are done thanks to the Message Passing Interface (MPI) library. PIC codes scale well to a large number of CPUs, today at least up to $\sim 10^6$ processes (see Fig. 11.5). These scaling plots are usually done under ideal conditions, and do not necessarily reflect problem-dependent loss of performance. In PIC, a poor load balancing severely slows down a simulation. If, for some reason, there is a concentration of particle in a few subdomains, only few processors will have to

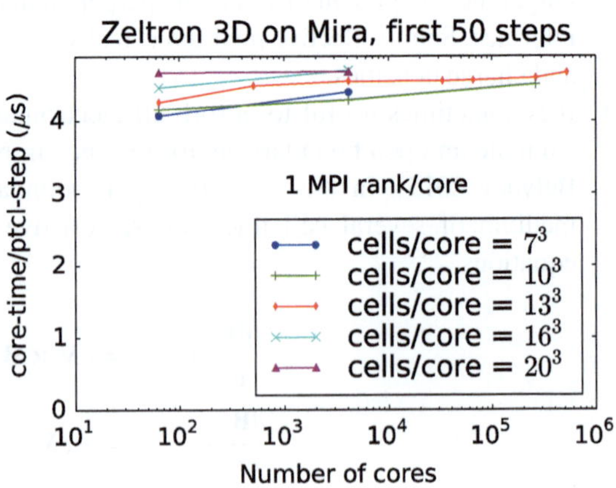

Fig. 11.5 Parallel scaling performance of the `zeltron` PIC code on the Mira supercomputer at the Argonne Leadership Computing Facility. Courtesy Greg Werner

push a lot of particles while the others remain idle.[4] The way how the domain is decomposed for a given setup can usual make a big difference. Hybrid codes combining MPI and OpenMP, variable particle weighting, or dynamical changes of the domain decomposition are other solutions to have better performances.

11.3 Application to Pulsar Wind Nebulae

In this section, we describe the main results obtained with PIC simulations on the efficiency of particle acceleration in PWNe. This section is divided into two parts: at first, we investigate particle acceleration at the termination shock of the pulsar wind, which is usually invoked to power the nebular *quiescent* emission (Sect. 11.3.1); then, we will discuss the origin of the gamma-ray *flares* detected from the Crab Nebula in the last few years (Sect. 11.3.2), focusing on the mechanisms that might explain such extreme particle acceleration events.

11.3.1 The Quiescent Emission

The Crab Nebula, our best laboratory for high energy astrophysics, has been observed over the entire electromagnetic spectrum from $\lesssim 100\,\mathrm{MHz}$ to $\gtrsim 1.5\,\mathrm{TeV}$ (see several contributions in this volume). Efficient acceleration of particles at the termination shock is required to explain its broadband spectrum. However, the flatness of the radio spectrum ($F_{\nu_r} \propto \nu^{-0.3}$, Bietenholz et al. 1997) is hard to reconcile with the steeper optical and X-ray slope ($F_{\nu_X} \propto \nu^{-1.1}$, Mori et al. 2004), unless the electron distribution is more complicated than a single power law. In fact, the radio band would require a distribution of emitting particles with a power-law slope $p = -d\log N/d\log\gamma \simeq 1.6$, whereas $p \gtrsim 2$ would be needed for the optical and X-ray emission. Even more fundamentally, how the pulsar wind termination shock can accelerate particles to the required "non-thermal" energies (i.e., well beyond the "thermal" peak of a Maxwellian distribution) is still an unsolved problem.

Particle acceleration in shocks is usually attributed to the Fermi process, where particles are energized by bouncing back and forth across the shock. Despite its importance, the Fermi process is still not understood from first principles. The highly nonlinear coupling between accelerated particles and magnetic turbulence—which is generated by the particles, and at the same time governs their acceleration—is extremely hard to incorporate in analytic models, and can

[4]Typically, pushing particles and depositing currents take 90% of the computing time (without communications), this is the reason why load-balancing is so critical in PIC.

be captured only with *ab initio* PIC simulations (for a review of the Fermi process in relativistic shocks, see Sironi et al. 2015a).

As we describe below, the efficiency of the Fermi process depends critically on the shock properties, e.g., composition, magnetization (i.e., the ratio σ between the Poynting flux and the kinetic energy flux of the pre-shock flow) and magnetic obliquity (i.e., the angle θ between the upstream magnetic field and the shock direction of propagation).[5] Pulsar winds are thought to be dominated by electron-positron pairs (Bucciantini et al. 2011). MHD models of PWNe require $\sigma \gtrsim 0.01 - 0.1$ in order to reproduce the morphology of the Crab jet/plume. Finally, polarization measurements indicate that the nebular magnetic field should be toroidal around the symmetry axis of the system, so that the termination shock is "perpendicular" (i.e., with the field orthogonal to the flow direction).

PIC simulations of perpendicular magnetized shocks show negligible particle acceleration (Gallant et al. 1992; Hoshino 2008; Sironi and Spitkovsky 2009, 2011b; Sironi et al. 2013). Here, due to the lack of significant self-generated turbulence, charged particles are forced to slide along the background field lines, whose orientation prohibits repeated crossings of the shock. This inhibits the Fermi process, and in fact the particle distribution behind perpendicular shocks is purely thermal.

In summary, PIC simulations have shown that the shock configurations which are apparently most relevant for PWNe (i.e., ultra-relativistic magnetized perpendicular shocks) do not naturally result in efficient particle acceleration. This is in sharp contrast with the pronounced non-thermal signatures of the quiescent emission of PWNe. However, one key ingredient of the PIC results summarized above is that the pre-shock magnetic field direction stays *uniform* throughout the timespan of the simulations. This is generally not the case in pulsar winds. If the rotational and magnetic axes of the central pulsar are misaligned, around the equatorial plane the wind consists of toroidal stripes of *alternating* magnetic polarity, separated by current sheets of hot plasma. It is still a subject of active research whether the alternating stripes will dissipate their energy into particle heat ahead of the termination shock, or whether the wind remains dominated by Poynting flux till the termination shock (Lyubarsky and Kirk 2001; Kirk and Skjæraasen 2003; Sironi and Spitkovsky 2011a). If the stripes are dissipated far ahead of the termination shock, the upstream flow is weakly magnetized and the pulsar wind reaches a terminal Lorentz factor (in the frame of the nebula) $\gamma_0 \sim L_{sd}/m_e c^2 \dot{N} \simeq 3.7 \times 10^4 L_{sd,38.5} \dot{N}_{40}^{-1}$, where $L_{sd} \equiv 3 \times 10^{38} L_{sd,38.5}$ ergs s^{-1} is the spin-down luminosity of the Crab, and $\dot{N} = 10^{40} \dot{N}_{40}$ s^{-1} is the particle flux entering the nebula, including the radio-emitting electrons (Bucciantini et al. 2011).

The two subsections below cover the two potential fates of the pulsar striped wind: at first, we investigate the physics of particle acceleration in a weakly

[5] In the limit $\gamma_0 \gg 1$ of ultra-relativistic shocks, as appropriate for PWNe, the efficiency of the Fermi process does not depend on the shock Lorentz factor γ_0 (Sironi and Spitkovsky 2009, 2011b; Sironi et al. 2013).

magnetized shock (i.e., assuming that the alternating stripes have dissipated their magnetic energy far ahead of the termination shock); then, we assume that the stripes persist until the termination shock.

11.3.1.1 The Termination Shock of a Weakly Magnetized Wind

Weakly magnetized ultra-relativistic shocks are mediated by electromagnetic plasma instabilities (the so-called Weibel instability, Weibel 1959; Medvedev and Loeb 1999; Gruzinov and Waxman 1999). These instabilities build up a magnetic barrier, up to a level[6] $\epsilon_B \sim 10^{-2} - 10^{-1}$, sufficient to deflect strongly the incoming particles and thus mediate the shock transition. The instability—triggered by a stream of shock-reflected particles propagating ahead of the shock—generates filamentary magnetic structures in the upstream region (Fig. 11.6), which in turn scatter the particles back and forth across the shock, mediating Fermi acceleration.

Such shocks do self-consistently accelerate particles up to nonthermal energies, via the Fermi process (Spitkovsky 2008a,b; Martins et al. 2009; Haugbølle 2011; Sironi et al. 2013). The accelerated particles populate in the downstream region a power-law tail $dN/d\gamma \propto \gamma^{-p}$ with a slope $p \sim 2.5$, that contains $\sim$ 3% of the particles and $\sim$ 10% of the flow energy.

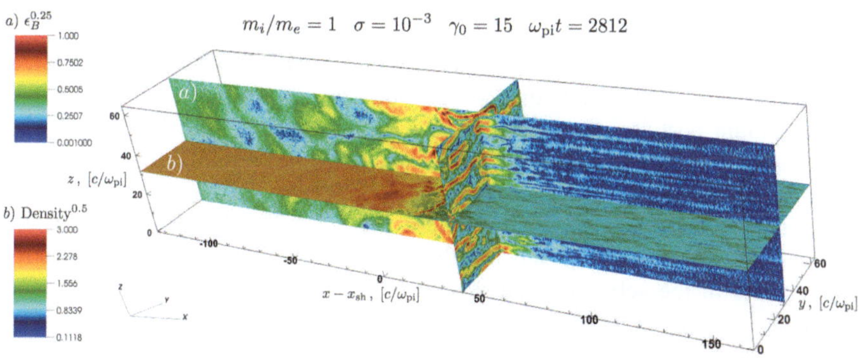

Fig. 11.6 Shock structure from the 3D PIC simulation of a $\sigma = 10^{-3}$ electron-positron shock with $\gamma_0 = 15$, from Sironi et al. (2013). The simulation is performed in the downstream frame and the shock propagates along $+\hat{x}$. We show the *xy* slice of the particle number density (normalized to the upstream density), and the *xz* and *yz* slices of the magnetic energy fraction ϵ_B. A stream of shock-accelerated particles propagates ahead of the shock, and their counter-streaming motion with respect to the incoming flow generates magnetic turbulence in the upstream via electromagnetic micro-instabilities. In turn, such waves provide the scattering required for particle acceleration

[6]The parameter ϵ_B denotes the magnetization of the turbulence, $\epsilon_B = \delta B^2/8\pi\gamma_0\rho_0 c^2$, where δB is the fluctuating magnetic field and ρ_0 is the mass density of the pre-shock flow. This should not be confused with the magnetization $\sigma = B_0^2/4\pi\gamma_0\rho_0 c^2$, which quantifies the strength of the pre-existing ordered upstream field B_0.

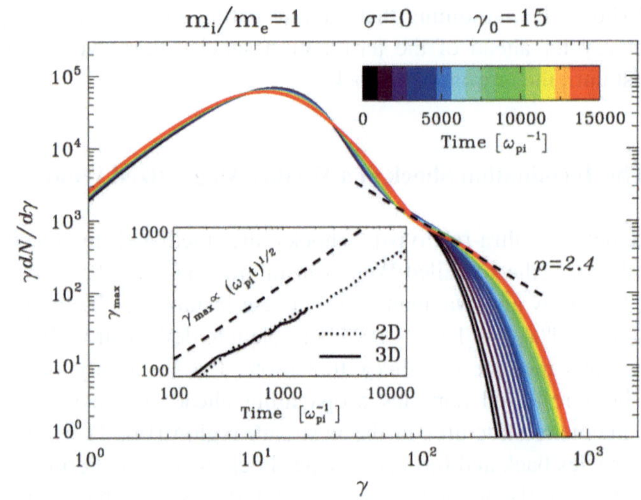

Fig. 11.7 Temporal evolution of the downstream particle spectrum, from the 2D simulation of a $\gamma_0 = 15$ electron-positron shock propagating into an unmagnetized flow (i.e., $\sigma = 0$), from Sironi et al. (2013). The evolution of the shock is followed from its birth (*black curve*) up to $\omega_{\rm p}t = 15000$ (*red curve*). The non-thermal tail approaches at late times a power law with a slope $p \sim 2.4$. Inset: temporal evolution of the maximum Lorentz factor, scaling as $\propto (\omega_{\rm p}t)^{1/2}$ (compare with the *black dashed line*) in both 2D (*dotted*) and 3D (*solid*)

The particle energy spectrum extends over time to higher and higher energies, as shown in Fig. 11.7. For electron-positron flows, as appropriate for pulsar winds, the maximum post-shock particle Lorentz factor increases with time as $\gamma_{max} \sim 0.5\,\gamma_0\,(\omega_{\rm p}t)^{1/2}$ (Sironi et al. 2013).[7] The plasma frequency $\omega_{\rm p}$ can be computed from the number density ahead of the termination shock, which is $n_{\rm TS} = \dot{N}/(4\pi R_{\rm TS}^2 c)$, assuming an isotropic particle flux. Here, $R_{\rm TS} \equiv 3 \times 10^{17} R_{\rm TS,17.5}$ cm is the termination shock radius. Balancing the acceleration rate with the synchrotron cooling rate in the self-generated Weibel fields, the maximum electron Lorentz factor is

$$\gamma_{sync,e} \simeq 3.5 \times 10^8 L_{sd,38.5}^{1/6} \dot{N}_{40}^{-1/3} \epsilon_{B,-2.5}^{-1/3} R_{\rm TS,17.5}^{1/3}\,. \tag{11.24}$$

A stronger constraint comes from the requirement that the diffusion length of the highest energy electrons be smaller than the termination shock radius (*i.e.* a confinement constraint). Alternatively, the acceleration time should be shorter than $R_{\rm TS}/c$, which yields the critical limit

$$\gamma_{conf,e} \simeq 1.9 \times 10^7 L_{sd,38.5}^{3/4} \dot{N}_{40}^{-1/2}\,, \tag{11.25}$$

[7]This scaling is shallower than the so-called (and commonly assumed) Bohm limit $\gamma_{max} \propto t$, and it naturally results from the small-scale nature of the Weibel turbulence generated in the shock layer (see Fig. 11.6).

which is generally more constraining than the cooling-limited Lorentz factor $\gamma_{sync,e}$. The corresponding synchrotron photons will have energies

$$h\nu_{conf,e} \simeq 0.17\, L^2_{sd,38.5} \dot{N}^{-1}_{40}\, \epsilon^{1/2}_{B,-2.5} R^{-1}_{\rm TS,17.5}\ \rm keV \tag{11.26}$$

which are apparently too small to explain the X-ray spectrum of the Crab, extending to energies beyond a few tens of MeV.

At face value, Fermi acceleration at the termination shock of PWNe is not a likely candidate for producing X-ray photons via the synchrotron process. Yet, the steady-state hard X-ray and gamma-ray spectra of PWNe do look like the consequences of Fermi acceleration—particle distributions with $p \simeq 2.4$ are a natural prediction of the Fermi process in ultra-relativistic shocks (Kirk et al. 2000; Achterberg et al. 2001; Keshet and Waxman 2005). In this regard, we argue that the wind termination shock might form in a macroscopically turbulent medium, with the outer scale of the turbulence driven by the large-scale shear flows in the nebula (Komissarov and Lyubarsky 2004; Del Zanna et al. 2004; Camus et al. 2009). If the large-scale motions drive a turbulent cascade to shorter wavelengths, back-scattering of the particles in this downstream turbulence, along with upstream reflection by the transverse magnetic field of the wind, might sustain Fermi acceleration to higher energies.

An alternative mechanism leading to particle acceleration to higher energies may be connected to the accelerator behind the recently discovered gamma-ray flares in the Crab Nebula (see Sect. 11.3.2). Runaway acceleration of electrons and positrons at reconnection X-lines, a linear accelerator, may inject energetic beams into the shock, with the mean energy per particle approaching the whole open field line voltage, $\gtrsim 10^{16}$ V in the Crab (Arons 2012), as required to explain the Crab GeV flares. This high-energy population can drive cyclotron turbulence when gyrating in the shock-compressed fields, and resonant absorption of the cyclotron harmonics can accelerate the electron-positron pairs in a broad spectrum, with maximum energy again comparable to the whole open field line voltage (Hoshino et al. 1992; Amato and Arons 2006).

11.3.1.2 The Termination Shock of a Strongly Magnetized Striped Wind

Assuming that the stripes survive until the termination shock, we now describe the physics of particle acceleration if the pre-shock flow carries a strong magnetic field of intensity B_0, oriented perpendicular to the shock direction of propagation and alternating with wavelength λ.[8] Although the magnetic field strength in the wind is always B_0, the wavelength-averaged field $\langle B_\phi \rangle_\lambda$ can vary from zero up to B_0, depending on the relative widths of the regions of positive and negative field (see the sketch in Fig. 11.8). In pulsar winds, one expects $\langle B_\phi \rangle_\lambda = 0$ only in the

[8]The wavelength λ of the striped wind equals $c\,P$, where P is the pulsar period.

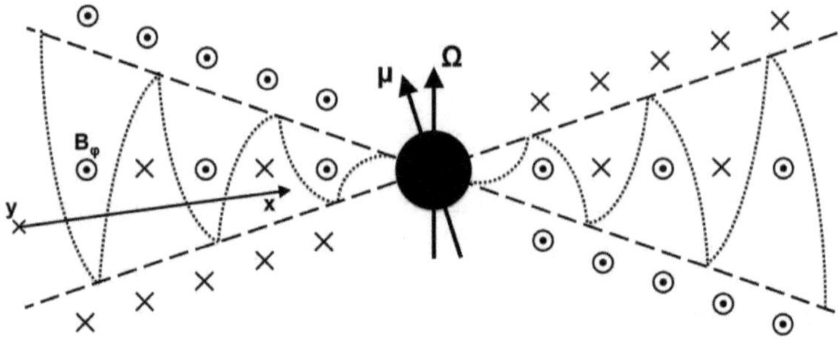

Fig. 11.8 Poloidal structure of the striped pulsar wind. The arrows denote the pulsar rotational axis (along $\boldsymbol{\Omega}$, vertical) and magnetic axis (along $\boldsymbol{\mu}$, inclined). Within the equatorial wedge bounded by the *dashed lines*, the wind consists of toroidal stripes of alternating polarity (see the reversals of B_ϕ), separated by current sheets (*dotted lines*). At latitudes higher than the inclination angle between $\boldsymbol{\Omega}$ and $\boldsymbol{\mu}$ (i.e., beyond the *dashed lines*), the field does not alternate. The simulation domain is in the *xy* plane, oriented as indicated

equatorial plane (where the stripes are symmetric), whereas $|\langle B_\phi\rangle_\lambda|/B_0 \to 1$ at high latitudes. As a proxy for latitude, we choose $\alpha = 2\langle B_\phi\rangle_\lambda/(B_0 + |\langle B_\phi\rangle_\lambda|)$, which varies between zero and unity.

At the termination shock, the compression of the flow forces the annihilation of nearby field lines, a process known as driven magnetic reconnection (Lyubarsky 2003; Sironi and Spitkovsky 2011a, 2012). As shown in Fig. 11.9, magnetic reconnection erases the striped structure of the flow (panel (a)), and transfers most of the energy stored in the magnetic fields (panel (d)) to the particles, whose distribution becomes much hotter behind the shock (see panel (f), for $x \lesssim 1000\, c/\omega_p$). As a result of field dissipation, the average particle energy increases by a factor of σ across the shock, regardless of the stripe width λ or the wind magnetization σ (as long as $\alpha \lesssim 0.1$). The reconnection process manifests itself as characteristic islands in density (panel (c)) and magnetic energy (panel (e)), separated by X-points where the magnetic field lines tear and reconnect.

The incoming particles are accelerated by the reconnection electric field at the X-points and, in the post-shock spectrum, they populate a broad distribution (red line in panel (g)), extending to much higher energies than expected in thermal equilibrium (dotted line). For the parameters studied in Fig. 11.9, the slope of the non-thermal tail is $p \simeq 1.5$ (dashed line in panel (g)), harder than what the Fermi process normally gives in relativistic shocks.[9] While efficient field dissipation (and so, efficient transfer of field energy to the particles) occurs irrespective of the wind properties (if $\alpha \lesssim 0.1$), the width of the downstream particle spectrum is sensitive to the stripe wavelength and the wind magnetization through the combination $\lambda/r_{L,hot}$,

[9]Hard particle spectra are found to be a generic by-product of magnetic reconnection in the relativistic regime appropriate for pulsar winds (e.g., Sironi and Spitkovsky 2014; Guo et al. 2014; Melzani et al. 2014; Werner et al. 2016; Sironi et al. 2015b, 2016).

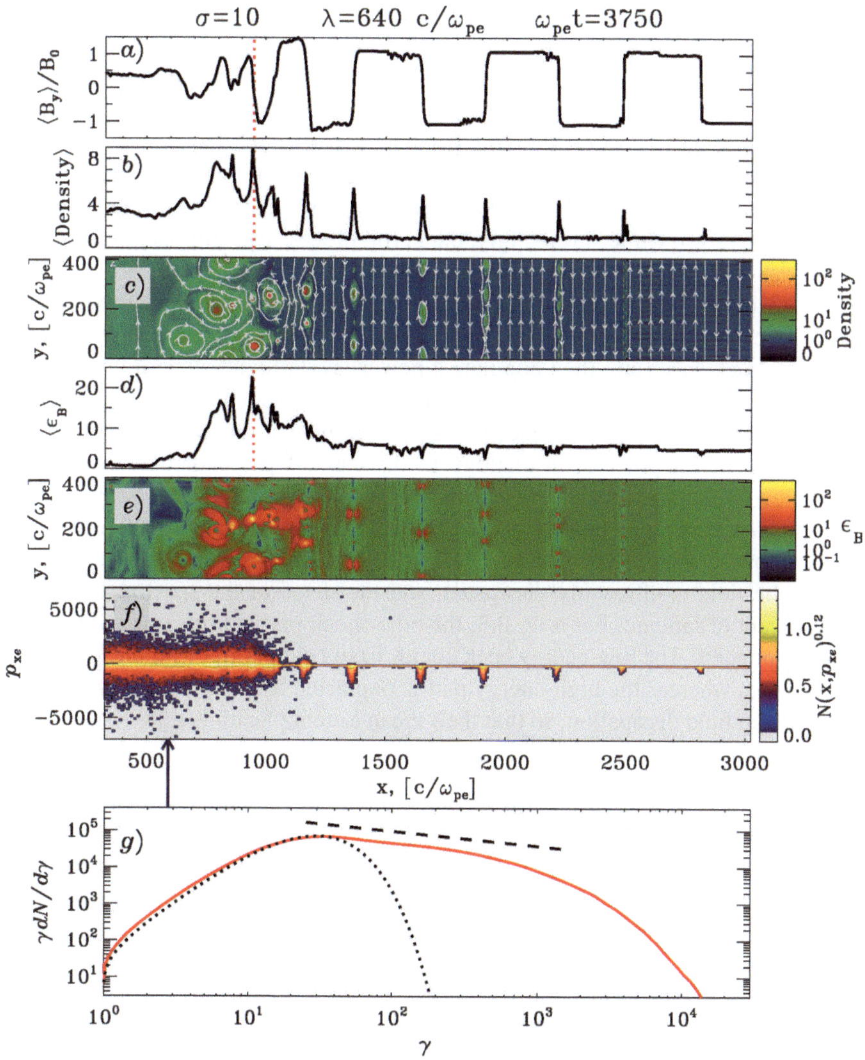

Fig. 11.9 2D PIC simulation of a relativistic shock propagating in a striped flow with magnetization $\sigma = 10$, $\alpha = 0.1$ and stripe wavelength $\lambda = 640\,c/\omega_p$, where c/ω_p is the so-called plasma skin depth, from Sironi and Spitkovsky (2011a). The shock is located at $x \sim 950\,c/\omega_{pe}$ (*vertical dotted red line*), and the incoming flow moves from right to left. At the shock, the striped structure of the magnetic field is erased (panel (**a**)), the flow compresses (density in panel (**b**)), and the field energy (panel (**d**)) is transferred to the particles (phase space in panel (**f**)). The microphysics of magnetic reconnection is revealed by the islands seen in the 2D plots of density and magnetic energy (panels (**c**) and (**e**), respectively) in a region around the shock. As a result of magnetic reconnection, the post-shock particle spectrum (*red line* in panel (**g**)) is much broader than a thermal distribution (*dotted line*), and it approaches a power-law tail with hard slope $p \simeq 1.5$ (*dashed line*)

namely the stripe wavelength measured in units of the post-shock particle Larmor radius (i.e., after dissipation has taken place, and the mean particle energy has increased by a factor of σ). A Maxwellian-like spectrum is obtained for $\lambda/r_{L,hot} \lesssim$ a few tens, whereas in the limit $\lambda/r_{L,hot} \gg 1$ the spectrum approaches a broad power-law tail of index $1 < p < 2$, extending from $\gamma_{\min} \simeq \gamma_0$ up to $\gamma_{\max} \simeq \gamma_0 \sigma^{1/(2-p)}$.

The particles are accelerated primarily by the reconnection electric field at the X-points, rather than by bouncing back and forth across the shock, as in the standard Fermi mechanism. Quite surprisingly, the Fermi process can still operate along the equatorial plane of the wind, where the stripes are quasi-symmetric ($\alpha \lesssim 0.01$). Here, the highest energy particles accelerated by the reconnection electric field can escape ahead of the shock, and be injected into a Fermi-like acceleration cycle. In the post-shock spectrum, they populate a power-law tail with slope $p \simeq 2.5$, that extends beyond the hard component produced by reconnection.

At higher latitudes, the presence of a non-negligible stripe-averaged field $\langle B_\phi \rangle_\lambda$ inhibits the Fermi process, in analogy to the case of perpendicular magnetized shocks (with uniform fields) discussed at the beginning of this section. The efficiency of particle acceleration via shock-driven reconnection is also affected, as we show in Fig. 11.10. For $\alpha \lesssim 0.1$ (i.e., relatively close to the equatorial plane), the field is efficiently dissipated, and the shape of the spectrum is nearly independent of latitude. For $\alpha \gtrsim 0.3$, the post-shock particle spectrum consists of two components. The low-energy peak comes from cold plasma with mean Lorentz factor $\sim \gamma_0$, whereas the high-energy part is populated by hot particles that gained energy from field dissipation, so that their mean Lorentz factor is now $\sim \gamma_0 \sigma$. As α

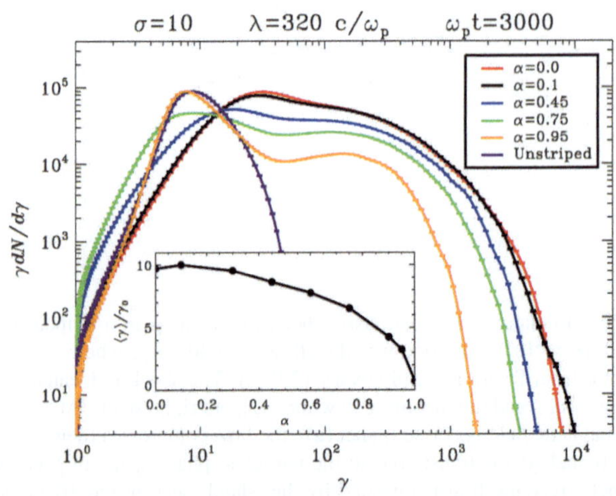

Fig. 11.10 Downstream particle spectrum at $\omega_p t = 3000$ for different values of the stripe-averaged field $\langle B_\phi \rangle_\lambda$ (or equivalently, of the parameter α), in a flow with $\lambda = 320\,c/\omega_p$ and $\sigma = 10$, from Sironi and Spitkovsky (2011a). The limit of an unstriped wind is shown for reference as a purple line. The *black line* in the subpanel shows the average downstream Lorentz factor as a function of α (with $\alpha = 1.0$ referring to the unstriped wind)

increases, the fraction of upstream Poynting flux available for dissipation decreases, which explains why the high-energy component in the spectra of Fig. 11.10 gets depopulated, at the expense of the low-energy part. The limit $|\alpha| \to 1$ (yellow curve for $\alpha = 0.95$) approaches the result expected for an unstriped wind (purple line).

Based on these findings, one could interpret the optical and X-ray signatures of the Crab, which require a particle spectrum with $p \simeq 2.5$, as synchrotron emission from the particles that are Fermi-accelerated close to the equatorial plane of the wind. In addition, the spectral index required for the radio spectrum of the Crab ($p \simeq 1.5$) could naturally result from the broad hard component of particles accelerated by shock-driven reconnection. However, the particle spectrum in the simulations approaches the hard tail required by the observations only when the combination $\lambda/r_{L,hot}$ exceeds a few tens (for smaller values, the spectrum is a narrow thermal-like distribution). At the termination shock of the pulsar wind ($R = R_{TS}$) we have

$$\frac{\lambda}{r_{L,hot}} \simeq 4\pi\kappa \frac{R_{LC}}{R_{TS}} , \tag{11.27}$$

where $R_{LC} = c/\Omega$ is the light cylinder radius ($\Omega = 2\pi/P$ is the pulsar rotational frequency), and κ is the so-called multiplicity in the wind (i.e., the ratio of the actual density to the Goldreich-Julian density, Goldreich and Julian (1969)). For the Crab, $R_{TS} \simeq 5 \times 10^8 R_{LC}$ (Hester et al. 2002) and most available models estimate $\kappa \simeq 10^4 - 10^6$ (Bucciantini et al. 2011). Based on our findings, the resulting value of $\lambda/(r_L\sigma) \lesssim 0.01$ would yield a Maxwellian-like spectrum, at odds with the wide flat spectrum required by observations. If radio-emitting electrons are accelerated at the termination shock of pulsar winds via magnetic reconnection, a revision of the existing theories of pulsar magnetospheres is required.

11.3.2 The Flaring Emission

In recent years, the *Fermi* and *AGILE* satellites have detected a number of hours-to week-long flares at GeV energies, which surprisingly falsify the widely-believed "standard candle" nature of the high-energy Crab emission. During these events the Crab nebula gamma-ray flux above 100 MeV exceeded its average value by a factor of several or higher (Tavani et al. 2011; Abdo et al. 2011; Buehler et al. 2012), while at other wavelengths nothing unusual was observed (Weisskopf et al. 2013). The observed gamma-ray flares happen with a cadence of $\sim$ 1 year (e.g., Buehler and Blandford 2014) and there are no associated pulsar timing glitches. Variability on timescales as short as a few hours has been reported. The peak isotropic luminosity is roughly 10^{36} ergs/s and the energy radiated is $\sim 10^{41}$ ergs. The flares and secular observations (Wilson-Hodge et al. 2011) demonstrate that the energy conversion is intermittent and that the mechanism can be locally cataclysmic.

The flare properties suggest that an extreme accelerator is at work. The typical decay time of the flaring episodes, which is attributed to synchrotron cooling, together with the $\sim$ GeV peak frequency, allows to solve simultaneously for the magnetic field strength $\sim$ 5 mG (as compared to the nebula-averaged $\sim 200\,\mu$ G) and for the extreme energy of the emitting particles $\sim$ PeV. From the $\sim$ 10 h rise time of the flares, one can estimate the size $\sim 10^{15}$ cm of the emission region. In order to accelerate up to PeV energies within this length, the accelerating electric field needs to be comparable to the inferred magnetic field (i.e., $E \sim B$).

Fermi acceleration at the termination shock of the Crab nebula fails to explain the observed GeV flares (Sironi et al. 2013). In contrast, the requirement that $E \sim B$ is naturally satisfied in reconnection layers, in the relativistic regime where the magnetic energy per particle exceeds its rest mass, or equivalently where the magnetization $\sigma = B_0^2/4\pi\rho_0 c^2 \gg 1$. The reconnection scenario would work best in the most magnetized regions of the nebula, i.e., near the poles and possibly in the jets (Cerutti et al. 2012; Lyubarsky 2012; Komissarov 2013; Mignone et al. 2013). Unfortunately, current gamma-ray telescopes do not have the angular resolution to pin down the precise location of the flares within the Nebula. Below, we discuss how PIC simulations help unveiling the role of magnetic reconnection in the Crab Nebula, as the underlying particle accelerator that powers the GeV flares.

11.3.2.1 Plane-Parallel Reconnection

In the simplest geometry of magnetic reconnection, the field lines are parallel to a pre-existing current sheet, with opposite polarity on the two sides of the current sheet. We shall call this setup as "plane-parallel reconnection" (see Kagan et al. 2015, for a review). Using 2D and 3D PIC simulations, it has been recently shown that most of the features of the Crab flares can be explained with relativistic plane-parallel reconnection (timescale, energetics, particle and photon spectra). The key arguments in favor of reconnection for the Crab flares are:

- The flare spectrum requires an electron power-law population with hard slope $p = -d\log N/d\log\gamma \lesssim 2$, which is not attainable in shocks, but it naturally results from relativistic reconnection (Sironi and Spitkovsky 2014; Guo et al. 2014; Melzani et al. 2014; Werner et al. 2016; Sironi et al. 2015b, 2016). As shown in Fig. 11.11, the power-law slope depends on the flow magnetization, being harder for higher σ ($p \sim 1.5$ for $\sigma = 50$, compare solid and dotted green lines). The slope is steeper for lower magnetizations ($p \sim 4$ for $\sigma = 1$, solid and dotted black lines), approaching the result of non-relativistic reconnection, yielding poor acceleration efficiencies (Drake et al. 2010).
- The $\sim$GeV peak energy of the flares is well above the classical synchrotron "burnoff" limit of $\sim 236\,\eta$ MeV (as measured in the fluid rest frame), which is obtained by balancing acceleration due to an electric field $E = \eta B$ with synchrotron cooling losses; unlike in shocks, where $\eta < 1$, in the reconnection layer one finds $\eta > 1$, thus boosting the synchrotron limit to the observed $\sim$ GeV peak. In a reconnection scenario, this requires the accelerating particles to stay confined within the reconnection layer, where $\eta > 1$. Uzdensky et al. (2011) showed analytically that as the particle energy increases, the trajectory gets more

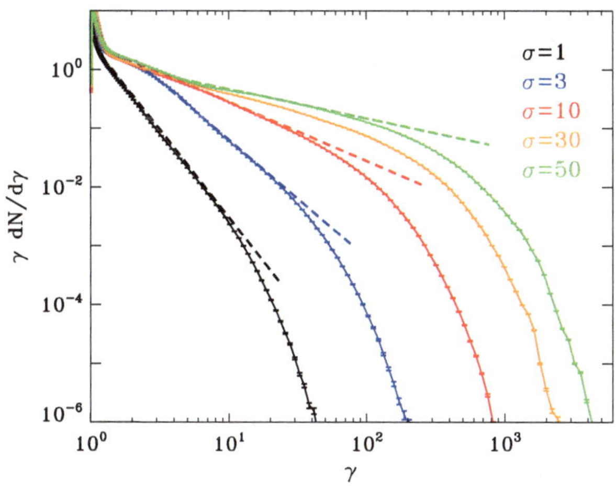

Fig. 11.11 Dependence of the spectrum on the magnetization, as indicated in the legend, from Sironi and Spitkovsky (2014). The *dotted lines* refer to power-law slopes of −4, −3, −2 and −1.5 (from *black to green*)

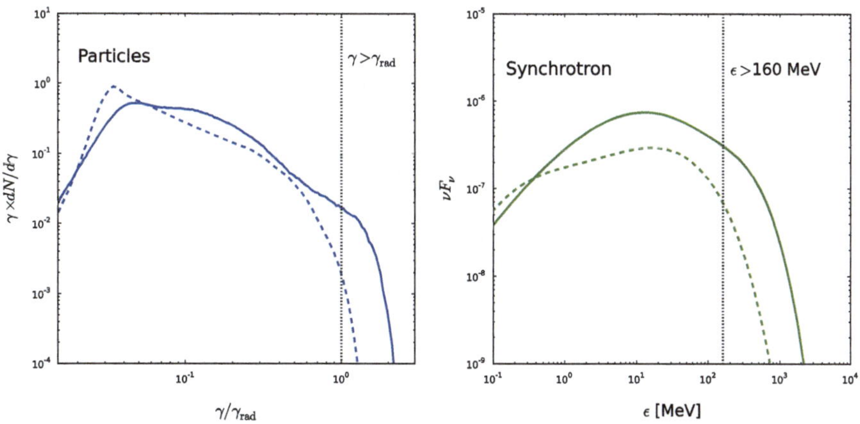

Fig. 11.12 Isotropically-averaged particle spectrum ($\gamma dN/d\gamma$, left panel) and synchrotron radiation energy distribution (νF_ν, *right panel*) in a 2D (*solid line*) and 3D (*dashed line*) PIC simulations of relativistic reconnection, including the effect of the radiation reaction force on the particles. The vertical *dotted lines* show the radiation-reaction limited energy of a particle if $E = B_0$ ($\gamma = \gamma_{\rm rad}$, *left*), and the corresponding maximum synchrotron photon energy ($\epsilon = 160$ MeV independent of E and B_0, *right*). Figure adapted from Cerutti et al. (2014a)

and more focused along the electric field, with vanishing cooling losses. This has now been confirmed with PIC simulations (Cerutti et al. 2013, 2014b). In particular, these studies demonstrated that reconnection can accelerate particles above the synchrotron radiation burn-off limit (Guilbert et al. 1983; de Jager et al. 1996) deep inside the reconnection layer where the electric field overcome the magnetic field (see Fig. 11.12). This result is crucial because it can explain

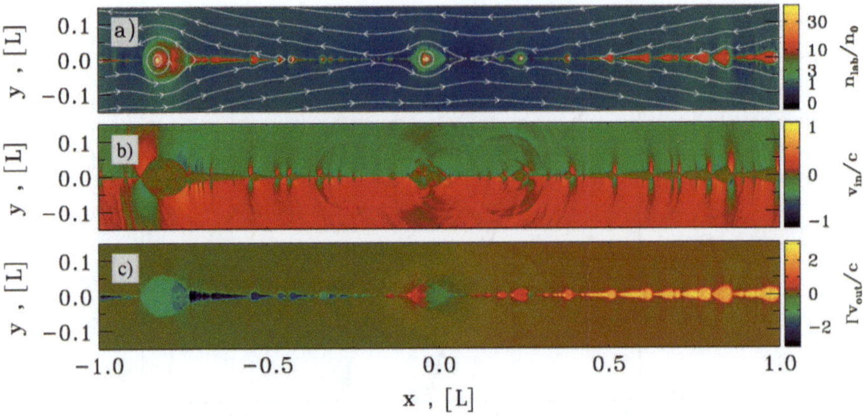

Fig. 11.13 The reconnection layer, from a 2D simulation with $\sigma = 10$, from Sironi et al. (2016). We present (**a**) particle density, in units of the number density far from the sheet, with overplotted magnetic field lines; (**b**) inflow velocity; and (**c**) outflow four-velocity, in units of the speed of light. The plasma enters the reconnection layer with $v_{\rm rec} \sim 0.15\,c$. The reconnection layer fragments into a series of magnetic islands (or plasmoids), moving away from the center of the current sheet at ultra-relativistic speeds

the emission of $>$ 100 MeV synchrotron radiation emitted during every Crab flare, which would be impossible to achieve in ideal MHD.

- The short rise time ($\sim$10 h) of the flaring episodes can naturally result from the inhomogeneity of the reconnection layer, which is fragmented into a chain of magnetic islands, or plasmoids, as shown in Fig. 11.13 (in 3D, these plasmoids appear as elongated magnetic flux ropes). The plasmoids are overdense (Fig. 11.13a), filled with energetic particles and confined by strong fields. The plasma flows into the reconnection layer at $v_{\rm rec} \simeq 0.15\,c$ for $\sigma = 10$ (Fig. 11.13b). The inflow speed is nearly independent of σ for larger magnetizations (Sironi et al. 2016), in agreement with analytical models (Lyubarsky 2005). After entering the sheet, the flow is advected out by the tension force of the reconnected field. The motion in the reconnection exhausts is ultra-relativistic (Fig. 11.13c), approaching a bulk four-velocity $\Gamma v_{\rm out} \sim \sqrt{\sigma}\,c$, in agreement with the theory (Lyubarsky 2005). The relativistic bulk motion of the plasmoids in the reconnection layer plays a critical role in enhancing—via Doppler boosting—their emission signatures.

Aside from bulk Doppler beaming, an energy-dependent "kinetic" beaming has also been proposed to explain the extreme time variability of the Crab flares (Cerutti et al. 2013, 2014b). In particular, while low-energy particles are nearly isotropic, at high energies ($\gamma \gtrsim \sigma$) the particles exhibit clear sign of anisotropy with two beams pointing roughly towards the $\pm x$-directions, i.e., along the reconnection exhausts. Hence, the beams are not necessarily pointing along the direction z of the reconnection electric field because the tension of the reconnected field lines pushes the particles away from the X-points in the form of a reconnection outflow towards the magnetic islands. Nonetheless, the direction of the beam of energetic particles is not static: it wiggles rapidly within the xz-plane, which results in rapid flares of energetic radiation when the beam crosses the line of sight of a distant observer. Since in the Crab the particles emitting > 100 MeV synchrotron radiation should be accelerated and radiating over a sub-Larmor timescale, one expects that the highest energy radiation should keep the imprint of the particle anisotropy (regardless of the acceleration process), while the low-energy radiation should be more isotropic (Cerutti et al. 2013, 2014b).

11.3.2.2 Explosive Reconnection

Despite its successes, plane-parallel reconnection does not seem to be fast enough to explain the short rise time of the Crab flares. In particular, recent PIC studies of relativistic reconnection have demonstrated that the reconnection rate (inflow velocity) in 3D simulations of plane-parallel reconnection is significantly lower than in 2D. For a reference magnetization $\sigma = 10$ the reconnection rate in 2D is $v_{rec}/c \sim 0.1$, whereas in 3D it is only $v_{rec}/c \sim 0.02$ (Sironi and Spitkovsky 2014). The slower reconnection rate leads to a weaker accelerating electric field. Moreover, for a given flare duration it translates into a smaller utilised magnetic energy.

To overcome this difficulty, it has recently been proposed that the Crab flares might result from explosive reconnection episodes, accompanying the relaxation of force-free equilibria on dynamical timescales (i.e., corresponding to an effective reconnection rate of $v_{rec}/c \sim 1$). In particular, Nalewajko et al. (2016), Lyutikov et al. (2016), and Yuan et al. (2016) have carried out PIC simulations of the relaxation of force-free equilibria in application to the Crab flares.

As a representative case, we consider the configuration of two Lundquist's force-free cylinders surrounded by uniform magnetic field (Lyutikov et al. 2016),

$$\mathbf{B}_L(r \leq r_\mathrm{j}) \propto J_1(r\alpha_0)\mathbf{e}_\phi + J_0(r\alpha_0)\mathbf{e}_z\,, \tag{11.28}$$

Here, J_0, J_1 are Bessel functions of zeroth and first order and the constant $\alpha_0 \simeq 3.8317$ is the first root of J_0. This solution is terminated at the first zero of J_1, which we denote as r_j and hence continued with $B_z = B_z(r_j)$ and $B_\phi = 0$ for $r > r_j$. Since the total current of the flux tube is zero, the azimuthal field vanishes at the boundary of the rope, and so the evolution is initially very slow (i.e., the initial configuration is dynamically stable). To speed things up, the ropes are pushed towards each other.

In Fig. 11.14, we present the 2D pattern of the out-of-plane field B_z (left column) and of the in-plane magnetic energy fraction $\epsilon_{B,\mathrm{in}} = (B_x^2 + B_y^2)/8\pi nmc^2$ (right column; with superimposed magnetic field lines), from a PIC simulation with $\sigma_\mathrm{in} = 43$ (only defined with the in-plane fields) and $r_\mathrm{j} = 61\, r_\mathrm{L,hot}$ (where $r_\mathrm{L,hot}$ is the Larmor radius of particles heated by reconnection). As the two magnetic ropes slowly approach, driven by the initial velocity push, reconnection is triggered in the plane $x = 0$, as indicated by the formation and subsequent ejection of small-scale plasmoids. As a result of reconnection, an increasing number of field lines,

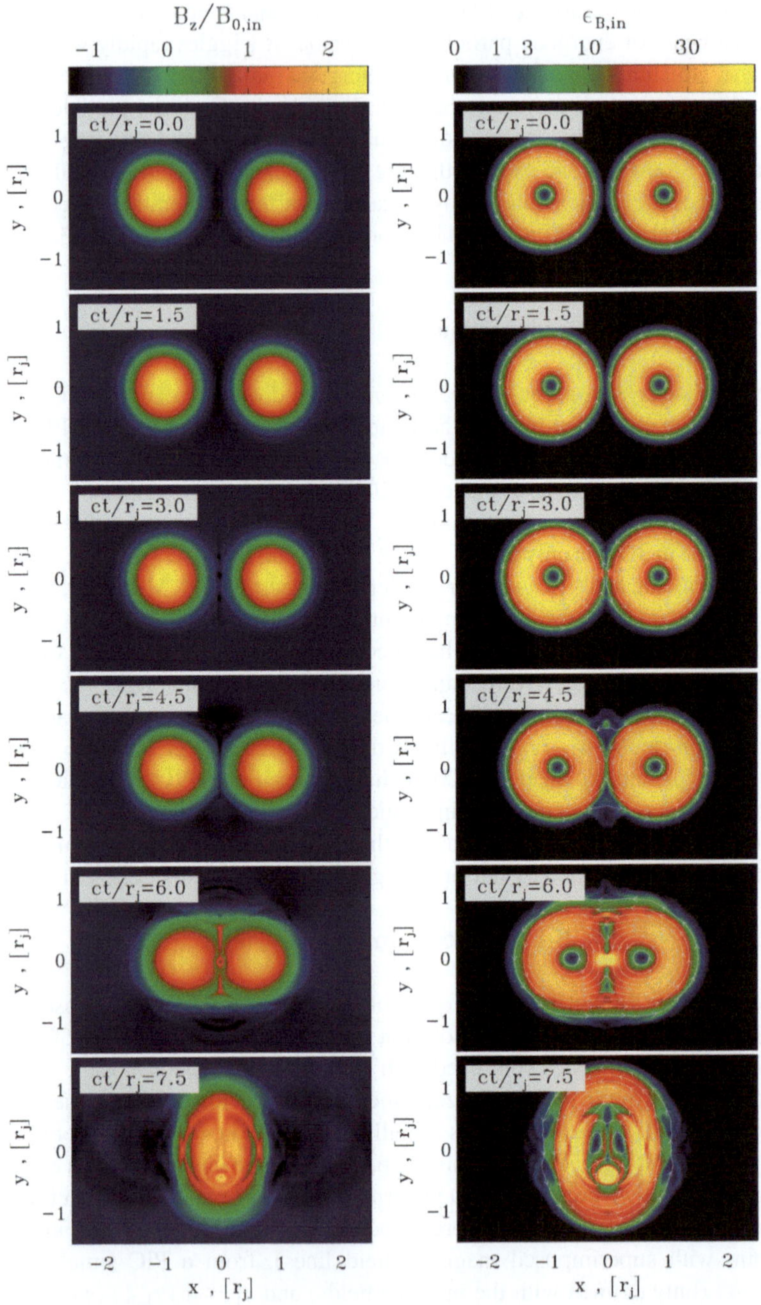

Fig. 11.14 Temporal evolution of 2D Lundquist ropes (time is measured in c/r_j and indicated in the *grey box* of each panel, increasing from *top to bottom*), from Lyutikov et al. (2016). The plot presents the 2D pattern of the out-of-plane field B_z (*left column*) and of the in-plane magnetic energy fraction $\epsilon_{B,\mathrm{in}} = (B_x^2 + B_y^2)/8\pi nmc^2$ (*right column*; with superimposed magnetic field lines), from a PIC simulation with $\sigma_{\mathrm{in}} = 43$ and $r_j = 61\, r_{\mathrm{L,hot}}$

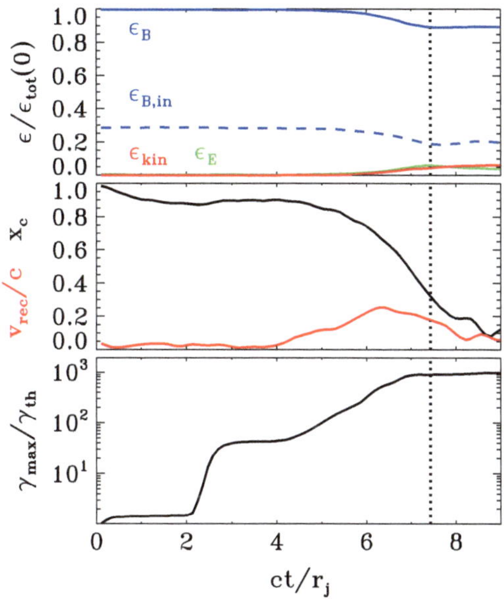

Fig. 11.15 Temporal evolution of various quantities, from a 2D PIC simulation of Lundquist ropes with $\sigma_{\rm in} = 43$ and $r_{\rm j} = 61\, r_{\rm L,hot}$ (the same as in Fig. 11.14), from Lyutikov et al. (2016). *Top panel*: fraction of energy in magnetic fields (*solid blue*), in-plane magnetic fields (*dashed blue*), electric fields (*green*) and particles (*red*; excluding the rest mass energy), in units of the total initial energy. *Middle panel*: reconnection rate $v_{\rm rec}/c$ (*red*), and location $x_{\rm c}$ of the core of the rightmost flux rope (*black*), in units of $r_{\rm j}$. *Bottom panel*: evolution of the maximum Lorentz factor $\gamma_{\rm max}$

that initially closed around one of the ropes, are now engulfing both magnetic islands. Their tension force causes the two ropes to approach and merge on a quick (dynamical) timescale, starting at $ct/r_{\rm j} \sim 4.5$ and ending at $ct/r_{\rm j} \sim 7.5$ (see that the distance of the rightmost island from the center rapidly decreases, as indicated by the black line in the middle panel of Fig. 11.15). The tension force drives the particles in the flux ropes toward the center, with a fast reconnection speed peaking at $v_{\rm rec}/c \sim 0.3$ (red line in the middle panel of Fig. 11.15).[10] The reconnection layer at $x = 0$ stretches up to a length of $\sim 2\, r_{\rm j}$, and secondary plasmoids are formed. In this phase of evolution, the fraction of initial energy released to the particles is small ($\epsilon_{\rm kin}/\epsilon_{\rm tot}(0) \sim 0.1$, top panel in Fig. 11.15), but the particles advected into the central X-point experience a dramatic episode of acceleration. As shown in the bottom panel of Fig. 11.15, the cutoff Lorentz factor $\gamma_{\rm max}$ of the particle spectrum presents a dramatic evolution, increasing up to $\gamma_{\rm max}/\gamma_{\rm th} \sim 10^3$ within a couple of dynamical times (here, $\gamma_{\rm th}$ is the initial "thermal" Lorentz factor).

[10] The reconnection rate is measured to be in the range $v_{\rm rec}/c \sim 0.2 - 0.5$, which increases with the magnetization and saturates at around 0.5 at high magnetization limit (Lyutikov et al. 2016).

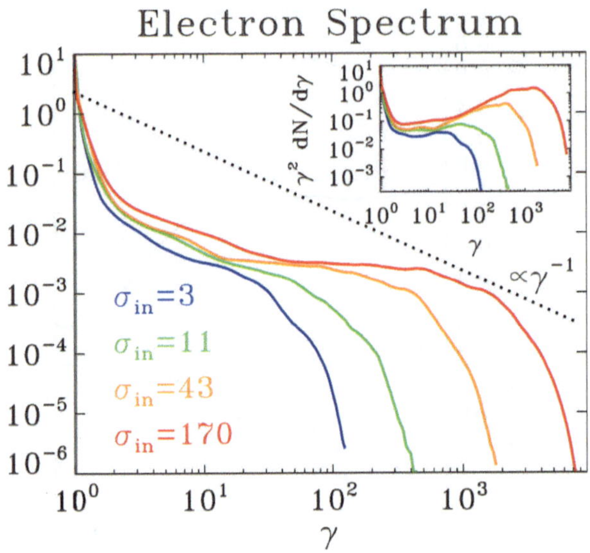

Fig. 11.16 Particle spectrum for a suite of PIC simulations of Lundquist ropes, from Lyutikov et al. (2016). We fix $r_j/r_{L,hot} = 61$ and we vary the magnetization σ_{in} from 3 to 170 (from *blue to red*, as indicated by the legend). The main plot shows $\gamma dN/d\gamma$ to emphasize the particle content, whereas the inset presents $\gamma^2 dN/d\gamma$ to highlight the energy census. The *dotted black line* is a power law $\gamma dN/d\gamma \propto \gamma^{-1}$, corresponding to equal energy content per decade (which would result in a flat distribution in the insets). The spectral hardness is strongly dependent on σ_{in}, with higher magnetizations giving harder spectra

This phase of extremely fast particle acceleration on a dynamical timescale is analogous to the relaxation of unstable "ABC" force-free structures discussed by Nalewajko et al. (2016) and Lyutikov et al. (2016), and it constitutes the most promising scenario to explain the Crab flares. The particle acceleration efficiency and the hardness of the power-law slope depend on the mean magnetization of the configuration, in a similar fashion as in plane-parallel reconnection scenarios. The particle spectrum gets harder as the mean magnetization increases (Fig. 11.16); both the non-thermal particle fraction and the maximum particle energy increase with the magnetization (Lyutikov et al. 2016; Nalewajko et al. 2016).

Since the highest energy particles are first accelerated in the current layers by the parallel electric field, they do not radiate much when they are inside the sheet, because the curvature of their trajectory is small. Most of the radiation is produced when particles are ejected from the current layers—their trajectories start to bend significantly in the ambient magnetic field (Yuan et al. 2016). Such a separation of acceleration site and radiative loss site facilitates acceleration beyond the synchrotron radiation reaction limit, as required by the Crab flares. Fast variability of the observed photon flux can be produced when compact plasmoids that contain high-energy particles are ejected from the ends of the current layers and get destroyed. These give beamed radiation. An observer sees high intensity radiation when the beam happens to be aligned with the line of sight. Such peaks in emission are accompanied by an increase in the polarization degree and rapid change of polarization angle in the high-energy band (Yuan et al. 2016). The variability timescale is determined by the spatial extent of the emitting structure, e.g. the plasmoids, thus can be much shorter than the light crossing time of the region that collapses.

11.4 Conclusions

In this chapter, we have discussed the role of PIC simulations in unveiling the origin of the emitting particles in PWNe. After describing the basics of the PIC technique, we have summarized its implications for the quiescent and flaring emission of the Crab Nebula, as a prototype of PWNe. A consensus seems to be emerging that, in addition to the standard scenario of particle acceleration via the Fermi process at the termination shock of the pulsar wind, magnetic reconnection in the wind, at the termination shock and in the nebula plays a major role in powering the multi-wavelength emission signatures of PWNe.

Acknowledgements LS acknowledges support from DoE DE-SC0016542 and NASA Fermi NNX16AR75G. BC acknowledges support from CNES and Labex OSUG@2020 (ANR10 LABX56).

References

Abdo, A.A., Ackermann, M., Ajello, M., Allafort, A., Baldini, L., Ballet, J., Barbiellini, G., Bastieri, D., Bechtol, K., Bellazzini, R., Berenji, B., Blandford, R.D., Bloom, E.D., Bonamente, E., Borgland, A.W., Bouvier, A., Brandt, T.J., Bregeon, J., Brez, A., Brigida, M., Bruel, P., Buehler, R., Buson, S., Caliandro, G.A., Cameron, R.A., Cannon, A., Caraveo, P.A., Casandjian, J.M., Çelik, Ö., Charles, E., Chekhtman, A., Cheung, C.C., Chiang, J., Ciprini, S., Claus, R., Cohen-Tanugi, J., Costamante, L., Cutini, S., D'Ammando, F., Dermer, C.D., de Angelis, A., de Luca, A., de Palma, F., Digel, S.W., do Couto e Silva, E., Drell, P.S., Drlica-Wagner, A., Dubois, R., Dumora, D., Favuzzi, C., Fegan, S.J., Ferrara, E.C., Focke, W.B., Fortin, P., Frailis, M., Fukazawa, Y., Funk, S., Fusco, P., Gargano, F., Gasparrini, D., Gehrels, N., Germani, S., Giglietto, N., Giordano, F., Giroletti, M., Glanzman, T., Godfrey, G., Grenier, I.A., Grondin, M.-H., Grove, J.E., Guiriec, S., Hadasch, D., Hanabata, Y., Harding, A.K., Hayashi, K., Hayashida, M., Hays, E., Horan, D., Itoh, R., Jóhannesson, G., Johnson, A.S., Johnson, T.J., Khangulyan, D., Kamae, T., Katagiri, H., Kataoka, J., Kerr, M., Knödlseder, J., Kuss, M., Lande, J., Latronico, L., Lee, S.-H., Lemoine-Goumard, M., Longo, F., Loparco, F., Lubrano, P., Madejski, G.M., Makeev, A., Marelli, M., Mazziotta, M.N., McEnery, J.E., Michelson, P.F., Mitthumsiri, W., Mizuno, T., Moiseev, A.A., Monte, C., Monzani, M.E., Morselli, A., Moskalenko, I.V., Murgia, S., Nakamori, T., Naumann-Godo, M., Nolan, P.L., Norris, J.P., Nuss, E., Ohsugi, T., Okumura, A., Omodei, N., Ormes, J.F., Ozaki, M., Paneque, D., Parent, D., Pelassa, V., Pepe, M., Pesce-Rollins, M., Pierbattista, M., Piron, F., Porter, T.A., Rainò, S., Rando, R., Ray, P.S., Razzano, M., Reimer, A., Reimer, O., Reposeur, T., Ritz, S., Romani, R.W., Sadrozinski, H.F.-W., Sanchez, D., Parkinson, P.M.S., Scargle, J.D.,

Schalk, T.L., Sgrò, C., Siskind, E.J., Smith, P.D., Spandre, G., Spinelli, P., Strickman, M.S., Suson, D.J., Takahashi, H., Takahashi, T., Tanaka, T., Thayer, J.B., Thompson, D.J., Tibaldo, L., Torres, D.F., Tosti, G., Tramacere, A., Troja, E., Uchiyama, Y., Vandenbroucke, J., Vasileiou, V., Vianello, G., Vitale, V., Wang, P., Wood, K.S., Yang, Z., Ziegler, M.: Gamma-ray flares from the Crab Nebula. Science **331**, 739 (2011). doi:10.1126/science.1199705

Achterberg, A., Gallant, Y.A., Kirk, J.G., Guthmann, A.W.: Particle acceleration by ultrarelativistic shocks: theory and simulations. Mon. Not. R. Astron. Soc. **328**, 393–408 (2001). doi:10.1046/j.1365-8711.2001.04851.x

Amato, E., Arons, J.: Heating and nonthermal particle acceleration in relativistic, transverse magnetosonic shock waves in proton-electron-positron plasmas. Astrophys. J. **653**, 325–338 (2006). doi:10.1086/508050

Arons, J.: Pulsar wind nebulae as cosmic pevatrons: a current sheet's tale. Space Sci. Rev. **173**, 341–367 (2012). doi:10.1007/s11214-012-9885-1

Belyaev, M.A.: PICsar: A 2.5D axisymmetric, relativistic, electromagnetic, particle in cell code with a radiation absorbing boundary. New Astron. **36**, 37–49 (2015). doi:10.1016/j.newast.2014.09.006

Berenger, J.-P.: A perfectly matched layer for the absorption of electromagnetic waves. J. Comput. Phys. **114**, 185–200 (1994). doi:10.1006/jcph.1994.1159

Berenger, J.-P.: Three-dimensional perfectly matched layer for the absorption of electromagnetic waves. J. Comput. Phys. **127**, 363–379 (1996). doi:10.1006/jcph.1996.0181

Bietenholz, M.F., Kassim, N., Frail, D.A., Perley, R.A., Erickson, W.C., Hajian, A.R.: The radio spectral index of the Crab Nebula. Astrophys. J. **490**, 291 (1997). doi:10.1086/304853

Birdsall, C.K., Langdon, A.B.: Plasma Physics via Computer Simulation, Taylor & Francis, New York (1991)

Bucciantini, N., Arons, J., Amato, E., Modelling spectral evolution of pulsar wind nebulae inside supernova remnants. Mon. Not. R. Astron. Soc. **410**, 381–398 (2011). doi:10.1111/j.1365-2966.2010.17449.x

Buehler, R., Blandford, R.: The surprising Crab pulsar and its Nebula: a review. Rep. Prog. Phys. **77**(6), 066901 (2014). doi:10.1088/0034-4885/77/6/066901

Buehler, R., Scargle, J.D., Blandford, R.D., Baldini, L., Baring, M.G., Belfiore, A., Charles, E., Chiang, J., D'Ammando, F., Dermer, C.D., Funk, S., Grove, J.E., Harding, A.K., Hays, E., Kerr, M., Massaro, F., Mazziotta, M.N., Romani, R.W., Saz Parkinson, P.M., Tennant, A.F., Weisskopf, M.C.: Gamma-ray activity in the Crab Nebula: the exceptional flare of 2011 April. Astrophys. J. **749**, 26 (2012). doi:10.1088/0004-637X/749/1/26

Camus, N.F., Komissarov, S.S., Bucciantini, N., Hughes, P.A.: Observations of 'wisps' in magnetohydrodynamic simulations of the Crab Nebula. Mon. Not. R. Astron. Soc. **400**, 1241–1246 (2009). doi:10.1111/j.1365-2966.2009.15550.x

Cerutti, B., Uzdensky, D.A., Begelman, M.C.: Extreme particle acceleration in magnetic reconnection layers: application to the gamma-ray flares in the Crab Nebula. Astrophys. J. **746**, 148 (2012). doi:10.1088/0004-637X/746/2/148

Cerutti, B., Werner, G.R., Uzdensky, D.A., Begelman, M.C.: Simulations of particle acceleration beyond the classical synchrotron burnoff limit in magnetic reconnection: an explanation of the Crab flares. Astrophys. J. **770**, 147 (2013). doi:10.1088/0004-637X/770/2/147

Cerutti, B., Werner, G.R., Uzdensky, D.A., Begelman, M.C.: Gamma-ray flares in the Crab Nebula: a case of relativistic reconnection? Phys. Plasmas **21**(5), 056501 (2014a). doi:10.1063/1.4872024

Cerutti, B., Werner, G.R., Uzdensky, D.A., Begelman, M.C.: Three-dimensional relativistic pair plasma reconnection with radiative feedback in the Crab Nebula. Astrophys. J. **782**, 104 (2014b). doi:10.1088/0004-637X/782/2/104

Cerutti, B., Philippov, A., Parfrey, K., Spitkovsky, A.: Particle acceleration in axisymmetric pulsar current sheets. Mon. Not. R. Astron. Soc. **448**, 606–619 (2015). doi:10.1093/mnras/stv042

Cerutti, B., Philippov, A.A., Spitkovsky, A.: Modelling high-energy pulsar light curves from first principles. Mon. Not. R. Astron. Soc. **457**, 2401–2414 (2016). doi:10.1093/mnras/stw124

Chen, A.Y., Beloborodov, A.M.: Electrodynamics of Axisymmetric Pulsar Magnetosphere with Electron-Positron Discharge: A Numerical Experiment. Astrophys. J. Lett. **795**, 22 (2014). doi:10.1088/2041-8205/795/1/L22

Cheng, C.Z., Knorr, G.: The integration of the Vlasov equation in configuration space. J. Comput. Phys. **22**, 330–351 (1976). doi:10.1016/0021-9991(76)90053-X

de Jager, O.C., Harding, A.K., Michelson, P.F., Nel, H.I., Nolan, P.L., Sreekumar, P., Thompson, D.J.: Gamma-ray observations of the Crab Nebula: a study of the synchro-compton spectrum. Astrophys. J. **457**, 253 (1996). doi:10.1086/176726

Del Zanna, L., Amato, E., Bucciantini, N.: Axially symmetric relativistic MHD simulations of pulsar wind nebulae in supernova remnants. On the origin of torus and jet-like features. Astron. Astrophys. **421**, 1063–1073 (2004). doi:10.1051/0004-6361:20035936

Drake, J.F., Opher, M., Swisdak, M., Chamoun, J.N.: A magnetic reconnection mechanism for the generation of anomalous cosmic rays. Astrophys. J. **709**, 963–974 (2010). doi:10.1088/0004-637X/709/2/963

Elkina, N.V., Büchner, J.: A new conservative unsplit method for the solution of the Vlasov equation. J. Comput. Phys. **213**, 862–875 (2006). doi:10.1016/j.jcp.2005.09.023

Esirkepov, T.Z.: Exact charge conservation scheme for particle-in-cell simulation with an arbitrary form-factor. Comput. Phys. Commun. **135**, 144–153 (2001). doi:10.1016/S0010-4655(00)00228-9

Gallant, Y.A., Hoshino, M., Langdon, A.B., Arons, J., Max, C.E.: Relativistic, perpendicular shocks in electron-positron plasmas. Astrophys. J. **391**, 73–101 (1992). doi:10.1086/171326

Goldreich, P., Julian, W.H.: Pulsar electrodynamics. Astrophys. J. **157**, 869 (1969). doi:10.1086/150119

Gruzinov, A., Waxman, E.: Gamma-ray burst afterglow: polarization and analytic light curves. Astrophys. J. **511**, 852–861 (1999). doi:10.1086/306720

Guilbert, P.W., Fabian, A.C., Rees, M.J.: Spectral and variability constraints on compact sources. Mon. Not. R. Astron. Soc. **205**, 593–603 (1983)

Guo, F., Li, H., Daughton, W., Liu, Y.-H.: Formation of hard power laws in the energetic particle spectra resulting from relativistic magnetic reconnection. Phys. Rev. Lett. **113**(15), 155005 (2014). doi:10.1103/PhysRevLett.113.155005

Haugbølle, T.: Three-dimensional modeling of relativistic collisionless ion-electron shocks. Astrophys. J. **739**, 42 (2011). doi:10.1088/2041-8205/739/2/L42

Hester, J.J., Mori, K., Burrows, D., Gallagher, J.S., Graham, J.R., Halverson, M., Kader, A., Michel, F.C., Scowen, P.: Hubble space telescope and chandra monitoring of the Crab synchrotron nebula. Astrophys. J. **577**, 49–52 (2002). doi:10.1086/344132

Hockney, R.W., Eastwood, J.W.: Computer Simulation Using Particles, Taylor & Francis, New York (1988)

Holland, R.: THREDS—a finite-difference time-domain EMP code in 3D spherical coordinates. IEEE Trans. Nucl. Sci. **30**, 4592–4595 (1983). doi:10.1109/TNS.1983.4333177

Hoshino, M.: Wakefield acceleration by radiation pressure in relativistic shock waves. Astrophys. J. **672**, 940–956 (2008). doi:10.1086/523665

Hoshino, M., Arons, J., Gallant, Y.A., Langdon, A.B.: Relativistic magnetosonic shock waves in synchrotron sources—shock structure and nonthermal acceleration of positrons. Astrophys. J. **390**, 454–479 (1992). doi:10.1086/171296

Kagan, D., Sironi, L. Cerutti, B., Giannios, D.: Relativistic magnetic reconnection in pair plasmas and its astrophysical applications. Space Sci. Rev. **191**, 545–573 (2015). doi:10.1007/s11214-014-0132-9

Kalapotharakos, C., Contopoulos, I.: Three-dimensional numerical simulations of the pulsar magnetosphere: preliminary results. Astron. Astrophys. **496**, 495–502 (2009). doi:10.1051/0004-6361:200810281

Keshet, U., Waxman, E.: Energy spectrum of particles accelerated in relativistic collisionless shocks. Phys. Rev. Lett. **94**(11), 111102 (2005). doi:10.1103/PhysRevLett.94.111102

Kirk, J.G., Skjæraasen, O.: Dissipation in poynting-flux-dominated flows: the σ-problem of the Crab pulsar wind. Astrophys. J. **591**, 366–379 (2003). doi:10.1086/375215

Kirk, J.G., Guthmann, A.W., Gallant, Y.A., Achterberg, A.: Particle acceleration at ultrarelativistic shocks: an eigenfunction method. Astrophys. J. **542**, 235–242 (2000). doi:10.1086/309533

Komissarov, S.S.: Magnetic dissipation in the Crab nebula. Mon. Not. R. Astron. Soc. **428**, 2459–2466 (2013). doi:10.1093/mnras/sts214

Komissarov, S.S., Lyubarsky, Y.E.: Synchrotron nebulae created by anisotropic magnetized pulsar winds. Mon. Not. R. Astron. Soc. **349**, 779–792 (2004). doi:10.1111/j.1365-2966.2004.07597.x

Lyubarsky, Y.E.: The termination shock in a striped pulsar wind. Mon. Not. R. Astron. Soc. **345**, 153–160 (2003). doi:10.1046/j.1365-8711.2003.06927.x

Lyubarsky, Y.E.: On the relativistic magnetic reconnection. Mon. Not. R. Astron. Soc. **358**, 113–119 (2005). doi:10.1111/j.1365-2966.2005.08767.x

Lyubarsky, Y.E.: Highly magnetized region in pulsar wind nebulae and origin of the Crab gamma-ray flares. Mon. Not. R. Astron. Soc. **427**, 1497–1502 (2012). doi:10.1111/j.1365-2966.2012.22097.x

Lyubarsky, Y., Kirk, J.G.: Reconnection in a striped pulsar wind. Astrophys. J. **547**, 437–448 (2001). doi:10.1086/318354

Lyutikov, M., Sironi, L., Komissarov, S., Porth, O.: Particle acceleration in explosive relativistic reconnection events and Crab Nebula gamma-ray flares. arXiv:1603.05731 (2016)

Marder, B.: A method for incorporating Gauss' law into electromagnetic PIC codes. J. Comput. Phys. **68**, 48–55 (1987). doi:10.1016/0021-9991(87)90043-X

Martins, S.F., Fonseca, R.A., Silva, L.O., Mori, W.B.: Ion dynamics and acceleration in relativistic shocks. Astrophys. J. **695**, 189–193 (2009). doi:10.1088/0004-637X/695/2/L189

Medvedev, M.V., Loeb, A.: Generation of magnetic fields in the relativistic shock of gamma-ray burst sources. Astrophys. J. **526**, 697–706 (1999). doi:10.1086/308038

Melzani, M., Walder, R., Folini, D., Winisdoerffer, C., Favre, J.M.: The energetics of relativistic magnetic reconnection: ion-electron repartition and particle distribution hardness. Astron. Astrophys. **570**, 112 (2014). doi:10.1051/0004-6361/201424193

Mignone, A., Striani, E., Tavani, M., Ferrari, A.: Modelling the kinked jet of the Crab Nebula. Mon. Not. R. Astron. Soc. **436**, 1102–1115 (2013). doi:10.1093/mnras/stt1632

Mori, K., Burrows, D.N., Hester, J.J., Pavlov, G.G., Shibata, S., Tsunemi, H.: Spatial variation of the x-ray spectrum of the Crab Nebula. Astrophys. J. **609**, 186–193 (2004). doi:10.1086/421011

Munz, C.-D., Omnes, P., Schneider, R., Sonnendrücker, E., Voß, U.: Divergence correction techniques for Maxwell solvers based on a hyperbolic model. J. Comput. Phys. **161**, 484–511 (2000). doi:10.1006/jcph.2000.6507

Nalewajko, K., Zrake, J., Yuan, Y., East, W.E., Blandford, R.D.: Kinetic simulations of the lowest-order unstable mode of relativistic magnetostatic equilibria. Astrophys. J. **826**, 115 (2016). doi:10.3847/0004-637X/826/2/115

Philippov, A.A., Spitkovsky, A., Cerutti, B.: Ab initio pulsar magnetosphere: three-dimensional particle-in-cell simulations of oblique pulsars. Astrophys. J. L. **801**, 19 (2015). doi:10.1088/2041-8205/801/1/L19

Sironi, L., Spitkovsky, A.: Particle acceleration in relativistic magnetized collisionless pair shocks: dependence of shock acceleration on magnetic obliquity. Astrophys. J. **698**, 1523–1549 (2009). doi:10.1088/0004-637X/698/2/1523

Sironi, L., Spitkovsky, A.: Acceleration of particles at the termination shock of a relativistic striped wind. Astrophys. J. **741**, 39 (2011a). doi:10.1088/0004-637X/741/1/39

Sironi, L., Spitkovsky, A.: Particle acceleration in relativistic magnetized collisionless electron-ion shocks. Astrophys. J. **726**, 75 (2011b). doi:10.1088/0004-637X/726/2/75

Sironi, L., Spitkovsky, A.: Particle-in-cell simulations of shock-driven reconnection in relativistic striped winds. Comput. Sci. Discovery **5**(1), 014014 (2012). doi:10.1088/1749-4699/5/1/014014

Sironi, L., Spitkovsky, A.: Relativistic reconnection: an efficient source of non-thermal particles. Astrophys. J. **783**, 21 (2014). doi:10.1088/2041-8205/783/1/L21

Sironi, L., Spitkovsky, A., Arons, J.: The maximum energy of accelerated particles in relativistic collisionless shocks. Astrophys. J. **771**, 54 (2013). doi:10.1088/0004-637X/771/1/54

Sironi, L., Keshet, U., Lemoine, M.: Relativistic shocks: particle acceleration and magnetization. Space Sci. Rev. **191**, 519–544 (2015a). doi:10.1007/s11214-015-0181-8

Sironi, L., Petropoulou, M., Giannios, D.: Relativistic jets shine through shocks or magnetic reconnection? Mon. Not. R. Astron. Soc. **450**, 183–191 (2015b). doi:10.1093/mnras/stv641

Sironi, L., Giannios, D., Petropoulou, M.: Plasmoids in relativistic reconnection, from birth to adulthood: first they grow, then they go. Mon. Not. R. Astron. Soc. **462**, 48–74 (2016). doi:10.1093/mnras/stw1620

Spitkovsky, A.: On the structure of relativistic collisionless shocks in electron-ion plasmas. Astrophys. J. **673**, 39–42 (2008a). doi:10.1086/527374

Spitkovsky, A.: Particle acceleration in relativistic collisionless shocks: Fermi process at last? Astrophys. J. **682**, 5–8 (2008b). doi:10.1086/590248

Tavani, M., Bulgarelli, A., Vittorini, V., Pellizzoni, A., Striani, E., Caraveo, P., Weisskopf, M.C., Tennant, A., Pucella, G., Trois, A., Costa, E., Evangelista, Y., Pittori, C., Verrecchia, F., Del Monte, E., Campana, R., Pilia, M., De Luca, A., Donnarumma, I., Horns, D., Ferrigno, C., Heinke, C.O., Trifoglio, M., Gianotti, F., Vercellone, S., Argan, A., Barbiellini, G., Cattaneo, P.W., Chen, A.W., Contessi, T., D'Ammando, F., DeParis, G., Di Cocco, G., Di Persio, G., Feroci, M., Ferrari, A., Galli, M., Giuliani, A., Giusti, M., Labanti, C., Lapshov, I., Lazzarotto, F., Lipari, P., Longo, F., Fuschino, F., Marisaldi, M., Mereghetti, S., Morelli, E., Moretti, E., Morselli, A., Pacciani, L., Perotti, F., Piano, G., Picozza, P., Prest, M., Rapisarda, M., Rappoldi, A., Rubini, A., Sabatini, S., Soffitta, P., Vallazza, E., Zambra, A., Zanello, D., Lucarelli, F., Santolamazza, P., Giommi, P., Salotti, L., Bignami, G.F.: Discovery of powerful gamma-ray flares from the Crab Nebula. Science **331**, 736 (2011). doi:10.1126/science.1200083

Timokhin, A.N., Arons, J.: Current flow and pair creation at low altitude in rotation-powered pulsars' force-free magnetospheres: space charge limited flow. Mon. Not. R. Astron. Soc. **429**, 20–54 (2013). doi:10.1093/mnras/sts298

Uzdensky, D.A., Cerutti, B., Begelman, M.C.: Reconnection-powered linear accelerator and gamma-ray flares in the Crab Nebula. Astrophys. J. **737**, 40 (2011). doi:10.1088/2041-8205/737/2/L40

Vay, J.-L.: Simulation of beams or plasmas crossing at relativistic velocity. Phys. Plasmas **15**(5), 056701 (2008). doi:10.1063/1.2837054

Villasenor, J., Buneman, O.: Rigorous charge conservation for local electromagnetic field solvers. Comput. Phys. Commun. **69**, 306–316 (1992). doi:10.1016/0010-4655(92)90169-Y

Weibel, E.S.: Spontaneously growing transverse waves in a plasma due to an anisotropic velocity distribution. Phys. Rev. Lett. **2**, 83–84 (1959). doi:10.1103/PhysRevLett.2.83

Weisskopf, M.C., Tennant, A.F., Arons, J., Blandford, R., Buehler, R., Caraveo, P., Cheung, C.C., Costa, E., de Luca, A., Ferrigno, C., Fu, H., Funk, S., Habermehl, M., Horns, D., Linford, J.D., Lobanov, A., Max, C., Mignani, R., O'Dell, S.L., Romani, R.W., Striani, E., Tavani, M., Taylor, G.B., Uchiyama, Y., Yuan, Y.: Chandra, Keck, and VLA observations of the Crab Nebula during the 2011-April gamma-ray flare. Astrophys. J. **765**, 56 (2013). doi:10.1088/0004-637X/765/1/56

Werner, G.R., Uzdensky, D.A., Cerutti, B., Nalewajko, K., Begelman, M.C.: The extent of power-law energy spectra in collisionless relativistic magnetic reconnection in pair plasmas. Astrophys. J. **816**, 8 (2016). doi:10.3847/2041-8205/816/1/L8

Wilson-Hodge, C.A., Cherry, M.L., Case, G.L., Baumgartner, W.H., Beklen, E., Narayana Bhat, P., Briggs, M.S., Camero-Arranz, A., Chaplin, V., Connaughton, V., Finger, M.H., Gehrels, N., Greiner, J., Jahoda, K., Jenke, P., Kippen, R.M., Kouveliotou, C., Krimm, H.A., Kuulkers, E., Lund, N., Meegan, C.A., Natalucci, L., Paciesas, W.S., Preece, R., Rodi, J.C., Shaposhnikov, N., Skinner, G.K., Swartz, D., von Kienlin, A., Diehl, R., Zhang, X.-L.: When a standard candle flickers. Astrophys. J. **727**, 40 (2011). doi:10.1088/2041-8205/727/2/L40

Yee, K.: Numerical solution of inital boundary value problems involving maxwell's equations in isotropic media. IEEE Trans. Antennas Propag. **14**, 302–307 (1966). doi:10.1109/TAP.1966.1138693

Yuan, Y., Nalewajko, K., Zrake, J., East, W.E., Blandford, R.D.: Kinetic study of radiation-reaction-limited particle acceleration during the relaxation of unstable force-free equilibria. Astrophys. J. **828**, 92 (2016). doi:10.3847/0004-637X/828/2/92

Chapter 12
Pulsar Wind Nebulae as a Source of Cosmic-Ray Electrons and Positrons

Kazumi Kashiyama

Abstract Cosmic-ray electron positron ($e^{\pm}$) observations by PAMELA, H.E.S.S, Fermi and AMS-02 have reported possible excesses of the fluxes with respect to the standard theoretical prediction in an energy range of from a few 10–100 GeV. Pulsar wind nebulae (PWNe) are a promising source of the excesses; multi-wavelength observations of PWNe have shown that $e^{\pm}$s are accelerated up to PeV energies in situ and the inferred acceleration rate can be sufficient to provide the observed $e^{\pm}$ flux. An uncertain point of this scenario is whether, how and when the accelerated $e^{\pm}$s can escape PWNe, which is also the key question for understanding the PWN physics. In this chapter, we first review the classical model of cosmic-ray $e^{\pm}$s and the observational history. Possible origins of the $e^{\pm}$ excess are generally discussed. Then, focusing on the PWN model, we consider the minimum requirements to be an $e^{\pm}$ source and the observational signatures. We show that $O(100)$ of pulsars with age of $< 10^7$ yrs and within a few kpc from the Earth or the nearby Geminga and Mongem pulsars alone can explain the observed $e^{\pm}$ excess. On the other hand, the young nearby Vela pulsar can contribute to $e^{\pm}$ cosmic rays with multiple-TeV energies. We discuss future prospects for testing the PWN scenarios and $e^{\pm}$ escape from the PWNe with using CALET, DAMPE, and CTA.

12.1 Introduction

Pulsar wind nebulae (PWNe) show broad non-thermal spectra from radio to gamma rays, which have been attributed to synchrotron and inverse Compton emission from ultra-relativistic electrons and positrons ($e^{\pm}$s) in situ. Such $e^{\pm}$s are considered to be produced in some regions in a pulsar magnetosphere, e.g., the so-called polar gap, carried out by a magnetized wind, and finally injected into a PWN through a termination shock and (re-)accelerated. In early stages of their evolution, PWNe are well confined in a supernovae ejecta and so as relativistic $e^{\pm}$s. They may start to escape from the PWN and be injected into the interstellar medium (ISM) when the

K. Kashiyama (✉)
Department of Physics, The University of Tokyo, Bunkyo, Tokyo 113-0033, Japan
e-mail: kashiyama@phys.s.u-tokyo.ac.jp

D.F. Torres (ed.), *Modelling Pulsar Wind Nebulae*, Astrophysics and Space Science Library 446, DOI 10.1007/978-3-319-63031-1_12

nebula is disrupted by a supernova reverse shock (i.e., reverberation) or the pulsar itself escapes from the supernova remnant (SNR) due to the kick velocity exerted by the supernova. Such systems can be an efficient producer of $e^{\pm}$ cosmic rays. For details of the above processes, readers are encouraged to see other chapters of this book.

Interestingly, recent observations by, e.g., PAMELA (Adriani et al. 2009), Fermi (Abdo et al. 2009; Ackermann et al. 2010), H.E.S.S. (Aharonian et al. 2008, 2009), and AMS-02 (Aguilar et al. 2013; Accardo et al. 2014; Aguilar et al. 2014a,b) show that the fluxes of $e^{\pm}$ cosmic rays with an energy from a few 10 GeV to $\sim$1 TeV exceed the conventional theoretical predictions, in which shock acceleration at SNRs and inelastic *pp* interactions in the ISM are the sources of $e^{\pm}$ cosmic rays. These observations may imply that an additional high-energy $e^{\pm}$ source exists, and PWNe are a promising candidate. A PWN contribution in the observed $e^{\pm}$ cosmic rays, if identified, can be used as a unique probe of $e^{\pm}$ acceleration, cooling, and propagation in the PWN, details of which are still unclear. Motivated by these, we consider PWNe as a source of cosmic-ray $e^{\pm}$ in this chapter.

The rest of this chapter is divided into two parts. The first part (Sect. 12.2) gives a general overview of cosmic-ray $e^{\pm}$s. In the latter part (Sect. 12.3), we consider PWNe as a source of cosmic-ray $e^{\pm}$s. We summarize the chapter in Sect. 12.4.

12.2 Cosmic-Ray $e^{\pm}$

In this section, we first introduce the classical theoretical framework of cosmic-ray $e^{\pm}$ production and their propagation through the Galaxy. A key prediction is that positron fraction (the ratio of positrons to electrons plus positrons) should decrease with energy. Then, we overview observational results of cosmic-ray $e^{\pm}$ with an energy of 10 GeV to $\sim$1 TeV in last decades. The observed positron fraction increases with energy contrary to the classical theory. Also the total flux of $e^{\pm}$ exceeds the standard theoretical prediction. We discuss possible interpretations of this positron "excess" or "anomaly" in general.

12.2.1 The Classical Model

Ginzburg and Syrovatsky (1961) and Hayakawa and Okuda (1962) recognized the importance of cosmic-ray $e^{\pm}$ in astrophysical contexts even before the first direct detection of cosmic-ray positrons. They first pointed out that cosmic-ray $e^{\pm}$ is a unique probe of physics of the ISM.

In the classical model, two independent channels of cosmic-ray $e^{\pm}$ production are considered. One is primary electrons accelerated by the Galactic SNRs and the other is secondary $e^{\pm}$s produced by inelastic *pp* interactions between cosmic-ray protons and the interstellar atoms. In this case, positrons are produced only as secondaries, and the e^+ injection rate depends on how many times parent cosmic-ray protons typically interact with the ISM atoms before diffusing out from the Galaxy. Importantly, this can be well calibrated by the flux ration of secondary-to-primary cosmic-ray hadrons, e.g., the boron-to-carbon ratio (see e.g., Berezinskii et al. 1990). The key is that how a charged particle propagates through the Galactic magnetic fields only depends on its rigidity, $R \equiv pc/Ze$, and the cross sections for production of secondary particles for each rigidity have been determined by collider experiments.

In a steady state, number density of a stable secondary cosmic-ray nucleus (i) formed from primaries (j) can be calculated from

$$\sum_{j\neq i} \sigma_{j\to i} v n_{\rm ISM} n_j \approx \left(\sigma_i v n_{\rm ISM} + \frac{1}{\tau_{\rm esc}}\right) n_i. \tag{12.1}$$

The left hand side corresponds to the formation rate of the secondary with $\sigma_{j\to i}$ being the cross sections for the spallation. On the other hand, the right hand side shows the loss rate where σ_i is the cross section for the fragmentation and $\tau_{\rm esc}$ is the effective escape time from the Galaxy. We can safely neglect the radiative cooling of the cosmic-ray nuclei. The escape time is usually parametrized by $\tau_{\rm esc} = X_{\rm esc}/\overline{m} n_{\rm ISM} v$, where $\overline{m}$ is the mean mass of nuclei in the ISM and $X_{\rm esc}$ is the so called grammage. From Eq. (12.1),

$$X_{\rm esc} \approx \overline{m} \left[\left(\sum_{j\neq i} \frac{n_j}{n_i} \sigma_{j\to i} - \sigma_i\right)\right]^{-1}. \tag{12.2}$$

The quantities in the right hand side of Eq. (12.2) have been obtained from terrestrial experiments and observations of cosmic-ray nuclei up to $\sim$1000 GV. Accordingly, the grammage $X_{\rm esc}$ depends only on rigidity and can be applied to all the types of cosmic rays. For example, the observed data by AMS-02 may be fitted as Blum et al. (2013)

$$X_{\rm esc} = 8.7 \text{ g cm}^{-2} \left(\frac{R}{10 \text{ GV}}\right)^{-0.4}. \tag{12.3}$$

In general, cosmic rays with a larger rigidity have a larger gyration radius and diffuse faster in the ISM, resulting in a smaller grammage and a shorter escape time. If one applies Eq. (12.3) to ultra-relativistic cosmic-ray $e^{\pm}$ (although the effect of radiative cooling is neglected so far),

$$\tau_{\rm esc} \sim 6.3 \times 10^5 \text{ yr} \left(\frac{\varepsilon_{\rm e}}{\rm TeV}\right)^{-0.4}. \tag{12.4}$$

Here we assume $\overline{m} = 1.402 m_{\rm H}$ and $n_{\rm ISM} = 1\ {\rm cm}^{-3}$. In the diffusion model, where the diffusion term $\nabla \cdot (D \nabla n_i)$ is considered instead of the leakage term $n_i/\tau_{\rm esc}$, the diffusion coefficient can be obtained based on a similar argument as above:

$$D \sim 10^{29}\ {\rm cm}^2\ {\rm s}^{-1} \left(\frac{\varepsilon_{\rm e}}{10\ {\rm GeV}}\right)^{0.4}. \tag{12.5}$$

Once $X_{\rm esc}$ and $\tau_{\rm esc}$ are obtained, the steady-state spectrum of proton cosmic rays can be described as

$$n_{\rm p} \approx Q_{\rm p} \tau_{\rm esc} \propto \varepsilon_{\rm p}^{-s-\delta}, \tag{12.6}$$

where $Q_{\rm p} \propto \varepsilon_{\rm p}^{-s}$ is the injection spectrum of the protons by SNRs. Accordingly, the injection rate of secondary cosmic-ray $e^{\pm}$ can be obtained as

$$Q_{{\rm e}^{\pm},{\rm sec}} \approx \frac{n_{\rm p}}{\sigma_{pp} v n_{\rm ISM}} F_{{\rm e}^{\pm}}(\varepsilon_{\rm e}/\varepsilon_{\rm p}) \propto \varepsilon_{\rm e}^{-s-\delta}. \tag{12.7}$$

Here σ_{pp} is the cross section for the hadronuclear interaction that is approximately energy independent in the inelastic regime, and $F_{{\rm e}^{\pm}}(\varepsilon_{\rm e}/\varepsilon_{\rm p})$ is the number of $e^{\pm}$ pairs[1] formed in an interaction and has a single peak at $\varepsilon_{\rm e}/\varepsilon_{\rm p} \approx 0.07$ and $F_{{\rm e}^{\pm}} \approx 4$ for 1 TeV $\leq \varepsilon_{\rm p} \leq$ 1 PeV. From Eqs. (12.6) and (12.7), the injection spectrum of the secondary $e^{\pm}$ should be similar to the observed spectrum of the parent protons.

We should note that gamma rays are also produced in hadronuclear interactions as decay product of neutral pions, which constitute the diffuse Galactic gamma-ray flux. Such a gamma-ray emission can be calculated in a similar way as above but with $F_{\gamma}(\varepsilon_{\gamma}/\varepsilon_{\rm p})$ instead of $F_{{\rm e}^{\pm}}(\varepsilon_{\rm e}/\varepsilon_{\rm p})$. Comparing it with the observed diffuse gamma-ray flux, one can calibrate the secondary $e^{\pm}$ injection rate.

After being injected, radiative energy loss via synchrotron and inverse Compton emission is crucial for high-energy $e^{\pm}$ cosmic rays. The energy loss rate is given by

$$\frac{d\varepsilon_{\rm e}}{dt} = -\frac{4}{3}\sigma_{\rm T} c \left(\frac{\varepsilon_{\rm e}}{m_{\rm e}c^2}\right)^2 \left[\frac{B^2}{8\pi} + \int d\varepsilon_{\rm ph} n_{\rm ph}(\varepsilon_{\rm ph}) f_{\rm KN}\left(\frac{4\varepsilon_{\rm e}\varepsilon_{\rm ph}}{m_{\rm e}^2 c^4}\right)\right]. \tag{12.8}$$

where $B \sim \mu{\rm G}$ is the ISM magnetic field strength, $n_{\rm ph}$ is the number density of the photon field basically consisting of stellar and dust emission and the cosmic microwave background, and $f_{\rm KN}$ is the Klein-Nishina suppression factor. Using the observed local values of the fields, Eq. (12.8) can be approximated as

$$\frac{d\varepsilon_{\rm e}}{dt} \approx -b\varepsilon_{\rm e}^2, \tag{12.9}$$

[1]More precisely, electrons are less abundant than positrons due to the charge conservation.

with $b = 10^{-16}\ \mathrm{GeV}^{-1}\mathrm{s}^{-1}$. Then, the cooling time of $e^{\pm}$ cosmic ray is

$$\tau_{\rm cool} \approx \frac{1}{b\varepsilon_{\rm e}} \sim 3\times 10^5\ \mathrm{yr}\left(\frac{\varepsilon_{\rm e}}{\rm TeV}\right)^{-1}. \tag{12.10}$$

Compared with Eq. (12.4), one can see that high-energy $e^{\pm}$ cosmic rays typically cool before escaping from the Galaxy. With the diffusion coefficient (Eq. 12.5), the maximum travel distance can be estimated as

$$d_{\rm cool} \approx \sqrt{D\tau_{\rm cool}} \sim 2\ \mathrm{kpc}\left(\frac{\varepsilon_{\rm e}}{100\ \mathrm{GeV}}\right)^{-0.3}, \tag{12.11}$$

which means that the sources of $e^{\pm}$ cosmic rays with $\varepsilon_{\rm e} > 100$ GeV should reside within a few kpc from the Earth. The number of young ($t_{\rm age} < 10^5$ yrs) and local ($d <$ kpc) SNRs is limited and accordingly the $e^{\pm}$ spectrum will have a cutoff at $e^{\pm} \sim$ TeV unless additional sources exist.

Due to the radiative cooling, the observed spectrum of cosmic-ray $e^{\pm}$ becomes softer than the injection spectrum. As for the primary electrons accelerated at SNRs,

$$n_{\rm e^-,prim} \approx Q_{e^-,\rm prim}\tau_{\rm cool} \propto \varepsilon_{\rm e}^{-s'-1}, \tag{12.12}$$

where we assume a steady power-law injection spectrum, $Q_{e^-,\rm prim} \propto \varepsilon_e^{-s'}$. Note that the observed energy flux of cosmic-ray electrons peaks at $\varepsilon_e \sim 10$ GeV and is roughly ~1% of that of cosmic-ray protons. The effect of the radiative cooling of the primary electrons via synchrotron and inverse Compton emission are calibrated by observing the Galactic diffuse ~GHz radio and ~MeV gamma-ray emission. Similarly to Eq, (12.12), the observed spectrum of the secondary $e^{\pm}$ becomes softer than the injection spectrum (Eq. 12.7);

$$n_{\rm e^{\pm},sec} \approx Q_{\rm e^{\pm},sec}\tau_{\rm cool} \propto \varepsilon_{\rm e}^{-s-\delta-1}. \tag{12.13}$$

Resultantly, the positron fraction becomes

$$\frac{n_{\rm e^+,sec}}{n_{\rm e^-,prim}} \propto \varepsilon_{\rm e}^{-s'+s-\delta}. \tag{12.14}$$

It is reasonable to assume that the injection spectra of primary protons and electrons have a similar power-law index, $s \approx s'$. Then, the positron fraction is predicted to decreases with energy as $\propto \varepsilon_{\rm e}^{-\delta}$.

The above arguments are based on a simple one-zone model. In fact, the Galaxy is far from uniform. Also, there are additional (re)acceleration and cooling processes of cosmic rays during the propagation. To incorporate these effects, one has to solve more general diffusion-convection equations. The model parameters, e.g., the diffusion coefficients, should be determined from observations of cosmic rays and Galactic diffuse emission in radio, x-ray, and gamma-ray bands. Some calculations with the GALPROP code[2] incorporating the above effects, also predict that the positron fraction should decrease with energy.

The above predictions of the classical scenario motivate dark matter (DM) search using high energy cosmic-ray $e^{\pm}$. In the weakly interacting massive particle (WIMP) scenario, the annihilation or decay of DM particle can provide an equal amount of electron and positron cosmic rays. If the DM contribution is significant at the Earth, the positron fraction should increase with ε_e up to the energy corresponding to the mass of the particle, $\varepsilon_e \sim$ TeV. In order to detect such DM signals definitely, we need to identify the background flux from the astrophysical sources as precise as possible.

12.2.2 Observations

Based on the above motivations, people have been trying to measure high energy $e^{\pm}$ in cosmic rays, especially the positron fraction. The first direct detection of cosmic-ray positrons was made by a group in Chicago in 1964 (de Shong et al. 1964). Subsequent measurements through 1960s and 1970s showed that the positron fraction decreases with energy up to $\sim$10 GeV (Fanselow et al. 1969; Daugherty et al. 1975). Balloon payloads through the 1980s measured the positron fraction to report an increase at around 10 GeV (Müller and Tang 1990). However, the effects of solar modulation is relevant below a few 10 GeV. Higher energy observations were being awaited.

In the middle of 1990s, HEAT (Barwick et al. 1995) and CAPRICE (Barbiellini et al. 1996) measured the cosmic-ray positron spectrum and fraction up to $\sim$50 GeV. These experiments showed that the positron fraction predominantly decreases with increasing energy while a small increase in up to $\sim$100 GeV was also announced. In the early 2000s, Alpha Magnetic Spectrometer (AMS)-01 (AMS-01 Collaboration et al. 2007) also reported the same kind of increase in the positron fraction. The significance was small at that times. The biggest obstacle was (and has been) the $\sim 10^{3-4}$ times larger proton contamination.

The situation drastically evolved after the late 2000s. The PAMELA satellite confirmed that the positron fraction increases with energy in the range of 10 to 100 GeV with a sufficient statistic (Adriani et al. 2009). On the other hand, the Fermi Large Area Telescope has measured the total flux of $e^{\pm}$s from $\sim$10 GeV up to $\sim$1 TeV, which can be fitted with a smooth power law $\sim \varepsilon_e^{-\gamma}$ with $\gamma \approx 3$ (Abdo et al. 2009; Ackermann et al. 2010).[3] The observed $e^{\pm}$ flux with Fermi exceeds the

[2]http://galprop.stanford.edu/.

[3]ATIC/PPB-BETS (Chang et al. 2008; Torii et al. 2008) suggest an sharp excess of the $e^{\pm}$ with a peak at 600 GeV, which is controversial to the results of Fermi/H.E.S.S/AMS-02 are controversial. We note that the statistical errors of ATIC are much larger than those of Fermi (Israel 2010) and so on.

theoretical prediction obtained by the GALPROP code and is likely connected to the PAMELA positron anomaly. Using a west-east effect of the Earth, Fermi can also determine the positron fraction, which showed that the anomaly extends up to ~200 GeV (Ackermann et al. 2012). Moreover, H.E.S.S. detected the $e^{\pm}$ total flux consistent with Fermi at $\varepsilon_e \sim$ a few 100 GeV while also identifying a steepening of the spectrum at $\varepsilon_e \sim 1$ TeV (Aharonian et al. 2008, 2009).

On May 16, 2011, AMS-02 was launched and has been operating successfully. AMS-02 has confirmed the previous results of PAMELA and Fermi at <100 GeV a much higher precision (Aguilar et al. 2013; Accardo et al. 2014; Aguilar et al. 2014a,b). More interestingly, it has reported that the positron spectrum exhibits a drop off at ~300 GeV although the uncertainties are large at this time. If confirmed, the cutoff feature will be the key to discriminate the scenarios. Recent observational results are summarized in Figs. 12.1 and 12.2.

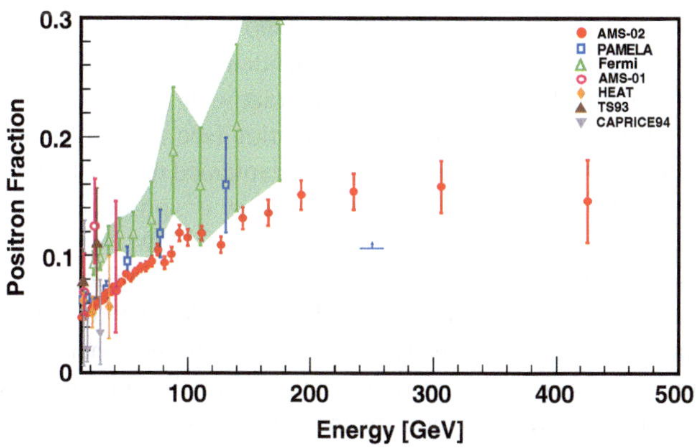

Fig. 12.1 The positron fraction above 10 GeV. The figure is taken from Accardo et al. (2014)

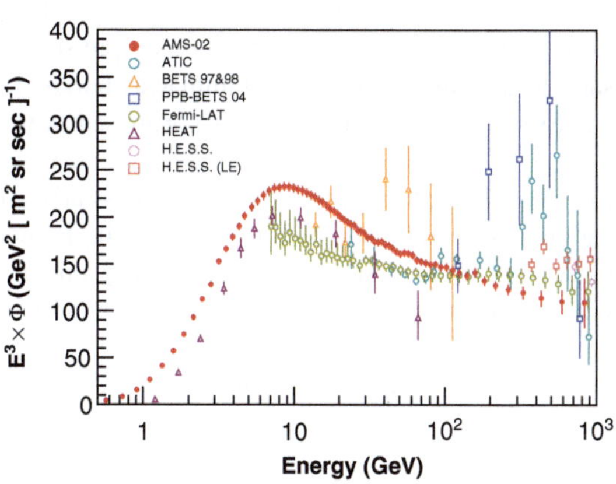

Fig. 12.2 The total flux of electron and positron cosmic rays. The figure is taken from Aguilar et al. (2014a)

In the near future, the TeV energy window will be explored. CALorimetric Electron Telescope (CALET) and Dark Matter Particle Explorer (DAMPE) were successfully launched in August and December 2015, respectively. These instruments will determine the electron spectrum up to $\sim$10 TeV with an energy resolution better than a few percent. In addition, the future Cherenkov Telescope Array (CTA) will be able to measure the cosmic-ray electron spectrum up to $\sim$15 TeV. Thanks to the large aperture, CTA could identify an anisotropy of the arrival direction of $e^{\pm}$ cosmic rays if a few sources dominates the flux.

12.2.3 The $e^{\pm}$ Excess and Possible Origins

As shown in Figs. 12.1 and 12.2, the observed fluxes of $e^{\pm}$ cosmic rays with > a few 10 GeV exceed the theoretical prediction. A simple solution is to modify the classical model, e.g., $e^{\pm}$ injection from SNRs and cosmic-ray propagation through the ISM. For example, young SNRs are surrounded by dense gas clouds. Hadronic cosmic rays accelerated at the SNR shock interact with such clouds and efficiently produce secondary $e^{\pm}$ via hadronuclear interaction (see e.g., Kohri et al. 2016 and references therein). Also, the $e^{\pm}$ excess can be explained merely by the propagation effects (see e.g., Blum et al. 2013) if the secondary $e^{\pm}$ production is enhanced or the cooling of $e^{\pm}$ cosmic rays is suppressed in the local ISM.

Another simple explanation for the $e^{\pm}$ excess is to consider new sources of cosmic-ray $e^{\pm}$. Decay or annihilation of DM particles are the most fascinating possibility and have been intensively investigated.[4] On the other hand, as we will discuss in the next section, neutron-star pulsars are the promising source since multi-wavelength observations indicate that ultra-relativistic $e^{\pm}$s do exist in the PWNe (e.g., Yüksel et al. 2009; Hooper et al. 2009; Malyshev et al. 2009; Grasso et al. 2009; Kawanaka et al. 2010; Heyl et al. 2010; Kawanaka et al. 2011; Hinton et al. 2011; Blasi and Amato 2011; Kisaka and Kawanaka 2012; Profumo 2012; Linden and Profumo 2013; Yin et al. 2013). Other astrophysical sources including white dwarf pulsars (Kashiyama et al. 2011), gamma-ray bursts (Ioka 2010), micro-quasars (Heinz and Sunyaev 2002) have been also proposed as the origin of the $e^{\pm}$ excess.

The difficulty of identifying the source of the $e^{\pm}$ excess arises from the fact that the required energy budget can be supplied by many of high-energy astrophysical objects. For comparison, let us first estimate the energy injection rate of proton cosmic rays. The observed energy density at their typical energy of $\varepsilon_{\rm p} \approx 10$ GeV is

$$u_{\rm p} \sim 1\ {\rm eV\ cm^{-3}}, \tag{12.15}$$

[4]In order to explain the $e^{\pm}$ excess, the annihilation cross section of the DM particle generally needs to be $\sim$100–1000 times larger than the thermal relic value by some unknown resonant processes.

which is comparable to the energy density of the ISM magnetic field, $u_{\rm B} \sim 1$ eV cm^{-3}. The total cosmic-ray energy in side the Milky way is

$$\mathcal{E}_{\rm p} \approx u_{\rm p} V_{\rm gal} \sim 3 \times 10^{55} \text{ erg gal}^{-1}. \tag{12.16}$$

where $V_{\rm gal} \sim 1000$ kpc^3 is the galactic volume. Given the escape time from the Galaxy, $\tau_{\rm esc} \sim 3 \times 10^6$ yrs for $\varepsilon_{\rm p} \sim 10$ GeV (Eq. 12.4), the required injection rate is

$$\mathcal{L}_{\rm p} \approx \frac{\mathcal{E}_{\rm p}}{\tau_{\rm esc}} \sim 10^{48} \text{ erg yr}^{-1} \text{ gal}^{-1}, \tag{12.17}$$

which is roughly $\sim$10% of the explosion energy of galactic SNe. On the other hand, the observed energy flux of $e^{\pm}$ cosmic rays at $\varepsilon_{\rm e} \sim 100$ GeV is less than $\sim$0.1% of proton cosmic rays, thus

$$\mathcal{E}_{e^{\pm}} < 0.001 \times \mathcal{E}_{\rm p} \sim 3 \times 10^{52} \text{ erg gal}^{-1}. \tag{12.18}$$

In the case of high energy $e^{\pm}$ cosmic rays with $\varepsilon_{\rm e} > 100$ GeV, the required injection rate can be determined by the cooling timescale (Eq. 12.10):

$$\mathcal{L}_{e^{\pm}} \approx \frac{\mathcal{E}_{e^{\pm}}}{\tau_{\rm cool}} < 10^{46} \text{ erg yr}^{-1} \text{ gal}^{-1} \sim 3 \times 10^{35} \text{ erg } s^{-1} \text{ kpc}^{-3}. \tag{12.19}$$

Although the estimate is crude, only less than $\sim$1% of the explosion energy of galactic SNe is required to explain the observed $e^{\pm}$ excess. One can check that the astrophysical sources listed above can satisfy this energy condition.

Thus, detailed investigation of the observed $e^{\pm}$ spectrum is necessary in order to discriminate the models. For example, DM models predict a sharp cutoff in the positron spectrum at an energy corresponding to the mass of the DM particle, which might be consistent with the possible cutoff at $\sim$300 GeV reported by AMS-02. We should note, however, that similar cutoff signature can be reproduced with other astrophysical sources, e.g., by considering an intrinsic break in the injection spectrum or an effectively instantaneous injection of high energy $e^{\pm}$s (Ioka 2010). In general, the $e^{\pm}$ spectrum from an astrophysical source is determined by how $e^{\pm}$s are accelerated and how and when they escape from the source. Conversely, those information can be obtained from high-energy $e^{\pm}$ observations, in principle.

Multi-messenger approach is also useful to discriminate models. In particular, the gamma-ray counterpart of the $e^{\pm}$ is important. Non-detection of diffuse gamma-ray emission from local dwarf galaxies using Fermi LAT set an upper limit for DM annihilation rate so that some DM models are excluded as the origin of the $e^{\pm}$ excess (e.g., López et al. 2016). Another useful messenger is antiproton. If the $e^{\pm}$ excess is originated from secondary $e^{\pm}$ produced by inelastic *pp* interaction, an excess in the antiproton fraction should be observed at energies above $\sim$10–100 GeV, which may be confirmed by AMS-02 (see Kohri et al. 2016 and references therein).

Lastly but importantly, the anisotropy of the arrival direction of $e^{\pm}$ cosmic rays can be detected if the number density of the source is relatively small. For example, an $O(1000)$ h observation using CTA could identify the anisotropy if a few sources dominate the observed $e^{\pm}$ flux with $\varepsilon_e \sim$ a few 100 GeV to $\sim$TeV (see e.g., Linden and Profumo 2013).

12.3 Pulsar Wind Nebulae as a Source of Cosmic-Ray $e^{\pm}$

In this section, we consider wind nebulae of neutron star pulsars as the sources of $e^{\pm}$ cosmic rays. The observed PWN spectra from the radio to gamma-ray bands can be typically explained by a superposition of synchrotron and inverse Compton emission from ultra-relativistic electrons (and positrons) with a broken power-law distribution. For example, in the case of the crab pulsar, the typical energy is $\sim$1–10 TeV (e.g., Tanaka and Takahara 2010). Such $e^{\pm}$s are considered to be produced in pair avalanches occurring in some gap regions in the pulsar magnetosphere, accelerated in the gap and wind region and finally injected in to the nebula region through the termination shock. In the early stage, the PWN and SNR structure can be stratified and ultra-relativistic $e^{\pm}$s are likely trapped in the PWN. They may escape into the ISM after the PWN is disrupted by the reverse shock of the SNR, or reverberation. How the reverberation proceeds and the $e^{\pm}$ escape sets in are uncertain and being actively studied using multi-wavelength observations. In principle, high energy $e^{\pm}$ cosmic rays can be a direct probe of these physical processes.

The reverberation occurs roughly when the SN ejecta starts to decelerate, i.e., the Sedov-Taylor time,

$$t_{\rm ST} \approx \left(\frac{3M_{\rm ej}}{4\pi \bar{m} n_{\rm ISM}}\right)^{1/3} \left(\frac{M_{\rm ej}}{2E_{\rm SN}}\right)^{1/2} \sim 0.71\ {\rm kyr} \left(\frac{E_{\rm SN}}{10^{51}\ {\rm erg}}\right)^{-1/2} \left(\frac{M_{\rm ej}}{5\,M_\odot}\right)^{5/6}, \tag{12.20}$$

where $E_{\rm SN}$ and $M_{\rm ej}$ is the explosion energy and ejecta mass of the SN, respectively. Thus, NSs with age older than $\sim$1 kyr are our primary target.

First, let us estimate the injection luminosity of $e^{\pm}$ cosmic rays from PWNe. The intrinsic power source of PWNe is the pulsar spin-down luminosity;

$$L_{\rm sd} \approx L_{\rm sd,i} \times \begin{cases} 1 & (t < t_{\rm sd}) \\ (t/t_{\rm sd})^{-2} & (t > t_{\rm sd}) \end{cases}. \tag{12.21}$$

The initial value can be estimated as Gruzinov (2005); Spitkovsky (2006); Tchekhovskoy et al. (2013)

$$L_{\rm sd,i} \approx \frac{B_{\rm dip}^2 (2\pi/P_{\rm i})^4 R_{\rm NS}^6}{4c^3}(1+\sin^2\chi)$$
$$\sim 2.7\times 10^{39}\,{\rm erg\,s^{-1}}\left(\frac{B_{\rm dip}}{10^{13}\,{\rm G}}\right)^2\left(\frac{P_{\rm i}}{100\,{\rm ms}}\right)^{-4}, \qquad (12.22)$$

where $B_{\rm dip}$ is the dipole magnetic field, $P_{\rm i}$ is the initial spin period, $R_{\rm NS} = 12$ km is the radius of the NS, and we set the angle between the magnetic and rotation axis as $\sin^2\chi = 2/3$. The spindown time is estimated as

$$t_{\rm sd} \approx \frac{E_{\rm rot,i}}{L_{\rm sd,i}} \sim 2.3\,{\rm kyr}\left(\frac{B_{\rm dip}}{10^{13}\,{\rm G}}\right)^{-2}\left(\frac{P_{\rm i}}{100\,{\rm ms}}\right)^{2}, \qquad (12.23)$$

where

$$E_{\rm rot,i} = 0.5\times I_{\rm NS}(2\pi/P_{\rm i})^2 \sim 1.9\times 10^{48}\,{\rm erg}\left(\frac{P_{\rm i}}{100\,{\rm ms}}\right)^{-2}, \qquad (12.24)$$

is the initial rotation energy with $I_{\rm NS} \approx 0.35\times M_{\rm NS}R_{\rm NS}^2$ and $M_{\rm NS} \sim 1.4\,M_\odot$ being the momentum of inertia and mass of NS, respectively. In the above estimates, we assume that the dipole magnetic fields is constant with time, but it can evolves with time in general. Also, the initial spin of NS is still highly uncertain. Population synthesis calculations including the magnetic field decay showed that typical initial dipole field is $\bar{B}_{\rm dip,i} \sim$ a few $\times 10^{13}$ G and the initial-spin distribution is a Gaussian with a peak at $\bar{P}_{\rm i} \sim$ a few $\times 100$ ms and standard deviation of $\sim$100 ms (Faucher-Giguère and Kaspi 2006; Popov et al. 2010). With such $\bar{B}_{\rm dip,i}$ and $\bar{P}_{\rm i}$, $t_{\rm ST}$ and $t_{\rm sd}$ are comparable. In this case, the injection of $e^{\pm}$s from a PWN is dominated by those at $t \sim t_{\rm sd} \sim t_{\rm ST}$ since the spindown luminosity decreases as $\propto t^{-2}$ for $t > t_{\rm sd}$. Then, the effective injection rate of $e^{\pm}$ from such a PWN with age $t_{\rm age}$ can be estimated as

$$\bar{L}_{e^\pm,\,{\rm PWN}} \approx \frac{\bar{E}_{\rm rot,i}}{t_{\rm age}} \sim 6.0\times 10^{36}\,{\rm erg\,s^{-1}}\left(\frac{\bar{P}_{\rm i}}{100\,{\rm ms}}\right)^{-2}\left(\frac{t_{\rm age}}{10\,{\rm kyr}}\right)^{-1}. \qquad (12.25)$$

The NS birth rate in our galaxy is $\mathcal{R}_{\rm NS} \sim 0.03\,{\rm yr^{-1}\,gal^{-1}}$ (Adams et al. 2013) [5] and powerful PWNe are formed in roughly $\sim$1/3 of them. With a galactic volume of $V_{\rm gal} \sim 1000\,{\rm kpc^3}$, the number density of NS pulsar with age of $t_{\rm age}$ becomes, on

[5]Note that $\mathcal{R}_{\rm NS}$ can be an order-of-magnitude higher in a starburst phase.

average,

$$n_{\rm PWN} \approx \frac{(\mathcal{R}_{\rm NS}/3)t_{\rm age}}{V_{\rm gal}} \sim 0.1\ {\rm kpc}^{-3}\left(\frac{t_{\rm age}}{10\ {\rm kyr}}\right). \tag{12.26}$$

Thus, the volume-averaged $e^{\pm}$ injection rate from PWNe is given as

$$\mathcal{L}_{e^{\pm},\,{\rm PWN}} \approx \int \bar{L}_{e^{\pm},\,{\rm PWN}} dn_{\rm PWN}$$
$$\sim 1.8 \times 10^{36}\ {\rm erg\ s^{-1}\ kpc^{-3}}\ \frac{C_{e^{\pm}}}{3}\left(\frac{\bar{P}_{\rm i}}{100\ {\rm ms}}\right)^{-2}, \tag{12.27}$$

where $C_{e^{\pm}} = \log(\tau_{\rm cool}/t_{\rm ST})$ is a correction factor of order unity. $C_{e^{\pm}} \sim 3.6$ and 2.6 for $\varepsilon_{\rm e} = 100$ GeV and 1 TeV, respectively. From Eqs. (12.19) and (12.27), one can see that PWNe can satisfy the minimum energy requirement for explaining the $e^{\pm}$ excess.

Eq. (12.27) is the volume and time averaged $e^{\pm}$ injection rate from PWNe. High-energy $e^{\pm}$ cosmic rays can only travel for $\tau_{\rm cool}$ (Eq. 12.10) and the corresponding travel distance is $d_{\rm cool}$ (Eq. 12.11). Accordingly, the number of potential sources of $e^{\pm}$ cosmic rays with an energy of $\varepsilon_{\rm e}$ can be given as

$$N_{\rm PWN,e^{\pm}} \approx n_{\rm PWN} \times \pi d_{\rm cool}^2 z_{\rm disk} \sim 1.3\left(\frac{t_{\rm age}}{10\ {\rm kyr}}\right)\left(\frac{\varepsilon_{\rm e}}{100\ {\rm GeV}}\right)^{-0.6}, \tag{12.28}$$

where $z_{\rm disk} \sim 1$ kpc is the local height of the galactic disk. For example, $N \sim (t_{\rm age}/10\ {\rm kyr})$ of multiple PWNe with age $t_{\rm age}$ contribute for $\varepsilon_{\rm e} \sim 100$ GeV. We should stress that $\bar{L}_{e^{\pm},\,{\rm PWN}} \times N_{\rm PWN,e^{\pm}} \propto t_{\rm age}^0$. More than 100 of old pulsars with $t_{\rm age} >$ Myr and a handful number of young and middle-aged pulsars with $t_{\rm age} <$ 100 kyr can give a comparable contribution to the total observed flux. Note that a pulsar with a typical kick velocity of ∼200–500 km s^{-1} (Lyne and Lorimer 1994) can escape from surrounding SNR at $t_{\rm age} \sim$ 10–100 kyr. Old pulsars with $t_{\rm age} >$ Myr are more likely observed as bow-shock PWNe. The bow-shock PWN scenario has an advantage that the escape of $e^{\pm}$ is rather straightforward. In particular, if the typical initial magnetic field of NS pulsar is as small as $B_{\rm dip,i} \sim 10^{12}$ G, the spindown time can be comparable to the exit time of the pulsar from the SNR, and the bulk of the rotation energy (Eq. 12.24) will be released after the exit. In this case, bow-shock PWNe alone could explain the $e^{\pm}$ excess (Blasi and Amato 2011).

From Eq. (12.28), we also expect that only at most a few of young and middle-aged pulsars can contribute to the $e^{\pm}$ flux at $\varepsilon_{\rm e} \sim$ TeV. In principle, nearby old pulsars can also inject ∼TeV $e^{\pm}$ cosmic rays, however the flux at the Earth is not likely significant. If the injection is dominated by those at $t \sim t_{\rm sd} \sim t_{\rm ST}$ as in the cases with $\bar{P}_{\rm i} \sim$ a few ×100 ms and $\bar{B}_{\rm dip,i} \sim$ a few $\times 10^{13}$ G, the TeV $e^{\pm}$s already lose their energies when $t_{\rm age} >$ Myr since $\tau_{\rm cool} < t_{\rm age}$. Thus, in order for an old pulsar to be a TeV $e^{\pm}$ source, $t_{\rm sd} \gg \tau_{\rm cool}$ so that the pulsar continues to inject fresh

TeV $e^{\pm}$s even at $t_{\rm age} >$ Myr. However, to meet this condition, the initial magnetic field needs to be extremely small as $B_{\rm dip,i} \ll 10^{12}$ G and such cases should be rare. From similar arguments, possibilities of astrophysical TeV $e^{\pm}$ sources are very limited, which is consistent with a spectral steepening observed at $\varepsilon_{\rm e} \sim$ TeV. In other wards, observations of multi-TeV $e^{\pm}$ cosmic ray may provide a clean window to see how $e^{\pm}$s are accelerated at and escape from a nearby younger PWN, e.g., Vela.

In summary, we have shown that PWNe satisfy the minimum energy requirement for explaining the excess of $e^{\pm}$ cosmic-ray flux at $\varepsilon_{\rm e} \sim 10$ GeV–TeV. More than 100 of relatively old pulsars or handful number of young and middle-aged pulsars can explain the flux up to $\varepsilon_{\rm e} \sim 100$ GeV while only at most a few of young and middle-aged pulsars can be the source $\sim$TeV $e^{\pm}$ cosmic rays. Up to these points, we have made several assumptions, but the conclusions are generally valid.

Nearby young and middle-aged pulsars, i.e., Vela, Geminga, and Monogem can meet the above criterions. Also, one can actually find $O(100)$ of old pulsars within a few kpc from the Earth in e.g., the ATNF pulsar catalogue.[6] Assuming these pulsars as sources, various authors calculated cosmic-ray $e^{\pm}$ spectra at the Earth and compared with the observed data. As expected,

- The $e^{\pm}$ excess can be explained by $O(100)$ of ATNF pulsars with age of $t_{\rm age} < 10^7$ yr and within a few kpc from the Earth (e.g., Grasso et al. 2009; Yin et al. 2013).
- In particular, only Geminga and Monogem together can be sufficient (e.g., Yüksel et al. 2009; Linden and Profumo 2013).
- Vela can contribute to $e^{\pm}$ cosmic rays with multi-TeV energies (e.g., Kawanaka et al. 2011; Hinton et al. 2011).

In these studies, the $e^{\pm}$ injection spectrum is often assumed to be exponential power law with a power-law index of $\Gamma \sim 2$ or smaller and a cutoff energy of $\varepsilon_{\rm e,\ cut} \sim$ TeV. This assumption is rather arbitrary at this stage. The next step will be to have a consistency check by combining multi-wavelength observations of PWNe, in particular with gamma rays, and cosmic-ray $e^{\pm}$ observations especially at higher energies. Before concluding this chapter, we discuss the future prospects.

The first scenario is relatively difficult to test. In this scenario, the observed $e^{\pm}$ spectrum will be a superposition of contributions from multiple dim pulsars. The information of each PWN will be smeared out especially at $\varepsilon_{\rm e} < 100$ GeV. Also it is not trivial to connect the $e^{\pm}$ observation to their gamma-ray emission, which is relatively dim too. If $\bar{P}_{\rm i} \sim$ a few $\times 100$ ms and $\bar{B}_{\rm dip,i} \sim$ a few $\times 10^{13}$ G are the typical initial parameters, the injection of $e^{\pm}$ cosmic rays is dominated by those at $t \sim t_{\rm sd} \sim t_{\rm ST}$ while the PWN gamma rays are emitted by $e^{\pm}$ injected at $t \sim t_{\rm age} \gg t_{\rm sd}$. Moreover, it is unlikely to detect an anisotropy of the arrival direction of $e^{\pm}$ since the number of contributing sources is much larger than unity.

[6]http://www.atnf.csiro.au/people/pulsar/psrcat/.

On the other hand, the second scenario could be tested in the near future. The distance from the Earth and age are estimated as $d \sim 250$ pc and $t_{\rm age} \sim 340$ kyr for Geminga (Faherty et al. 2007) and $d \sim 290$ pc and $t_{\rm age} \sim 110$ kyr for Monogem (Brisken et al. 2003) . Importantly, Milagro detected an extended TeV gamma-ray emission from Geminga (Abdo et al. 2007), which was recently confirmed by HAWC (Abeysekara et al. 2017). The nebula of Geminga pulsar is considered to be a bow shock PWN and the extended gamma-ray "halo" can be a tracer of how nebula electrons are diffusing into the surrounding media. CTA will be able to determine the detailed profile of the halo (de Oña-Wilhelmi et al. 2013), with which one could directly connect the $e^{\pm}$ injection at a bow-shock PWN and the observed $e^{\pm}$ cosmic rays. Also note that, if the $e^{\pm}$ flux at $\varepsilon_{\rm e} \sim$ TeV is dominated by Geminga or Monogem or both, an anisotropy of the arrival direction of $e^{\pm}$ cosmic ray will be detectable by CTA (Linden and Profumo 2013).

The last possibility is also testable and relevant for understanding how accelerated cosmic rays escape into the ISM. The Vera pulsar is at $d \sim 300$ pc from the Earth (Dodson et al. 2003) and the age is $t_{\rm age} \sim 11$ kyr (Manchester et al. 2005). Both the morphological and spectral properties are complex. These can be interpreted as a result of the interaction between the PWN and SNR reverse shock and that the $e^{\pm}$ escape has been just started. Thanks to the proximity and young age, the Vera pulsar can be the source of $e^{\pm}$ cosmic rays up to $\varepsilon_{\rm e} \sim$ a few 10 TeV. CALET, DAMPE, and CTA will be able to detect such multi-TeV $e^{\pm}$s. Details of the spectrum reflect the escape process (Kawanaka et al. 2011; Hinton et al. 2011), which is the missing link in the origin of cosmic rays.

12.4 Summary

In this chapter, we have considered PWNe as a source of cosmic-ray $e^{\pm}$s. Cosmic-ray $e^{\pm}$ observations with PAMELA, H.E.S.S, Fermi and AMS-02 have revealed that the fluxes exceed the standard theoretical prediction in an energy range of from a few 10 GeV to TeV. PWNe are the promising source of the $e^{\pm}$ excess given the multi-wavelength observations of PWNe indicating TeV-PeV $e^{\pm}$ acceleration in situ. We have discussed the minimum requirements for PWNe and showed that $O(100)$ of pulsars with age of $<10^7$ yrs and within a few kpc from the Earth or the nearby Geminga and Mongem pulsars alone can explain the observed $e^{\pm}$ excess, and also that the young nearby Vela pulsar can contribute to $e^{\pm}$ cosmic rays with multiple-TeV energies. Future observations using CALET, DAMPE, and CTA can test the PWN scenarios and probe how and when $e^{\pm}$s escape the PWNe.

Acknowledgements The author thank Norita Kawanaka and Kohta Murase for discussion. KK is supported by JST CREST and NASA through Einstein Postdoctoral Fellowship grant number PF4-150123 awarded by the Chandra X-ray Center, which is operated by the Smithsonian Astrophysical Observatory for NASA under contract NAS8-03060.

References

Abdo, A.A., et al.: Astrophys. J. **664**, L91 (2007)
Abdo, A.A., et al.: Phys. Rev. Lett. **102**, 181101 (2009)
Abeysekara, A.U., et al.: arXiv, arXiv:1702.02992 (2017)
Accardo, L., et al.: Phys. Rev. Lett. **113**, 121101 (2014)
Ackermann, M., et al.: Phys. Rev. D **82**, 092004 (2010)
Ackermann, M., et al.: Phys. Rev. Lett. **108**, 011103 (2012)
Adams, S.M., Kochanek, C.S., Beacom, J.F., Vagins, M.R., Stanek, K.Z.: Astrophys. J. **778**, 164 (2013)
Adriani, O., et al.: Nature **458** 607 (2009)
Aguilar, M., et al.: Phys. Rev. Lett. **110**, 141102 (2013)
Aguilar, M., et al.: Phys. Rev. Lett. **113**, 121102 (2014a)
Aguilar, M., et al.: Phys. Rev. Lett. **113**, 221102 (2014b)
Aharonian, F., et al.: Phys. Rev. Lett. **101**, 261104 (2008)
Aharonian, F., et al.: Astron. Astrophys. **508**, 561 (2009)
AMS-01 Collaboration, et al.: Phys. Lett. B **646**, 145 (2007)
Barbiellini G., et al.: Astron. Astrophys. **309**, L15 (1996)
Barwick, S.W., et al., Phys. Rev. Lett. **75**, 390 (1995)
Berezinskii, V.S., Bulanov, S.V., Dogiel, V.A., Ptuskin, V.S.: In: Ginzburg, V.L. (ed.) Astrophysics of Cosmic Rays. North-Holland, Amsterdam (1990)
Blasi, P., Amato, E.: High-Energy Emission from Pulsars and Their Systems: Proceedings of the First Session of the Sant Cugat Forum on Astrophysics. Astrophysics and Space Science Proceedings, p. 624. Springer, Heidelberg (2011). ISBN 978-3-642-17250-2
Blum, K., Katz, B., Waxman, E.: Phys. Rev. Lett. **111**, 211101 (2013)
Brisken, W.F., Thorsett, S.E., Golden, A., Goss, W.M.: Astrophys. J. **593**, L89 (2003)
Chang, J., et al.: Nature **456**, 362 (2008)
Daugherty, J.K., Hartman, R.C., Schmidt, P.J.: Astrophys. J. **198**, 493 (1975)
de Oña-Wilhelmi E., et al.: Astropart. Phys. **43**, 287 (2013)
de Shong, J.A., Hildebrand, R.H., Meyer, P.: Phys. Rev. Lett. **12**, 3 (1964)
Dodson, R., Legge, D., Reynolds, J.E., McCulloch, P.M.: Astrophys. J. **596**, 1137 (2003)
Faherty, J., Walter, F.M., Anderson, J.: Astrophys. Space Sci. **308**, 225 (2007)
Fanselow, J.L., Hartman, R.C., Hildebrad, R.H., Meyer, P.: Astrophys. J. 158, 771 (1969)
Faucher-Giguère, C.-A., Kaspi, V.M.: Astrophys. J. **643**, 332 (2006)
Ginzburg, V.L., Syrovatsky, S.I.: Prog. Theor. Phys. Suppl. **20**, 1 (1961)
Grasso, D., Profumo, S., Strong, A.W., et al.: Astropart. Phys. **32**, 140 (2009)
Gruzinov, A.: Phys. Rev. Lett. **94**, 021101 (2005)
Hayakawa, S., Okuda, H.: Prog. Theor. Phys. **28**, 517 (1962)
Heinz, S., Sunyaev, R.: Astron. Astrophys. **390**, 751 (2002)
Heyl, J.S., Gill, R., Hernquist, L.: Mon. Not. R. Astron. Soc. **406**, L25 (2010)
Hinton, J.A., Funk, S., Parsons, R.D., Ohm, S.: Astrophys. J. **743**, L7 (2011)
Hooper, D., Blasi, P., Dario Serpico, P.: J. Cosmol. Astropart. Phys. **1**, 025 (2009)
Ioka, K.: Prog. Theor. Phys. **123**, 743 (2010)
Israel, M.H.: Bull. Am. Astron. Soc. **41**, 698 (2010). American Astronomical Society, HEAD Meeting #11, id.26.09
Kashiyama, K., Ioka, K., Kawanaka, N.: Phys. Rev. D **83**, 023002 (2011)
Kawanaka, N., Ioka, K., Nojiri, M.M.: Astrophys. J. **710**, 958 (2010)
Kawanaka, N., Ioka, K., Ohir,a Y., Kashiyama, K.: Astrophys. J. **729**, 93 (2011)
Kisaka, S., Kawanaka, N.: Mon. Not. R. Astron. Soc. **421**, 3543 (2012)
Kohri, K., Ioka, K., Fujita, Y., Yamazaki, R.: Prog. Theor. Exp. Phys. **2016**, 021E01 (2016)
Linden, T., Profumo, S.: Astrophys. J. **772**, 18 (2013)
López, A., Savage, C., Spolyar, D., Adams, D.Q.: J. Cosmol. Astropart. Phys. **3**, 033 (2016)
Lyne, A.G., Lorimer, D.R.: Nature **369**, 127 (1994)

Malyshev, D., Cholis, I., Gelfand, J.: Phys. Rev. D **80**, 063005 (2009)
Manchester, R.N., Hobbs, G.B., Teoh, A., Hobbs, M.: Astron. J. **129**, 1993 (2005)
Müller, D., Tang, K.K.: In: Proceedings of the 21st International Cosmic Ray Conference, vol. 3, p. 249 (1990)
Popov, S.B., Pons, J.A., Miralles, J.A., Boldin, P.A., Posselt, B.: Mon. Not. R. Astron. Soc. **401**, 2675 (2010)
Profumo, S.: Centr. Eur. J. Phys. **10**, 1 (2012)
Spitkovsky, A.: Astrophys. J. **648**, L51 (2006)
Tanaka, S.J., Takahara, F.: Astrophys. J. **715**, 1248 (2010)
Tchekhovskoy, A., Spitkovsky, A., Li, J.G.: Mon. Not. R. Astron. Soc. **435**, L1 (2013)
Torii, S., et al.: arXiv, arXiv:0809.0760 (2008)
Yin, P.-F., Yu, Z.-H., Yuan, Q., Bi, X.-J.: Phys. Rev. D **88**, 023001 (2013)
Yüksel, H., Kistler, M.D., Stanev, T.: Phys. Rev. Lett. **103**, 051101 (2009)

Chapter 13
Winds in Transitional Millisecond Pulsars

Slavko Bogdanov

Abstract Transitional millisecond pulsars are a newly recognized population of neutron stars in compact binaries that undergo metamorphoses between a disk-free rotation-powered pulsar dominated state and an accretion-disk-dominated X-ray binary state. In the non-accreting state, the action of the pulsar wind can result in heating of the close stellar companion and the occurrence of an X-ray (and possibly γ-ray) luminous shock. In the accreting state, there is growing evidence that the MSP wind is not completely extinguished by the accretion flow and may be partly responsible for the peculiar behavior observed across the electromagnetic spectrum. Herein, I present an overview of our current understanding of transitional MSP systems, with an emphasis on the role of the rotation-powered pulsar wind on the phenomenology of the radio pulsar and accreting states.

13.1 Introduction

Rotation-powered (radio) millisecond pulsars (MSPs) are a subset of neutron stars distinct from that population of "ordinary" pulsars (Alpar et al. 1982). They are characterized by rapid spins, comparatively low magnetic fields ($\sim 10^{8-9}$ Gauss), and old age ($\sim 1 - 10$ Gyr). The majority of MSPs are found in binaries, consistent with the favored formation scenario based on spin-up by accretion of matter and angular momentum in a low-mass X-ray binary (Alpar et al. 1982). Pulsar wind nebulae and bow shocks driven by MSPs appear to be exceedingly rare. Of the over 300 MSPs discovered to date,[1] only three, PSRs J0437−4715 (Bell et al. 1995; Rangelov et al. 2016), B1957+20 (Stappers et al. 2003), and J2124−3358 (Hui and

[1] See the ATNF Pulsar Catalogue at http://www.atnf.csiro.au/people/pulsar/psrcat/.

S. Bogdanov (✉)
Columbia University, 550 West 120th Street, New York, NY 10027, USA
e-mail: slavko@astro.columbia.edu

D.F. Torres (ed.), *Modelling Pulsar Wind Nebulae*, Astrophysics and Space Science Library 446, DOI 10.1007/978-3-319-63031-1_13

Becker 2006), are known to exhibit Hα and/or X-ray bow shocks (Brownsberger and Romani 2014). The absence of extended nebular emission around most MSPs suggest that the combination of low ambient density, systematically lower space velocities compared to normal pulsars, and/or unfavorable wind geometry are not conducive to generation of luminous wind nebulae or bow shocks. Nevertheless, a number of binary MSPs provide alternative means of probing the interaction of the relativistic outflow from the pulsar with the environment of the pulsar.

In compact MSP binaries (with orbital periods of order a few hours), the companion star can intercept an appreciable fraction of the pulsar wind. For example, it is well-established that in eclipsing "black widow" pulsars like the original PSR B1957+20 (Fruchter et al. 1988), the frequency-dependent radio eclipses are caused by the energetic pulsar wind stripping off material from the very low mass ($\sim$0.03 $M_{\odot}$) companion (Phinney et al. 1988), producing an extended envelope of material around the secondary star. In addition, certain optical variability associated with the orbital period is attributed to heating of the face of the tidally-locked secondary by the pulsar wind (Reynolds et al. 2007; Romani et al. 2012). Similar manifestations of the pulsar wind are found in the recently recognized variety of eclipsing MSPs, nicknamed "redbacks" (Roberts 2013). These systems differ from black widows because they contain more massive ($\gtrsim$0.2 $M_{\odot}$) non-degenerate stars, which are nearly or fully Roche-lobe filling. Redbacks were first discovered in globular clusters, in particular PSR J1740–5340 (D'Amico et al. 2001; Bogdanov et al. 2010) in NGC 6397 and PSR J0024–7204W (Edmonds et al. 2002; Bogdanov et al. 2005) in 47 Tuc. The evidence for Roche-lobe filling companions, possibly bloated by the external energy supplied by the pulsar wind, led to speculations that such systems represent the missing links between X-ray binaries and radio MSPs. This was in part because the optical and X-ray properties of these systems appeared similar to that of the canonical accreting millisecond X-ray pulsar (AMXP) SAX J1808.4–3658 in quiescence[2] (Campana et al. 2004).

This suspected connection between MSPs in low-mass X-ray binaries (LMXBs) and redbacks was definitively confirmed in 2013 with the detection of an accretion-induced thermonuclear outburst from PSR J1824–2452I in the globular cluster M28 (Papitto et al. 2013). The source exhibited a Type I X-ray burst in March 2013 reaching a peak 0.5–10 keV outburst luminosity of $\sim 5 \times 10^{36}$ erg s^{-1}. X-ray pulsations at 254 Hz were discovered during two *XMM-Newton* observations taken on 2013 April 3 and 13. Remarkably, the spin frequency and orbital period were identical to that of a previously known redback radio MSP, PSR J1824−2452I. The object was detected again as a radio MSP after the 2013 outburst. Re-examination

[2]The implication of this likely connection is that in deep quiescence the pulsar winds in AMXP systems may reactivate. At present, this conjecture is difficult confirm due to the large distances of most AMXPs.

of archival *Chandra* data further revealed that the binary had been in a low-luminosity X-ray binary state in 2008 as well (Linares et al. 2014). It is interesting to note that in archival *Hubble Space Telescope* data from 2009 and 2010 PSR J1824−2452I/IGR J18245−2452 exhibited a brighter (by ∼2 magnitudes) and slightly bluer counterpart in August 2009 compared to April 2009 and April 2010 (Pallanca et al. 2013). This is likely associated with another switch between an accretion disk state (present in August 2009) and a radio pulsar wind dominated state, indicating that the accretion disk has reappeared at least three times since 2008.

Prior compelling evidence for a similar transformation was revealed with the identification of the first redback outside of a globular cluster, PSR J1023+0038. Optical observations from 2001 showed an accretion disk (Wang et al. 2009), which was absent at the time of the radio pulsar discovery in 2009 (Archibald et al. 2009). During its disk episode in 2001, PSR J1023+0038 appeared optically blue and bright in addition to showing strong, double-peaked H and He emission lines (Bond et al. 2002; Szkody et al. 2003), which are commonly seen in accreting systems. This suggests that the companion star overflowed its Roche lobe, overpowered the pulsar wind and mass transfer occurred via a disk surrounding the MSP. In contrast, spectra obtained starting in early 2003 showed substantially lower optical flux and an ordinary-looking G-type spectrum (Thorstensen and Armstrong 2005), implying the absence of an accretion disk. In 2013 June, intensive radio monitoring uncovered that PSR J1023+0038 had ceased emitting radio pulsations. Followup multi-wavelength observations found that the accretion disk had reappeared, which has been continuously present since.

A search for similar behavior among *Fermi* LAT sources revealed that the X-ray binary and candidate high-energy γ-ray source XSS J12270–4859 (Hill et al. 2011; de Martino et al. 2010) had undergone a sudden decline in X-ray and optical brightness sometime in November or December of 2012 (Bassa et al. 2014; Bogdanov et al. 2014). The system presently hosts an active 1.69 ms radio and γ-ray pulsar (Roy et al. 2015).

It has become evident that these so-called transitional MSPs can sporadically alternate between two clearly distinct states (Fig. 13.1), each lasting months to years, for reasons that remain to be understood. In the context of pulsar winds, this behavior makes them particularly appealing as they offer a rare opportunity to probe the wind of the same pulsar under vastly different circumstances. In what follows, I provide an overview of our current understanding of how the pulsar wind in transitional MSPs interacts with its immediate environment in both the accreting and non-accreting states and the insight we can gain regarding the physics of pulsar winds, relativistic shocks, and accretion onto magnetized objects.

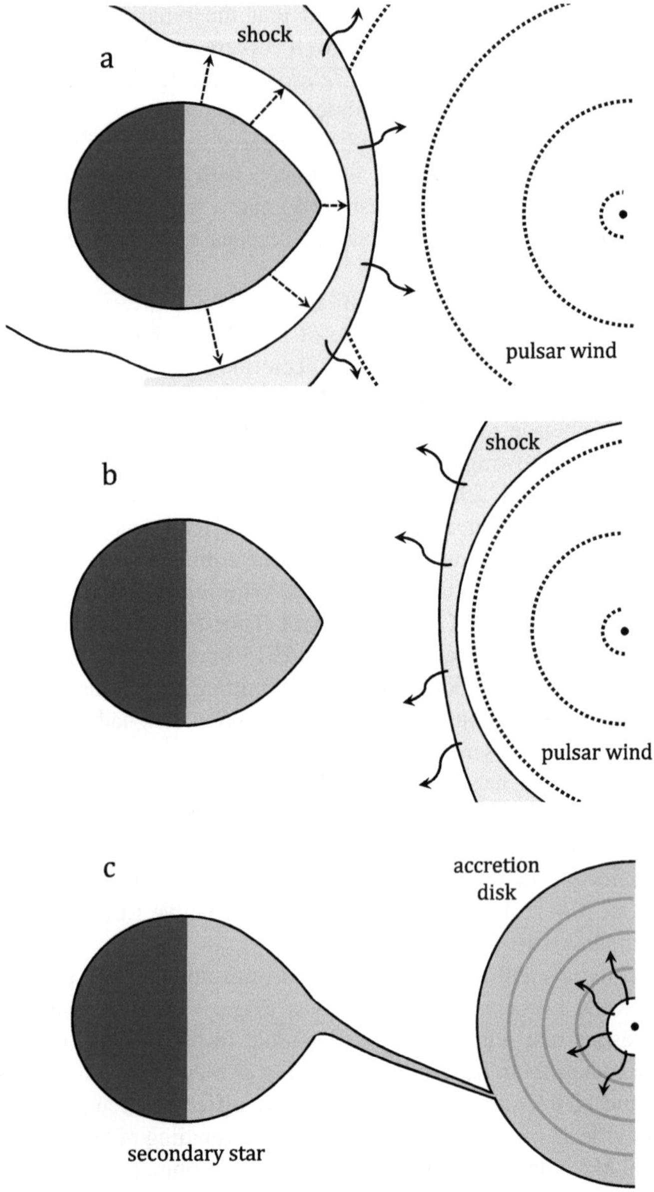

Fig. 13.1 Schematic illustrations (not to scale) of a transitional MSP binary in the disk-free, radio pulsar state, where the pulsar wind drives an intra-binary shock as it interacts with material emanating from the secondary star for the two possible cases of (**a**) the shock bowing around the secondary star and (**b**) the shock bowing around the pulsar. Panel (**c**) shows the accretion-disk dominated state where matter overflowing from the secondary star accumulates in an accretion disk extending down to the neutron star

13.2 Phenomenology of the Non-accreting State

The growing sample of redbacks in the field of the Galaxy allows us to establish several ubiquitous observational features. In addition to radio eclipses around superior conjunction also seen in black widows, redback MSPs undergo random and irregular frequency-dependent eclipses at all orbital phases (Archibald et al. 2013; Camilo et al. 2016). In addition, redbacks universally show optical modulations at the orbital period, due to either ellipsoidal variations (owing to the changing view of the tidally deformed secondary star) or illumination of the face of the companion either directly by the pulsar wind or indirectly through the intrabinary shock.

The interaction between the outflows from both stars can generate a standing intra-binary shock (Harding and Gaisser 1990; Arons and Tavani 1993), as apparent from X-ray radiation that is modulated at the binary period. Redbacks typically exhibit non-thermal X-ray emission with luminosities in the range $\approx 10^{31-33}$ erg s^{-1}, with a pronounced orbital phase dependence of the flux (see Fig. 13.2 for two illustrative examples), due to the changing view of the emitting region as the stars orbit each other (Bogdanov et al. 2005; Archibald et al. 2010; Bogdanov et al. 2010, 2011, 2014; Hui et al. 2015; Salvetti et al. 2015). Specifically, the flux declines by a factor of $\sim 2-3$ at superior conjunction, i.e. when the secondary star is between the observer and the pulsar (de Martino et al. 2015). This behavior can be interpreted as

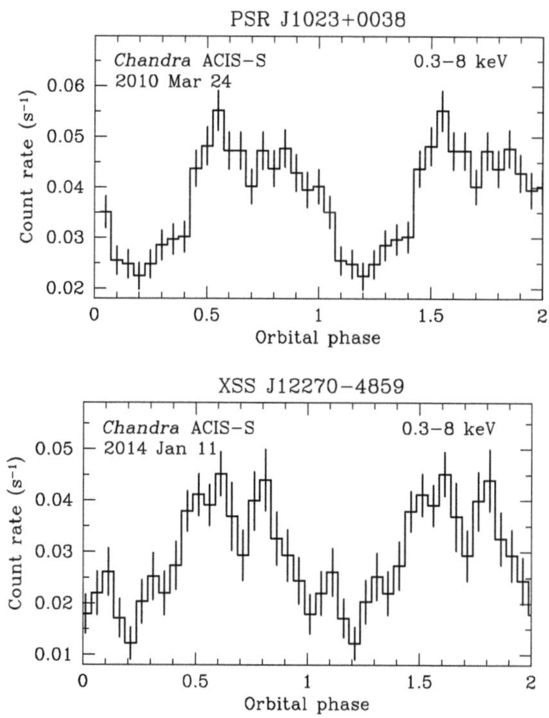

Fig. 13.2 *Chandra* ACIS-S 0.3–8 keV folded light curves of PSR J1023+0038 (*top*) and XSS J12270−4859 (*bottom*) in their non-accreting states (Bogdanov et al. 2011, 2014), illustrating the orbital modulation of the pulsar wind driven intrabinary shock-emission. The decline in X-ray flux at superior conjunction ($\phi_b \approx 0.25$ based on the radio pulsar timing convention) and the "horned" peaks appear to be universal features in redbacks. Two binary orbital cycles are shown for clarity

being due to a partial geometric occultation of the synchrotron X-ray emitting region by the companion (Bogdanov et al. 2005) if the shock is close to the surface of the star. However, the maximum of the X-ray light curves near inferior conjunction often has one or two sharp peaks, which suggest that relativistic beaming is the dominant effect responsible for the X-ray variability. Recently, efforts have been undertaken to develop detailed semi-analytical models in order to reproduce the observational features associated with the intrabinary shock and wind irradiation of the companion (Romani and Sanchez 2016; Wadiasingh et al. 2017). These models qualitatively capture the essential features of the X-ray and radio data, in particular the X-ray light curves and the radio eclipse durations, although there appear to be additional unmodeled physical effects.

One important element in redbacks that needs to be investigated in more detail is the role of the magnetic field of the secondary star in confining the pulsar wind and shaping the intrabinary shock. The presence of quasi-cyclic orbital period variations in redbacks are likely the result of tidal dissipation in the companion star (Applegate 1992). As a consequence the star is expected to be much more magnetically active compared to an isolated star main sequence of the same mass. Observational support for magnetic activity has come from optical studies (Romani et al. 2012; van Staden and Antoniadis 2016), which have revealed instances of strong flaring activity as well as large star spots. Additionally, observations indicate that the observed heating effect may not be due to direct exposure to the pulsar wind; rather, the asymmetries in the heating light curve are an indication of indirect heating from the shock (Romani 2015; van Staden and Antoniadis 2016). For a standing shock that is closer to the pulsar, as suggested by modeling X-ray light curves (Romani and Sanchez 2016; Wadiasingh et al. 2017), X-rays/γ-rays generated by the shock alone do not supply sufficient power to account for the heating of the face of the secondary star. One possible explanation involves the existence of a mechanism that preferentially channels the post-shock flow to the surface of the companion such as magnetic "tubes" that connect to the stellar surface.

Like most MSPs, redbacks are typically bright γ-ray sources, which has allowed targeted radio searches of *Fermi* sources to uncover the majority of these systems (Abdo et al. 2013). However, this emission is predominantly in the form of pulsed γ-rays produced by relativistic particles accelerated in the pulsar magnetosphere. There is tantalizing evidence that a portion of the emission $>100\,\mathrm{MeV}$ may actually originate from the same pulsar-wind driven intrabinary shock that produces synchrotron X-rays (Takata et al. 2014; Li et al. 2014). Theoretical models have also predicted acceleration of particles up to TeV energies in the intrabinary shock via an inverse Compton mechanism (Harding and Gaisser 1990; Arons and Tavani 1993). Although observations of PSR J1023+0038 in the radio pulsar state by VERITAS Aliu et al. (2016) have resulted in a non-detection of very high energy γ-rays above $>100\,\mathrm{GeV}$, they provide a useful limit of $\geq 2\,\mathrm{G}$ on the magnetic field in the shock.

Shortly after the discovery of the black widow pulsar, it was recognized that the observed multi-wavelength radiation from such systems can be a unique diagnostic of the largely unknown physics of winds from recycled pulsars and particle acceleration in collisionless shocks (Harding and Gaisser 1990; Kulkarni et al.

1992; Arons and Tavani 1993). For typical redbacks, the separation between the pulsar and a shock is $\approx 10^{11}$ cm or less, whereas for the Crab pulsar the termination shock occurs at $\sim 10^{14-15}$ cm. As such, the intra-binary shock probes a different physical regime of the pulsar wind compared to large-scale wind nebulae associated with other pulsars. Therefore, further detailed modeling of the wind interactions and attendant intrabinary shocks in redback binaries can greatly aid in establishing where the transition occurs from a magnetic-flux-dominated to a particle-dominated pulsar wind.

13.3 Phenomenology of the Accreting State

In its last accreting state, PSR J1824−2452I/IGR J18245−2452 underwent a Type I X-ray burst, such as those commonly observed from "classical" AMXPs (Patruno and Watts 2012). However, such outbursts appear to be exceptionally rare among transitional MSPs, based on the lack of a luminous outburst from the nearby ($\lesssim$3 kpc) non-accreting redbacks or accreting transitional pulsars since the advent of the existing sensitive transient monitors (*Swift* BAT, *INTEGRAL*, and *MAXI*). Evidently, accreting transitional MSPs spend most of the time in a low luminosity ($\sim 10^{32-33}$ ergs s^{-1}) state, which in LMXBs is regarded as quiescence. In this state, the multi-wavelength behavior is unlike anything seen in other accreting systems.

Perhaps the most surprising aspects of the state transition of PSR J1023+0038 from a radio pulsar state to an accretion-disk-dominated state was the factor of $\sim$6.5 enhancement in the high-energy γ-ray emission seen with *Fermi* LAT (Stappers et al. 2014; Deller et al. 2015). A similar but less pronounced flux change was also found for the state transition of XSS J12270−4859 from a prolonged LMXB state to a radio pulsar state (Torres et al. 2017). While radio MSPs are commonly observed as pulsed sources by *Fermi* LAT, γ-ray emission was not expected to be associated with neutron star LMXBs.

In X-rays, all known accreting transitional MSPs exhibit an extraordinary variability pattern. In 2008, IGR J18245−2452 experienced very rapid (<500 s) transitions between two clearly distinguishable high and low X-ray states that differ in luminosity by nearly an order of magnitude, from 5.6×10^{32} erg s^{-1} up to 3.8×10^{33} erg s^{-1} (Papitto et al. 2013; Linares et al. 2014). XSS J12270−4859 and PSR J1023+0038 are known to exhibit similar aperiodic "moding" between two X-ray flux states that last for seconds to hours (see Fig. 13.3). This behavior is seen with *NuSTAR* in hard X-rays as well, up to at least 79 keV, with no evidence for a spectral break or turnover (Tendulkar et al. 2014; Bogdanov 2016). This unique observational signature of accreting transitional pulsars in X-rays has enabled the identification of two new candidate transitional MSPs systems, 3FGL J1544.6−1125 (Bogdanov and Halpern 2015) and 3FGL J0427.9−6704 (Strader et al. 2016). It is important to note that this phenomenology is completely distinct from that seen in the X-ray dipper variety of LMXBs (White and Swank 1982), where the dips are periodic at the binary orbit and are strongly energy-

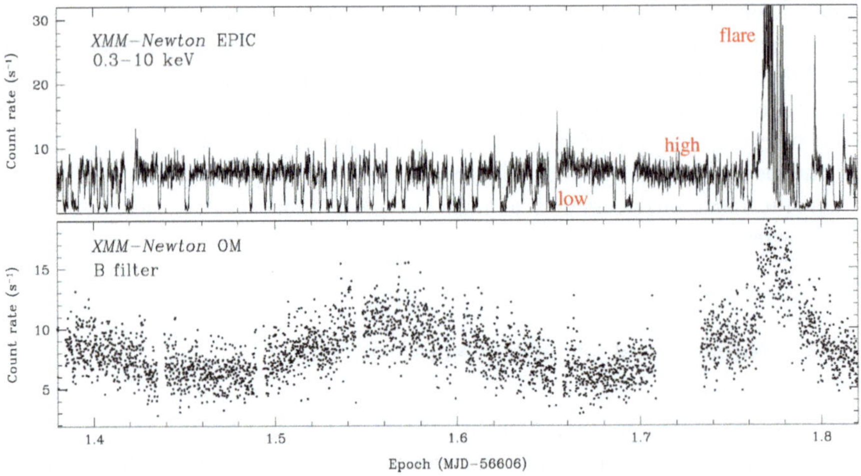

Fig. 13.3 A representative portion of an *XMM-Newton* EPIC time series of PSR J1023+0038 obtained on 2013 November 10. The X-ray variability (*top panel*) is characterized by rapid changes between two clearly separated flux modes ("high" and "low"), as well as occasional strong enhancements in flux ("flare"). The optical time series (*bottom panel*) shows broad sinusoidal modulation at the orbital period, rapid flickering typical of accreting binaries, and intense flares associated with the X-ray ones. The gaps in the optical light curve are due to interruptions in exposure

dependent. In PSR J1023+0038 and XSS J12270−4859, occasional intense flares of unknown origin are also seen (reaching up to a few $\times 10^{34}$ erg s^{-1}), while no strong flaring is seen in IGR J18245−2452 during the 200-ks *Chandra* exposure in 2008.

The most remarkable finding from recent studies has certainly been the detection of coherent X-ray pulsations in the high mode at a luminosity of just $\sim 10^{33}$ erg s^{-1} in both PSR J1023+0038 and XSS J12270−4859 (Archibald et al. 2015; Papitto et al. 2015). This implies that low-level accretion ($\approx 10^{-5} - 10^{-4}$ of the Eddington luminosity) is taking place through an unknown mechanism. Unlike AMXPs, which exhibit profile and luminosity variation in the accretion powered X-ray pulsations, the pulsed emission from PSR J1023+0038 has remained virtually constant in flux and in shape/amplitude for an interval spanning years. The detection of X-ray pulsations from PSR J1023+0038 has enabled an investigation of the spin behavior spanning over 2 years (Jaodand et al. 2016). In particular, *XMM-Newton* monitoring has provided a measurement of the spin-down rate of the pulsar in its accretion-powered state: the neutron star is spinning down only $26.8 \pm 0.4\%$ faster compared to the disk-free radio pulsar state (Archibald et al. 2013).

Ultra-violet, optical, and infrared photometric data obtained for PSR J1023+0038 as well as XSS J12270−4859 also show remarkable features: (1) rapid flickering (on timescales of seconds); (2) orbital modulations, likely the result of heating of the face of the secondary star; and (3) intense flares, some (but not all) of which are coincident with the X-ray flares (de Martino et al. 2013; Bogdanov et al. 2015); (4) occasional bimodality in brightness that is reminiscent of the X-ray mode switching (Bogdanov and Halpern 2015; Shahbaz et al. 2015).

At radio frequencies, the pulsar emission appears to be extinguished or heavily obscured when an accretion disk is present (Stappers et al. 2014; Bogdanov et al. 2015). At the same time, unpulsed but highly-variable (on timescales as short as minutes) flat-spectrum radio emission is seen during the accreting state (Hill et al. 2011; Deller et al. 2015), indicative of synchrotron originating in an outflow from the system, possibly in the form of a compact, partially self-absorbed jet. The radio continuum emission from transitional MSP is much brighter than expected given their X-ray luminosity; the radio luminosity greatly exceeds extrapolations made from observations of more vigorously accreting neutron star systems. This implies that the systems are in a jet/outflow-dominated state, contrary to prior predictions for neutron star LMXBs at very low accretion rates (Migliari and Fender 2006; Migliari et al. 2011). Moreover, rather than being two or more orders of magnitude fainter in the radio than black hole LMXBs with an equivalent X-ray luminosity, transitional MSPs appear to be slightly less than one order of magnitude fainter. There is compelling evidence that this source class follows an X-ray/radio luminosity scaling relationship ($L_X - L_R$) that appears to be distinct from other neutron star LMXBs, and which may follow a parallel track to the black-hole binaries, with the same $L_R \propto L_X^{0.6}$.

The wealth of multi-wavelength information collected for transitional MSPs has established the following observational characteristics that any viable theoretical model needs to explain self-consistently: (1) rapid X-ray switching between two discrete luminosity levels that differ by a factor of $6 - 10$ with possibly matching UV/optical mode switching; (2) accretion onto the neutron star polar caps at a single fixed luminosity ($\approx 10^{33}$ ergs s^{-1}); (3) sporadic X-ray, optical, and radio flaring; (4) unpulsed high-energy γ-ray emission; (5) unusually bright and variable, flat-spectrum radio continuum emission. While certain "propeller" (Lovelace et al. 1999) and "trapped" disk (D'Angelo and Spruit 2012) accretion models bear qualitative resemblance to the observational data, none can adequately account for several aspects of the rich phenomenology of these objects, which is an indication of key missing physics.

A key uncertainty regarding accreting transitional MSPs is whether the pulsar wind remains active after the accretion disk moves in or whether it is completely quenched by the inflowing material. This ambiguity has led to two general categories of interpretations, which differ principally in the assumption of an active or inactive wind. Based on the currently available observational evidence, the extraordinary X-ray mode switching between the two discrete flux levels may be plausibly interpreted as being due to rapid transitions between magnetospheric accretion and X-ray/γ-ray pulsar wind shock emission regimes (Linares et al. 2014; Takata et al. 2014; Coti Zelati et al. 2014) with the two different non-thermal X-ray emission mechanisms coincidentally producing the same power-law spectrum. Alternatively, emptying and refilling of the inner accretion disk due to the action of a propeller mechanism, with some accretion taking place, may account for the rapid variability (Papitto et al. 2014; Papitto and Torres 2015), while the observed γ-rays are produced by accelerated particles in the turbulent region where the accretion disk meets the rapidly rotating pulsar magnetosphere. In either case, the typical luminosities in the disk state for these systems are at quiescent levels (10^{33-34} erg s^{-1}), much lower than the 10^{35-37} erg s^{-1} typical of actively accreting LMXBs. This is an indication that these systems are in a jet/outflow-dominated regime with the bulk of infalling matter being prevented from accreting onto the neutron star surface. However, in such a scenario it is puzzling how the large amount of ejected material can maintain a low luminosity.

It should be noted that the lack of radio pulsations in accreting MSPs alone is not an indication of an extinguished pulsar wind. For instance, the radio emission mechanism may still operate, but may be undetectable in the radio band due to free-free absorption by a large amount of intervening ionized material surrounding the neutron star. Furthermore, while the electron-positron pair cascade production process needed for magnetospheric emission can be disrupted by "pollution" from accreting material, the relativistic primary particle and Poynting flux should still be present (Parfrey et al. 2016a). The low implied accretion rate onto the neutron star surface in PSR J1023+0038 and XSS J12270−4859 indicates that the inflow is not over the entire open field region. This, in turn, suggests that a large fraction of the pulsar open field region remains unobstructed for a wind outflow. Indeed, the finding for PSR J1023+0038 is that the pulsar spin-down is mostly unaffected by the accretion flow and interaction of the magnetosphere with the accretion disk (Jaodand et al. 2016), which strongly favors an active pulsar wind in the accreting state. Clearly, further investigation is required to conclusively establish if this interpretation is correct. Resolving this essential mystery may provide valuable insight into the little-understood low-luminosity ("quiescent") regime in neutron star X-ray binaries.

13.4 Open Questions and Future Prospects

The wealth of observational information gathered on transitional MSPs so far reveals fascinating new phenomenology of accreting pulsars. However, it also raises a host of fundamental questions about the underlying physical processes at play. What precipitates the formation of an accretion disk in these otherwise pulsar wind-dominated systems? What system parameters determine the propensity of certain redbacks to transition between the two states, e.g., pulsar period, spin-down power, and magnetic field strength, orbital period/separation, companion mass and magnetic activity, etc. (Linares 2014)? What is the cause for the stochastic switching between two clearly separated X-ray luminosity levels? Is the pulsar wind still active and if so does it play a crucial role in these processes?

A number of multi-wavelength investigations of the known transitional MSPs are currently in progress aimed at addressing these essential questions. In addition, surveys employing complementary approaches (Strader et al. 2016; Halpern et al.

2017; Torres et al. 2017) are underway to uncover new systems in both the accreting and non-accreting states that can serve as additional testbeds. In parallel, it is crucial to develop sophisticated theoretical models and numerical simulations that consider both accretion processes and pulsar electrodynamics (Parfrey et al. 2016,a). These efforts on the observational and theoretical fronts have important implications for understanding multiple aspects of close binary evolution, accretion onto magnetized objects, jet formation physics, and rotation-powered pulsar winds.

References

Abdo, A.A., Ajello, M., Allafort, A., Baldini, L., Ballet, J., Barbiellini, G., Baring, M.G., Bastieri, D., Belfiore, A., Bellazzini, R., et al.: The second Fermi large area telescope catalog of gamma-ray pulsars. Astrophys. J. Suppl. Ser. **208**, 17 (2013)

Aliu, E., Archambault, S., Archer, A., Benbow, W., Bird, R., Biteau, J., Buchovecky, M., Buckley, J.H., Bugaev, V., Byrum, K., Cardenzana, J.V., Cerruti, M., Chen, X., Ciupik, L., Connolly, M.P., Cui, W., Dickinson, H.J., Eisch, J.D., Falcone, A., Feng, Q., Finley, J.P., Fleischhack, H., Flinders, A., Fortin, P., Fortson, L., Furniss, A., Gillanders, G.H., Griffin, S., Grube, J., Gyuk, G., Hütten, M., Håkansson, N., Holder, J., Humensky, T.B., Johnson, C.A., Kaaret, P., Kar, P., Kelley-Hoskins, N., Kertzman, M., Kieda, D., Krause, M., Lang, M.J., Loo, A., Maier, G., McArthur, S., McCann, A., Meagher, K., Moriarty, P., Mukherjee, R., Nguyen, T., Nieto, D., O'Faoláin de Bhróithe, A., Ong, R.A., Otte, A.N., Pandel, D., Park, N., Pelassa, V., Petrashyk, A., Pohl, M., Popkow, A., Pueschel, E., Quinn, J., Ragan, K., Reynolds, P.T., Richards, G.T., Roache, E., Rulten, C., Santander, M., Sembroski, G.H., Shahinyan, K., Smith, A.W., Staszak, D., Telezhinsky, I., Tucci, J.V., Tyler, J., Varlotta, A., Vincent, S., Wakely, S.P., Weiner, O.M., Weinstein, A., Wilhelm, A., Williams, D.A., Zitzer, B., Chernyakova, M., Roberts, M.S.E.: A search for very high energy gamma rays from the missing link binary pulsar J1023+0038 with VERITAS. Astrophys. J. **831**, 193 (2016)

Alpar, M.A., Cheng, A.F., Ruderman, M.A., Shaham, J.: A new class of radio pulsars. Nature **300**, 728–730 (1982)

Applegate, J.H.: A mechanism for orbital period modulation in close binaries. Astrophys. J. **385**, 621–629 (1992)

Archibald, A.M., Stairs, I.H., Ransom, S.M., Kaspi, V.M., Kondratiev, V.I., Lorimer, D.R., McLaughlin, M.A., Boyles, J., Hessels, J.W.T., Lynch, R., van Leeuwen, J., Roberts, M.S.E., Jenet, F., Champion, D.J., Rosen, R., Barlow, B.N., Dunlap, B.H., Remillard, R.A.: A radio pulsar/X-ray binary link. Science **324**, 1411 (2009)

Archibald, A.M., Kaspi, V.M., Bogdanov, S., Hessels, J.W.T., Stairs, I.H., Ransom, S.M., McLaughlin, M.A.: X-ray variability and evidence for pulsations from the unique radio pulsar/X-ray binary transition object FIRST J102347.6+003841. Astrophys. J. **722**, 88–95 (2010)

Archibald, A.M., Kaspi, V.M., Hessels, J.W.T., Stappers, B., Janssen, G., Lyne, A.: Long-term radio timing observations of the transition millisecond pulsar PSR~J1023+0038. ArXiv e-prints, Nov 2013

Archibald, A.M., Bogdanov, S., Patruno, A., Hessels, J.W.T., Deller, A.T., Bassa, C., Janssen, G.H., Kaspi, V.M., Lyne, A.G., Stappers, B.W., Tendulkar, S.P., D'Angelo, C.R., Wijnands, R.: Accretion-powered pulsations in an apparently quiescent neutron star binary. Astrophys. J. **807**, 62 (2015)

Arons, J., Tavani, M.: High-energy emission from the eclipsing millisecond pulsar PSR 1957+20. Astrophys. J. **403**, 249–255 (1993)

Bassa, C.G., Patruno, A., Hessels, J.W.T., Keane, E.F., Monard, B., Mahony, E.K., Bogdanov, S., Corbel, S., Edwards, P.G., Archibald, A.M., Janssen, G.H., Stappers, B.W., Tendulkar, S.: A state change in the low-mass X-ray binary XSS J12270-4859. Mon. Not. R. Astron. Soc. **441**, 1825–1830 (2014)

Bell, J.F., Bailes, M., Manchester, R.N., Weisberg, J.M., Lyne, A.G.: The proper motion and wind nebula of the nearby millisecond pulsar J0437-4715. Astrophys. J. **440**, L81–L83 (1995)

Bogdanov, S.: A NuSTAR observation of the gamma-ray-emitting X-ray binary and transitional millisecond pulsar candidate 1RXS J154439.4-112820. Astrophys. J. **826**, 28 (2016)

Bogdanov, S., Halpern, J.P.: Identification of the high-energy gamma-ray source 3FGL J1544.6-1125 as a transitional millisecond pulsar binary in an accreting state. Astrophys. J. **803**, L27 (2015)

Bogdanov, S., Grindlay, J.E., van den Berg, M.: An X-ray variable millisecond pulsar in the globular cluster 47 tucanae: closing the link to low-mass X-ray binaries. Astrophys. J. **630**, 1029–1036 (2005)

Bogdanov, S., van den Berg, M., Heinke, C.O., Cohn, H.N., Lugger, P.M., Grindlay, J.E.: A Chandra X-ray observatory study of PSR J1740-5340 and candidate millisecond pulsars in the globular cluster NGC 6397. Astrophys. J. **709**, 241–250 (2010)

Bogdanov, S., van den Berg, M., Servillat, M., Heinke, C.O., Grindlay, J.E., Stairs, I.H., Ransom, S.M., Freire, P.C.C., Bégin, S., Becker, W.: Chandra X-ray observations of 12 ms pulsars in the globular cluster M28. Astrophys. J. **730**, 81 (2011)

Bogdanov, S., Archibald, A.M., Hessels, J.W.T., Kaspi, V.M., Lorimer, D., McLaughlin, M.A., Ransom, S.M., Stairs, I.H.: A Chandra X-ray observation of the binary millisecond pulsar PSR J1023+0038. Astrophys. J. **742**, 97 (2011)

Bogdanov, S., Patruno, A., Archibald, A.M., Bassa, C., Hessels, J.W.T., Janssen, G.H., Stappers, B.W.: X-ray observations of XSS J12270-4859 in a new low state: a transformation to a disk-free rotation-powered pulsar binary. Astrophys. J. **789**, 40 (2014)

Bogdanov, S., Archibald, A.M., Bassa, C., Deller, A.T., Halpern, J.P., Heald, G., Hessels, J.W.T., Janssen, G.H., Lyne, A.G., Moldón, J., Paragi, Z., Patruno, A., Perera, B.B.P., Stappers, B.W., Tendulkar, S.P., D'Angelo, C.R., Wijnands, R.: Coordinated X-ray, ultraviolet, optical, and radio observations of the PSR J1023+0038 system in a low-mass X-ray binary state. Astrophys. J. **806**, 148 (2015)

Bond, H.E., White, R.L., Becker, R.H., O'Brien, M.S.: FIRST J102347.6+003841: the first radio-selected cataclysmic variable. Publ. Astron. Soc. Pac. **114**, 1359–1363 (2002)

Brownsberger, S., Romani, R.W.: A survey for Hα pulsar bow shocks. Astrophys. J. **784**, 154 (2014)

Camilo, F., Reynolds, J.E., Ransom, S.M., Halpern, J.P., Bogdanov, S., Kerr, M., Ray, P.S., Cordes, J.M., Sarkissian, J., Barr, E.D., Ferrara, E.C.: Discovery of a millisecond pulsar in the 5.4 day binary 3FGL J1417.5-4402: observing the late phase of pulsar recycling. Astrophys. J. **820**, 6 (2016)

Campana, S., D'Avanzo, P., Casares, J., Covino, S., Israel, G., Marconi, G., Hynes, R., Charles, P., Stella, L.: Indirect evidence of an active radio pulsar in the quiescent state of the transient millisecond pulsar SAX J1808.4-3658. Astrophys. J. **614**, L49–L52 (2004)

Coti Zelati, F., Baglio, M.C., Campana, S., D'Avanzo, P., Goldoni, P., Masetti, N., Muñoz-Darias, T., Covino, S., Fender, R.P., Jiménez Bailón, E., Otí-Floranes, H., Palazzi, E., Ramón-Fox, F.G.: Engulfing a radio pulsar: the case of PSR J1023+0038. Mon. Not. R. Astron. Soc. **444**, 1783–1792 (2014)

D'Amico, N., Possenti, A., Manchester, R.N., Sarkissian, J., Lyne, A.G., Camilo, F.: An eclipsing millisecond pulsar with a possible main-sequence companion in NGC 6397. Astrophys. J. **561**, L89–L92 (2001)

D'Angelo, C.R., Spruit, H.C.: Accretion discs trapped near corotation. Mon. Not. R. Astron. Soc. **420**, 416–429 (2012)

Deller, A.T., Moldon, J., Miller-Jones, J.C.A., Patruno, A., Hessels, J.W.T., Archibald, A.M., Paragi, Z., Heald, G., Vilchez, N.: Radio imaging observations of PSR J1023+0038 in an LMXB state. Astrophys. J. **809**, 13 (2015) de Martino, D., Falanga, M., Bonnet-Bidaud, J.-M., Belloni, T., Mouchet, M., Masetti, N., Andruchow, I., Cellone, S.A., Mukai, K., Matt, G.: The intriguing nature of the high-energy gamma ray source XSS J12270-4859. Astron. Astrophys. **515**, A25 (2010)

de Martino, D., Belloni, T., Falanga, M., Papitto, A., Motta, S., Pellizzoni, A., Evangelista, Y., Piano, G., Masetti, N., Bonnet-Bidaud, J.-M., Mouchet, M., Mukai, K., Possenti, A.: X-ray follow-ups of XSS J12270-4859: a low-mass X-ray binary with gamma-ray Fermi-LAT association. Astron. Astrophys. **550**, A89 (2013)

de Martino, D., Papitto, A., Belloni, T., Burgay, M., De Ona Wilhelmi, E., Li, J., Pellizzoni, A., Possenti, A., Rea, N., Torres, D.F.: Multiwavelength observations of the transitional millisecond pulsar binary XSS J12270-4859. Mon. Not. R. Astron. Soc. **454**, 2190–2198 (2015)

Edmonds, P.D., Gilliland, R.L., Camilo, F., Heinke, C.O., Grindlay, J.E.: A millisecond pulsar optical counterpart with large-amplitude variability in the globular cluster 47 tucanae. Astrophys. J. **579**, 741–751 (2002)

Fruchter, A.S., Stinebring, D.R., Taylor, J.H.: A millisecond pulsar in an eclipsing binary. Nature **333**, 237–239 (1988)

Halpern, J.P., Bogdanov, S., Thorstensen, J.R.: X-ray and optical study of the gamma-ray source 3FGL J0838.8−2829: identification of a candidate millisecond pulsar binary and an asynchronous polar. ArXiv e-prints, Jan 2017

Harding, A.K., Gaisser, T.K.: Acceleration by pulsar winds in binary systems. Astrophys. J. **358**, 561–574 (1990)

Hill, A.B., Szostek, A., Corbel, S., Camilo, F., Corbet, R.H.D., Dubois, R., Dubus, G., Edwards, P.G., Ferrara, E.C., Kerr, M., Koerding, E., Kozieł, D., Stawarz, Ł.: The bright unidentifiedγ-ray source 1FGL J1227.9-4852: can it be associated with a low-mass X-ray binary? Mon. Not. R. Astron. Soc. **415**, 235–243 (2011)

Hui, C.Y., Becker, W.: Searches for diffuse X-ray emission around millisecond pulsars: an X-ray nebula associated with PSR J2124-3358. Astron. Astrophys. **448**, L13–L17 (2006)

Hui, C.Y., Hu, C.P., Park, S.M., Takata, J., Li, K.L., Tam, P.H.T., Lin, L.C.C., Kong, A.K.H., Cheng, K.S., Kim, C.: Exploring the intrabinary shock from the redback millisecond pulsar PSR J2129-0429. Astrophys. J. **801**, L27 (2015)

Jaodand, A., Archibald, A.M., Hessels, J.W.T., Bogdanov, S., D'Angelo, C.R., Patruno, A., Bassa, C., Deller, A.T.: Timing observations of PSR J1023+0038 during a low-mass X-ray binary state. Astrophys. J. **830**, 122 (2016)

Kulkarni, S.R., Phinney, E.S., Evans, C.R., Hasinger, G.: X-ray detection of the eclipsing millisecond pulsar PSR1957+20. Nature **359**, 300–302 (1992)

Li, K.L., Kong, A.K.H., Takata, J., Cheng, K.S., Tam, P.H.T., Hui, C.Y., Jin, R.: NuSTAR observations and broadband spectral energy distribution modeling of the millisecond pulsar binary PSR J1023+0038. Astrophys. J. **797**, 111 (2014)

Linares, M.: X-ray states of redback millisecond pulsars. Astrophys. J. **795**, 72 (2014)

Linares, M., Bahramian, A., Heinke, C., Wijnands, R., Patruno, A., Altamirano, D., Homan, J., Bogdanov, S., Pooley, D.: The neutron star transient and millisecond pulsar in M28: from sub-luminous accretion to rotation-powered quiescence. Mon. Not. R. Astron. Soc. **438**, 251–261 (2014)

Lovelace, R.V.E., Romanova, M.M., Bisnovatyi-Kogan, G.S.: Magnetic propeller outflows. Astrophys. J. **514**, 368–372 (1999)

Migliari, S., Fender, R.P.: Jets in neutron star X-ray binaries: a comparison with black holes. Mon. Not. R. Astron. Soc. **366**, 79–91 (2006)

Migliari, S., Miller-Jones, J.C.A., Russell, D.M.: The influence of spin on jet power in neutron star X-ray binaries. Mon. Not. R. Astron. Soc. **415**, 2407–2416 (2011)

Pallanca, C., Dalessandro, E., Ferraro, F.R., Lanzoni, B., Beccari, G.: The optical counterpart to the X-ray transient IGR J1824-24525 in the globular cluster M28. Astrophys. J. **773**, 122 (2013)

Papitto, A., Torres, D.F.: A propeller model for the sub-luminous state of the transitional millisecond pulsar PSR J1023+0038. Astrophys. J. **807**, 33 (2015)

Papitto, A., Ferrigno, C., Bozzo, E., Rea, N., Pavan, L., Burderi, L., Burgay, M., Campana, S., di Salvo, T., Falanga, M., Filipović, M.D., Freire, P.C.C., Hessels, J.W.T., Possenti, A., Ransom,

S.M., Riggio, A., Romano, P., Sarkissian, J.M., Stairs, I.H., Stella, L., Torres, D.F., Wieringa, M.H., Wong, G.F.: Swings between rotation and accretion power in a binary millisecond pulsar. Nature **501**, 517–520 (2013)

Papitto, A., Torres, D.F., Li, J.: A propeller scenario for the gamma-ray emission of low-mass X-ray binaries: the case of XSS J12270-4859. Mon. Not. R. Astron. Soc. **438**, 2105–2116 (2014)

Papitto, A., de Martino, D., Belloni, T.M., Burgay, M., Pellizzoni, A., Possenti, A., Torres, D.F.: X-ray coherent pulsations during a sub-luminous accretion disc state of the transitional millisecond pulsar XSS J12270-4859. Mon. Not. R. Astron. Soc. **449**, L26–L30 (2015)

Parfrey, K., Spitkovsky, A., Beloborodov, A.M.: Simulations of the magnetospheres of accreting millisecond pulsars. ArXiv e-prints, Aug 2016

Parfrey, K., Spitkovsky, A., Beloborodov, A.M.: Torque enhancement, spin equilibrium, and jet power from disk-induced opening of pulsar magnetic fields. Astrophys. J. **822**, 33 (2016)

Patruno, A., Watts, A.L.: Accreting millisecond X-ray pulsars. ArXiv e-prints, June 2012

Phinney, E.S., Evans, C.R., Blandford, R.D., Kulkarni, S.R.: Ablating dwarf model for eclipsing millisecond pulsar 1957 + 20. Nature **333**, 832–834 (1988)

Rangelov, B., Pavlov, G.G., Kargaltsev, O., Durant, M., Bykov, A.M., Krassilchtchikov, A.: First detection of a pulsar bow shock nebula in Far-UV: PSR J0437-4715. Astrophys. J. **831**, 129 (2016)

Reynolds, M.T., Callanan, P.J., Fruchter, A.S., Torres, M.A.P., Beer, M.E., Gibbons, R.A.: The light curve of the companion to PSR B1957+20. Mon. Not. R. Astron. Soc. **379**, 1117–1122 (2007)

Roberts, M.S.E.: Surrounded by spiders! New black widows and redbacks in the Galactic field. In: van Leeuwen, J. (ed.) Neutron Stars and Pulsars: Challenges and Opportunities After 80 Years. IAU Symposium, vol. 291, pp. 127–132. Cambridge University Press, Cambridge (2013)

Romani, R.W.: A likely millisecond pulsar binary counterpart for Fermi source 2FGL J2039.6-5620. Astrophys. J. **812**, L24 (2015)

Romani, R.W., Sanchez, N.: Intra-binary shock heating of black widow companions. Astrophys. J. **828**, 7 (2016)

Romani, R.W., Filippenko, A.V., Silverman, J.M., Cenko, S.B., Greiner, J., Rau, A., Elliott, J., Pletsch, H.J.: PSR J1311-3430: a heavyweight neutron star with a flyweight helium companion. Astrophys. J. **760**, L36 (2012)

Roy, J., Ray, P.S., Bhattacharyya, B., Stappers, B., Chengalur, J.N., Deneva, J., Camilo, F., Johnson, T.J., Wolff, M., Hessels, J.W.T., Bassa, C.G., Keane, E.F., Ferrara, E.C., Harding, A.K., Wood, K.S.: Discovery of Psr J1227-4853: a transition from a low-mass X-ray binary to a redback millisecond pulsar. Astrophys. J. **800**, L12 (2015)

Salvetti, D., Mignani, R.P., De Luca, A., Delvaux, C., Pallanca, C., Belfiore, A., Marelli, M., Breeveld, A.A., Greiner, J., Becker, W., Pizzocaro, D.: Multi-wavelength observations of 3FGL J2039.6-5618: a candidate redback millisecond pulsar. Astrophys. J. **814**, 88 (2015)

Shahbaz, T., Linares, M., Nevado, S.P., Rodríguez-Gil, P., Casares, J., Dhillon, V.S., Marsh, T.R., Littlefair, S., Leckngam, A., Poshyachinda, S.: The binary millisecond pulsar PSR J1023+0038 during its accretion state – I. Optical variability. Mon. Not. R. Astron. Soc. **453**, 3461–3473 (2015)

Stappers, B.W., Gaensler, B.M., Kaspi, V.M., van der Klis, M., Lewin, W.H.G.: An X-ray nebula associated with the millisecond pulsar B1957+20. Science **299**, 1372 (2003)

Stappers, B.W., Archibald, A.M., Hessels, J.W.T., Bassa, C.G., Bogdanov, S., Janssen, G.H., Kaspi, V.M., Lyne, A.G., Patruno, A., Tendulkar, S., Hill, A.B., Glanzman, T.: A state change in the missing link binary pulsar system PSR J1023+0038. Astrophys. J. **790**, 39 (2014)

Strader, J., Li, K.-L., Chomiuk, L., Heinke, C.O., Udalski, A., Peacock, M., Shishkovsky, L., Tremou, E.: A new γ-ray loud, eclipsing low-mass X-ray binary. Astrophys. J. **831**, 89 (2016)

Szkody, P., Fraser, O., Silvestri, N., Henden, A., Anderson, S.F., Frith, J., Lawton, B., Owens, E., Raymond, S., Schmidt, G., Wolfe, M., Bochanski, J., Covey, K., Harris, H., Hawley, S., Knapp, G.R., Margon, B., Voges, W., Walkowicz, L., Brinkmann, J., Lamb, D.Q.: Cataclysmic variables from the sloan digital sky survey. II. The second year. Astron. J. **126**, 1499–1514 (2003)

Takata, J., Li, K.L., Leung, G.C.K., Kong, A.K.H., Tam, P.H.T., Hui, C.Y., Wu, E.M.H., Xing, Y., Cao, Y., Tang, S., Wang, Z., Cheng, K.S.: Multi-wavelength emissions from the millisecond pulsar binary PSR J1023+0038 during an accretion active state. Astrophys. J. **785**, 131 (2014)

Tendulkar, S.P., Yang, C., An, H., Kaspi, V.M., Archibald, A.M., Bassa, C., Bellm, E., Bogdanov, S., Harrison, F.A., Hessels, J.W.T., Janssen, G.H., Lyne, A.G., Patruno, A., Stappers, B., Stern, D., Tomsick, J.A., Boggs, S.E., Chakrabarty, D., Christensen, F.E., Craig, W.W., Hailey, C.A., Zhang, W.: NuSTAR observations of the state transition of millisecond pulsar binary PSR J1023+0038. Astrophys. J. **791**, 77 (2014)

Thorstensen, J.R., Armstrong, E.: Is First J102347.6+003841 really a cataclysmic binary? Astron. J. **130**, 759–766 (2005)

Torres, D.F., Ji, L., Li, J., Papitto, A., Rea, N., de Oña Wilhelmi, E., Zhang, S.: A search for transitions between states in redbacks and black widows using seven years of Fermi-LAT observations. Astrophys. J. **836**, 68 (2017)

van Staden, A.D., Antoniadis, J.: An active, asynchronous companion to a redback millisecond pulsar. Astrophys. J. **833**, L12 (2016)

Wadiasingh, Z., Harding, A.K., Venter, C., Böttcher, M., Baring, M.G.: Constraining relativistic bow shock properties in rotation-powered millisecond pulsar binaries. ArXiv e-prints, Mar 2017

Wang, Z., Archibald, A.M., Thorstensen, J.R., Kaspi, V.M., Lorimer, D.R., Stairs, I., Ransom, S.M.: SDSS J102347.6+003841: a millisecond radio pulsar binary that had a hot disk during 2000-2001. Astrophys. J. **703**, 2017–2023 (2009)

White, N.E., Swank, J.H.: The discovery of 50 minute periodic absorption events from 4U 1915–05. Astrophys. J. **253**, L61–L66 (1982)

Index

D.F. Torres (ed.), *Modelling Pulsar Wind Nebulae*, Astrophysics and Space Science Library 446, DOI 10.1007/978-3-319-63031-1

MIX
Papier aus verantwortungsvollen Quellen
Paper from responsible sources
FSC® C105338

If you have any concerns about our products, you can contact us on
ProductSafety@springernature.com

In case Publisher is established outside the EU, the EU authorized representative is:
Springer Nature Customer Service Center GmbH
Europaplatz 3, 69115 Heidelberg, Germany

Printed by Libri Plureos GmbH
in Hamburg, Germany